Lecture Notes in Mechanical Engineering

ABCM Series on Mechanical Sciences and Engineering

Series Editors

Ricardo Diego Torres, PUC-PR, Curitiba, Brazil

Marcello Augusto Faraco de Medeiros, USP, Sao Carlos, Brazil

Marco Bittencourt, UNICAMP, Campinas, Brazil

This series publishes selected papers as well as full proceedings of events organized and/or promoted by the Brazilian Society for Mechanical Sciences and Engineering (ABCM) on an international level. These include the International Congress of Mechanical Engineering (COBEM) and the International Symposium on Dynamic Problems of Mechanics (DINAME), among others.

Valter Estevão Beal · Armando Sá Ribeiro Junior
Editors

Advances in Manufacturing Engineering Research

Selected Works from COBEF 2025

ABCM
Associação Brasileira de
Engenharia e Ciências Mecânicas

Springer

Editors
Valter Estevão Beal
Universidade SENAI CIMATEC
Salvador, Bahia, Brazil

Armando Sá Ribeiro Junior
Universidade Federal da Bahia
Salvador, Bahia, Brazil

ISSN 2195-4356 ISSN 2195-4364 (electronic)
Lecture Notes in Mechanical Engineering
ISSN 2662-3021 ISSN 2662-303X (electronic)
ABCM Series on Mechanical Sciences and Engineering
ISBN 978-3-032-09857-3 ISBN 978-3-032-09858-0 (eBook)
https://doi.org/10.1007/978-3-032-09858-0

This Springer imprint is published by the registered company Springer Nature Switzerland AG
The registered company address is: Gewerbestrasse 11, 6330 Cham, Switzerland

Preface

The "Brazilian Manufacturing Engineering Congress—COBEF" is promoted by the Brazilian Society of Mechanical Sciences and Engineering (ABCM) and is an itinerant event that occurs every two years, representative of all the country.

Since its first edition, held in Curitiba/PR in 2001, COBEF has already established itself as the most important event in Manufacturing Engineering in Brazil, occurring in different locations in the country, namely Uberlândia/MG (2003), Joinville/SC (2005), Águas de São Pedro/SP (2007), Belo Horizonte/MG (2009), Caxias do Sul/RS (2011), Penedo/RJ (2013), Salvador/BA (2015), Joinville/SC (2017), São Carlos/SP (2019), Curitiba/PR (online 2021), and Brasilia (2023).

The 13th edition took place from May 7th to 9th, 2025, in Salvador, the state capital of Bahia, Brazil. After 10 years, SENAI CIMATEC University hosted COBEF for the second time. Nevertheless, because of restorations on its infrastructure, the congress place was nearby SENAI CIMATEC, on the Grand Hotel Stella Maris convention center facilities.

The organizing committee was Dr. Rodrigo Santiago Coelho (SENAI CIMATEC) as Chair, Dra. Josiane Dantas Viana Barbosa (SENAI CIMATEC) as Co-chair, Dra. Lilian Lefol Nani Guarieiro (SENAI CIMATEC) as Treasurer, Dr. Gilmar Ferreira Batalha (USP) as responsible for the Scientific Committee, Dr. Silmar Baptista Nunes (SENAI CIMATEC) as links and actions with industries, Dr. Valter Estevão Beal (SENAI CIMATEC) and Dr. Armando Sá Ribeiro Jr. (UFBA) as Editors, Dr. Guilherme Oliveira de Souza (SENAI CIMATEC) and Dra. Auristela Carla de Miranda (UFRB) as Secretaries.

The realization and organization were from SENAI CIMATEC University together with ABCM. The congress was supported by CAPES, CNPq, FAPESB, and the Bahia State Government. Companies that also provided support for the event were ESAB, Zeiss, Instron, Termomecânica, Versatus, Altair, and Gleeble.

The COBEF 2025 Congress brought together 286 participants from 19 Brazilian states and the Federal District, as well as representatives from other countries. The event featured a final program comprising the presentation of 259 technical works across eleven areas of knowledge: Additive Manufacturing; Control Systems; Materials Characterization and Processing; Metallurgical Processes; Metrology;

Processes with Material Removal; Product Engineering; Project and Management; Robotics, Sensors and Actuators; Tribology; and Union and Assembly Processes.

As an important part of the COBEF 2025, 5 keynote lectures were presented. The keynote lectures were delivered by distinguished experts: Prof. Dr. Jesper Henri Hattel (Technical University of Denmark—DTU) on Multiphysics Modeling of Manufacturing Processes; Prof. Dr. André Costa e Silva (EEIMVR-UFF and IBQN) on Computational Thermodynamics as a Tool for the Development of Metallic Materials and Their Manufacturing Processes; Prof. Dr. Sastry Kandukuri (DNV—Norway) on Combining Innovation and Quality Assurance: The Role of Standards in Advancing Additive Manufacturing in the Energy, Oil and Gas, and High-Risk Industries; Prof. Dr. Maria Elizete Kunkel (UNIFESP) on Challenges and Opportunities of Advanced Additive Manufacturing Processes Applied to Prostheses and Orthoses; and Dr.-Ing. Sören Müller (Technical University of Berlin) on Beyond Traditional Forming—Challenges and Opportunities.

The event also marked the launch of the book *Tecnologias e aplicações da manufatura aditiva*, authored by Prof. Dr. Neri Volpato, and included a sponsored lecture by ESAB, delivered by Adriano Marinho, entitled "ESAB Digital Solutions: Welding Connectivity and Monitoring." As a special activity, participants visited the facilities of SENAI CIMATEC Headquarters in Salvador and SENAI CIMATEC PARK in Camaçari.

Papers that were approved and presented in the 13th Brazilian Manufacturing Engineering Congress—COBEF were invited to resubmit and were carefully selected to take part in this book. The first selection occurred with the congress session hosts that awarded six papers as the best of the congress. The other papers were selected based on relevance of the subject, methodology, results and innovation. In total, 22 papers were selected, some translated, many refined and organized in this book. Each chapter of the book is one of these selected papers.

This book has 22 chapters. Chapters "Low-Cost and In-line Monitoring System for Pressure Estimation in a FFF 3D Printhead"–"Effects of Electrolyte Temperature on Jet ECD of Copper Monolayers in Electrochemical Additive Manufacturing" are from Additive Manufacturing area. Chapters "Permanent Mold Casting of A356 Alloy to Simulate Critical Parts of Lightweight Wheels in Laboratory Scale"–"Development and Manufacturing of a Motor for Mini Rockets with an Apogee of 200 Meters" are, respectively, from Casting, Composites and Design for Manufacturing. Chapters "Influence of Cryogenic Cooling on Surface Roughness and Cutting Forces During Milling of ASTM A890 GR 6A Super Duplex Stainless Steel"–"Chap. Influence of Dressing Depth on the Roughness and Surface Texture of Aeronautical Alloy During Grinding" cover works of Machining or Material Removal Process. Chapter "Chap. Determination of Flow Curves for AISI 430 and AISI 430 DDQ Steel" is about Materials and "Application of Active Learning Methodologies and Skills Development in the Metrology Laboratory of FEM UNICAMP" discusses Metrology and active learning strategies. The last three chapters, "Study of Residual Stresses in Laser Cutting of High Strength Welded Tubes for Use in Vehicle Impact"– "Analysis of Corrosion Resistance in Salt Solution of Welded Joints of AISI 409 Ferritic Stainless Steel" are from Welding field.

With contributions from leading experts in the field, this book is essential reading for anyone looking to stay up to date on the latest trends and developments in manufacturing engineering. The book provides valuable insights and information to academic researchers, practicing engineers, and undergraduate and graduate students.

Editors of the Advances in Manufacturing Engineering Research: Selected works from COBEF 2025 and Chair of the congress.

Salvador, Bahia, Brazil

Valter Estevão Beal—Editor
Armando Sá Ribeiro Junior—Editor
Rodrigo Santiago Coelho—Chair

Contents

Low-Cost and In-line Monitoring System for Pressure Estimation in a FFF 3D Printhead

Dávila Moreira Lopes Silva, Luís Fernando Borges Caldeira, Jorge Lino Alves, and Zilda Castro Silveira

Abstract This work presents the design and implementation of a low-cost (€80) modular in-line monitoring system for estimating pressure at the extrusion nozzle in the FFF desktop 3D printhead (Prusa i3 MK3S model). One-dimensional analytical–numerical models were implemented, considering boundary conditions and geometric constraints of this extrusion printhead type. Two tension springs models 565 and 599 (Fanamol®, Portugal), were configured in parallel, in order to mimic the behavior, the force transducers. Calibration tests were conducted according to ISO 376 (International Organization for Standardization (2011) ISO 376:2011: Metallic materials—calibration of force-proving instruments used for the verification of uniaxial testing machines. ISO, Geneva, Switzerland), using a universal testing machine Multitest 2.5 (Mecmesin®, United Kingdom). The rheological parameters of Polylactic Acid (PLA) ExtralFill Natural (Fillamentum®, Czech Republic), the system's calibration material, were characterized using a parallel plate rheometer (TA Instruments®, United States of America), with an elongation rate of 5% in the linear regime, at melting temperatures near 200 °C. The Cox-Merz transformation was applied to estimate the rheological properties of the material, particularly the shear rates under steady shear flow. Pseudoplastic behavior during the plasticization process was investigated using the coefficients from the Power Law and Carreau-Yasuda models. The preliminary experimental design considered a hot-end temperature of 200 °C and extrusion speeds from 50 mm/min to 200 mm/min for pressure estimation. The experimental results indicated that the pressures obtained at the extrusion nozzle ranged from 1.27 to 3.32 MPa, for speeds of 50 to 200 mm/min. These values were higher than the results obtained from the numerical model, which presented values ranging from 0.85 to 1.35 MPa. In-line pressure monitoring during the plasticization process in the FFF 3D printhead contributes to the estimation of

D. M. L. Silva (✉) · L. F. B. Caldeira · Z. C. Silveira
Department of Mechanical Engineering, São Carlos School of Engineering, University of São Paulo, Av. Trabalhador São-Carlense 400, São Carlos 13566-590, SP, Brazil
e-mail: davilamlopes@gmail.com

J. L. Alves
LAETA, INEGI, Faculty of Engineering, University of Porto, Rua Dr. Roberto Frias, Porto 4200–465, Portugal

V. E. Beal and A. S. Ribeiro Junior (eds.), *Advances in Manufacturing Engineering Research*, Lecture Notes in Mechanical Engineering,
https://doi.org/10.1007/978-3-032-09858-0_1

the material's thermomechanical and rheological properties during flow, making it a crucial step, especially in the context of data collection for active control techniques and machine learning applications.

Keywords Fused filament fabrication · Additive manufacturing · Polylactic acid (PLA) · Spring system · Stiffness constant

1 Introduction

Additive manufacturing (AM) is a common term used to describe a group of techniques that build objects by adding material layer by layer. This approach is different from traditional manufacturing methods, which usually involve removing or shaping material. AM creates three-dimensional objects by depositing, joining, or transforming small volumes of raw material, called voxels. In a single process, it defines both the shape and the properties of the final product (Thompson et al. 2016; Bose et al. 2018).

The most common pressure-generating mechanisms include the movement of a plunger, the displacement of the feedstock in filament form, and the drag and/or backflow forces generated by the rotation of single or twin screws (Gibson et al. 2021; Gonzalez-Gutierrez et al. 2018).

In the field of extrusion-based additive manufacturing, Fused Deposition Modeling (FDM) is the most widely used technique, with its patent expiring around 2008 (Crump 1989). The popularity of this technique can be attributed to the simplicity of the principles involved in material processing and the low-cost of the mechanical and electromechanical components used in the equipment (Gibson et al. 2021).

In this context, the increasing range of both commercial and research-grade filaments for the FDM technique has enabled the production of prototypes and parts with enhanced functionality and durability, supporting advances in fields such as 4D printing, cellular structures (lattices), and the fabrication of hybrid injection molds, among others. Furthermore, the development of filaments from polymer-based raw materials incorporating fillers and/or additives that provide customized physicochemical properties to the final parts—such as bactericidal, conductive, or biocompatible functions—has been extensively investigated in the literature (Roberson et al. 2014; Torrado et al. 2015; Gardan 2018; Khalid et al. 2022).

In current research on the Fused Filament Fabrication (FFF) technique, monitoring process parameters during material flow, especially when developing advanced materials, is a growing field aimed at better understanding the thermomechanical environment during printing (Read et al. 2024). In-line measurements of temperature and pressure during extrusion, combined with analytical and numerical models (even simple one-dimensional ones), help predict material flow and are thus important tools for controlling the extrusion process.

Kattinger et al. (2023) and Chung et al. (2024) used X-ray computed tomography to investigate the flow inside the nozzle of a 3D FFF printer. In Kattinger et al. (2023), to enable adequate contrast for the X-ray technique, the filaments studied were loaded with tungsten powder. The experiment supported the validation of a solid–fluid transition model as well as the estimation of the extrusion force. Experimental data were compared with the Carreau–Yasuda model (Bretas and D'ávila 2005), showing good agreement only for extrusion speeds up to 5 mm/s. The main limitation of this technique is the need to add tungsten powder, which may alter the material's rheology. Go et al. (2017) and Serdeczny et al. (2020) implemented a load cell setup in contact with the filament, positioned between the extrusion motor and the hot-end in direct drive 3D printers, aiming to measure the feeding force, which can be indirectly correlated with the pressure drop inside the 3D printer nozzle. Serdeczny et al. (2020) evaluated the fit of experimental data against three models for pressure development inside the 3D printer nozzle: the traditional model by Bellini et al. (2005) and more recent models by Phan et al. (2018) and Osswald et al. (2018). The model by Osswald et al. (2018) showed the poorest fit to the experimental data; however, both Phan et al. (2018) and Bellini et al. (2005) only showed good agreement for speeds up to approximately 2 mm/s. Coogan et al. (2019) developed an in-line temperature and pressure monitoring system to evaluate flow properties of thermoplastic materials inside the nozzle of an FFF 3D printer. In this study, a pressure transmission column was connected to an LCL-010 Omega load cell (Norwalk®, CT). For calibrating the pressure transmission system, the modified nozzle was connected to a capillary rheometer (not specified), allowing direct measurement of nozzle pressure. However, no design details of the pressure transmission column or manufacturing information were provided, nor were assembly challenges addressed.

In this context, this work presents the preliminary design and development of a low-cost in-line monitoring system for estimating the pressure at the extrusion nozzle of an FFF desktop 3D printhead (Prusa i3 MK3S model). Two tension springs, models 565 and 599 (Fanamol®, Portugal), were configured in parallel to mimic the behavior of force transducers, and their displacement measurements were obtained through image processing. One-dimensional analytical–numerical models were implemented, taking into account the boundary conditions and geometric constraints of this type of extrusion printhead.

2 Materials and Methods

The proposed procedure was divided into five stages: (a) rheological characterization of the feedstock (filament) and flow rate estimation at the extrusion nozzle; (b) modeling of material flow in the calibrated nozzle of the FFF 3D printhead; (c) characterization of the mechanical properties of tension springs; (d) conceptual design of the monitoring system; and (e) the exploratory prototype development (functional testing).

Table 1 Main properties of natural ExtralFill PLA (Fillamentum®, Czech Republic)

Properties	Typical value	Method
Density	1,24 g/cm^3	ASTM D792
Tensile Strength	60 MPa	ASTM D882
MFI (Melt Flow Index)	6g/10min (210°C, 216 kg)	ASTM D1238

Source Fillamentum®, Czech Republic

2.1 Rheological Characterization of the Feedstock and Flow Rate Characterization of the AM Equipment

Natural Extrafill Polylactic Acid (PLA) filament (Fillamentum®, Czech Republic) was used in this study as the calibration material. The main properties of the filament are summarized in Table 1. The rheological properties of the PLA were characterized using a parallel plate rheometer (TA Instruments®, USA), applying a 5% strain rate within the linear regime at a temperature of 200 °C. The Cox-Merz transform was employed to estimate the material properties under steady shear conditions, thereby enabling the determination of the power-law coefficients and the Carreau–Yasuda parameters, which were the pseudoplasticity models investigated.

The Prusa i3 MK3S additive manufacturing equipment (Prusa®, Czech Republic) was used for this study. A flow rate characterization test of the equipment was required. For this purpose, the mass of filaments extruded at extrusion speeds ranging from 50 mm/min to 450 mm/min was measured using a scientific scale. The extrusion time was set to one minute, and five replicates were performed for each test condition. This procedure allowed the experimental estimation of the equipment's flow rate at different extrusion speeds.

2.2 Modeling of Material Flow at the Extrusion Nozzle of FFF 3D Printhead

Mathematically, the variation of viscosity (η) in melted polymers, considering the shear rate parameter ($\dot{\gamma}$), can be modeled by the Power Law, as shown in Eq. 1. The parameter (k) is called the consistency factor, and the parameter (n) represents the Power Law index (Bretas and D'ávila 2005; Rauwendaal 2014). In an isothermal and one-dimensional flow, the pressure drop (Δp) can be estimated using Eq. 2 (Michaeli, 2003).

$$\eta = k\dot{\gamma}^{n-1} \tag{1}$$

$$Q = K'\varphi\Delta p^{m} \tag{2}$$

In Eq. (2), K' is the conductance factor of the die geometry, and φ and m are the reciprocal parameters of the Power Law indices, k and n, respectively. Equations 2 and 3 were applied to determine K' according to the different nozzle geometries, namely cylinder and truncated cone. Table 2 presents the geometries as well as the equations for the conductance factor K' (michaeli 2003).

For the estimation of the pressure drop in the die, a routine was developed using MATLAB®.

The material was also modeled as a pseudoplastic fluid following the isothermal Carreau–Yasuda model, as described in Eq. 5. Pressure development was numerically estimated using the PolyFlow software (Ansys®, USA) due to analytical limitations caused by the model's nonlinearities. The flow was considered one-dimensional, isothermal, and with no slip at the wall. The boundary conditions adopted were based on (Nzebuka et al. 2022), considering the radial symmetry of the nozzle; the geometry used is shown in Fig. 1. The generated mesh consisted of 108,417 predominantly hexahedral elements. Numerical modeling is not the focus of this study; therefore, an analytical approach was applied to estimate the pressure development using the Carreau–Yasuda model.

In Eq. (5), η^0 is the viscosity at low shear rates, and η^∞ is the viscosity at high shear rates. The parameter λ represents the characteristic time, α is the coefficient that adjusts the smoothness of the transition between low and high shear rate regimes, and n is the pseudoplasticity index.

$$\eta = \eta^\infty + (\eta^o - \eta^\infty)[1 + \left(\lambda\dot{\gamma}\right)\alpha]^{n-1/\alpha} \tag{5}$$

Table 2 Conductance for the considered geometries

	Geometry	Conductance	
Cylinder	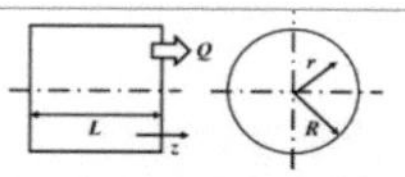	$K' = \frac{\pi R^{m+3}}{2^m(m+3)}\left(\frac{1}{L}\right)^m$	(3)
Conical frustum	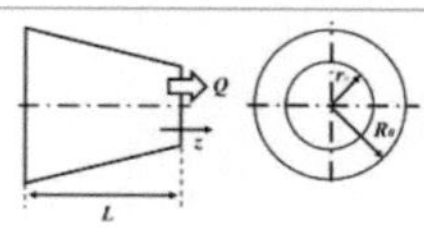	$K' = \frac{\pi}{2^m(m+3)}\left(\frac{3\left(\frac{R_o}{r_o}-1\right)}{mL\left(1-\left(\frac{r_o}{R_o}\right)^{\frac{3}{m}}\right)}\right)^m$	(4)

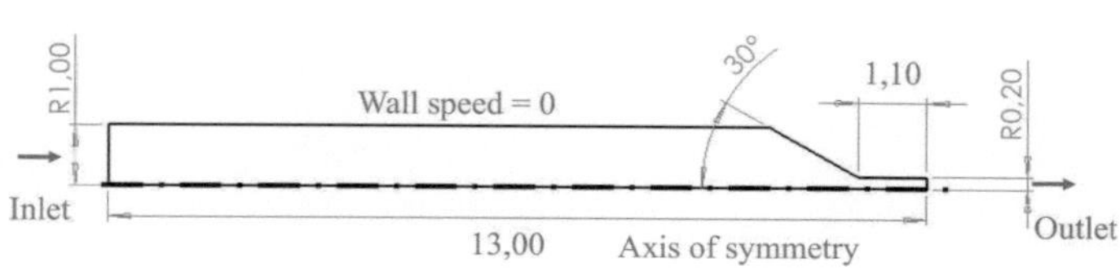

Fig. 1 Simplified geometry of the 3D printing nozzle

The flow modeling results were fundamental for determining the requirements for the force-sensing springs used in this study and for mathematically estimating the pressure developed in the 3D printing nozzle.

2.3 Mechanical Characterization of the Springs

The springs were calibrated through tensile tests performed on a Multitest 2.5 universal testing machine (Mecmesin®, United Kingdom) following the guidelines of ISO 376 (2011). The test was programmed to apply a maximum displacement of 4 mm over 1 min, and five experiments were conducted for each spring analyzed. The validation of the image processing system was carried out simultaneously with the spring calibration tests. Images were acquired using an Exmor-R camera (Sony®, Japan). Subsequently, the images were analyzed using ImageJ software to verify spring displacement and compare it with the calibration data. Figure 2 illustrates the tests and the use of ImageJ software.

After the rheological characterization of the material and the characterization of the springs used, it was possible to create the simplified free body diagram of the extrusion nozzle, as shown in Fig. 3a. Equation 6 presents the static equilibrium free body equation, where F_{bp} is the force due to pressure development, F_{el} is the elastic force given by the spring stiffness, and P is the weight of the mechanical system. In Eq. 7, ΔP_1, $\Delta P_2 e \Delta P_3$ represent the pressure development in the three main regions of the 3D printing nozzle (see Fig. 3b), and A_{fil} is the filament cross-sectional area. In Eq. 8, c is the stiffness constant and x is the spring displacement.

$$F_{\text{bp}} + P - 2F_{\text{el}} = 0 \tag{6}$$

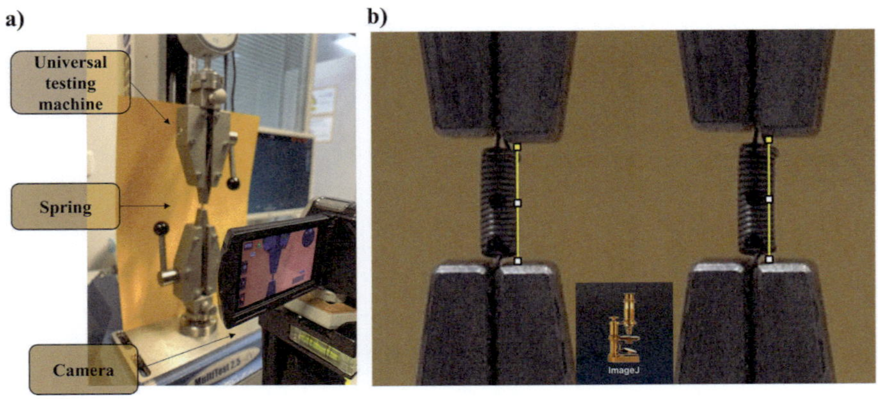

Fig. 2 Experimental procedure for spring calibration and validation of the image acquisition system

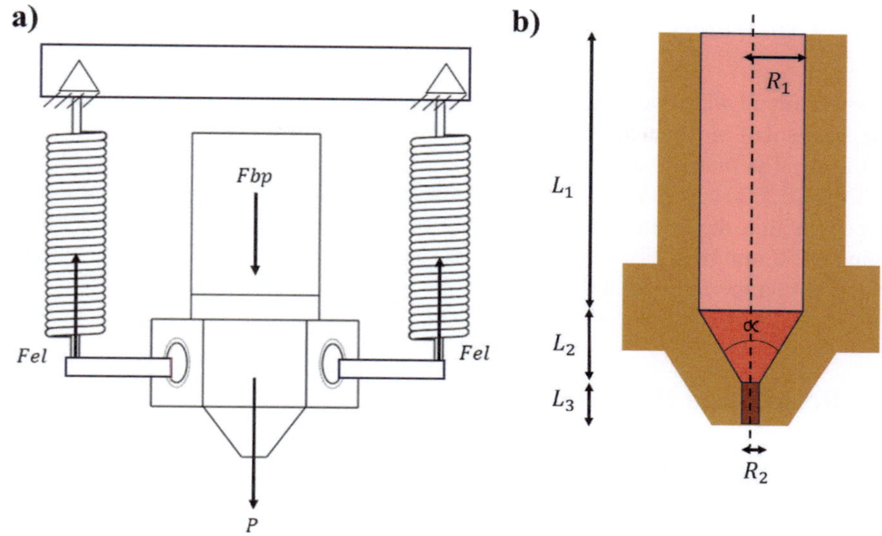

Fig. 3 Illustration of the 3D printing nozzle: **a** Free body diagram of the force transduction system and **b** Representation of the three main regions of pressure development

$$F_{\mathrm{bp}} = (\Delta P_1 + \Delta P_2 + \Delta P_3)A_{\mathrm{fil}} \tag{7}$$

$$F_{\mathrm{el}} = cx \tag{8}$$

Using Eqs. 6, 7, and 8, the pressure developed at the 3D printing nozzle can be experimentally estimated, considering that F_{bp} is determined by Eq. 7 and F_{el} by Eq. 8.

2.4 Conceptual Design of the Monitoring System

The conceptual design of the pressure monitoring system includes the following main geometric requirements: (a) a region for positioning the force transducers in the heating block of the hot-end; and (b) a nozzle adapted to allow the assembly of the force transducers and to enable a measurable displacement of the springs. The main constraints of the monitoring system are related to the importance of maintaining similarity in weight and dimensions between the modified hot-end and the original Prusa i3 MK3S hot-end (Prusa®, Czech Republic), which has an approximate weight of 60 g and maximum dimensions fitting within a 25 mm × 65 mm × 25 mm space. The three-dimensional geometric model was developed using the software CREO 8.0.0 (PTC, United States of America).

2.5 Exploratory Prototype and Functional Testing

The prototype was manufactured at the Mechanical Engineering Workshops of the Department of Mechanical Engineering, Faculty of Engineering, University of Porto, Portugal. The material used for both the nozzle and the modified heating block was ASME C22000 copper alloy. Functional tests were carried out considering extrusion speeds within the linear flow rate regime of the printhead, ranging from 50 mm/min to 250 mm/min. Five experiments were conducted for each condition, with an extrusion temperature of 200 °C.

3 Results and Discussion

3.1 Rheological Characterization of the Feedstock and Flow Rate Characterization of the AM Equipment

Material characterization and mass flow rate curves of the AM equipment are presented in Fig. 4. As shown, the fluid viscosity is presented for shear rates ranging from 10^{-2} to 10^{2} s^{-1}. Since the rheological tests were performed using a parallel plate geometry, it becomes necessary to extrapolate the data for higher shear rate regions. Figure 4b, in turn, shows the mass flow rate behavior (kg/s) in the hot-end of the Prusa i3 MK3S printer (Prusa®, Czech Republic), with flow rates ranging from 2 10^{-6} a 2 10^{-5} kg/s for extrusion speeds between 50 mm/min and 450 mm/min. In addition, good agreement is observed between the model data (Mod) and the experimental data at different temperatures (200 °C, 205 °C, and 210 °C) for extrusion speeds up to 250 mm/min. The temperatures of 205 °C and 210 °C were tested in order to assess whether linearity in the flow rate could be observed at any extrusion temperature, which was not the case.

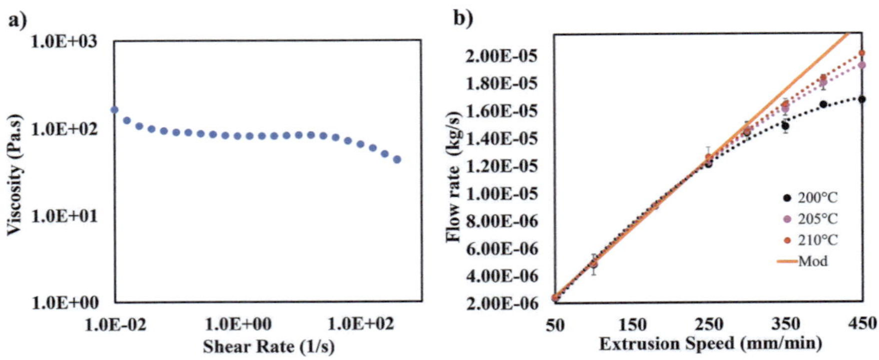

Fig. 4 Characterization curves: **a** Rheological behavior, **b** Mass flow rate

3.2 Modeling of Material Flow at the Extrusion Nozzle of FFF 3D Printhead

Pseudoplasticity coefficients were estimate from Eqs. 1 and 5 for higher shear rates, the graph in Fig. 4a was extrapolated as shown in Fig. 5.

Figure 5 shows the experimental approximations used to determine the pseudoplasticity coefficients of PLA ExtralFill Natural (Fillamentum®, Czech Republic). Based on the Power Law coefficients, it was possible to analytically estimate the volumetric flow rate (feeding rate), feeding speed, and the pressure development within the extrusion nozzle under various printing conditions, including layer height and printing speed, as illustrated in Fig. 6.

Figure 6b shows the pressure development field considering the printing conditions. The results indicate that the pressure reached 2 MPa at a printing speed of 300 mm/s and a maximum layer height of 0.4 mm (Fig. 6c). However, given the typical printing conditions in desktop printers, printing speeds are generally below 150 mm/s, and for a 0.4 mm nozzle, the layer height is usually 0.2 mm, reaching pressures close to 1.3 MPa.

Figure 7 presents numerical pressure estimates for extrusion speeds ranging from 50 mm/min to 250 mm/min based on the Carreau–Yasuda model. Analyzing Fig. 7, it can be observed that the Carreau–Yasuda model predicts pressure development values ranging from 0.25 MPa to 0.70 MPa for extrusion speeds between 50 mm/min and 250 mm/min. Additionally, as expected, the pressure development is more intense in the most restrictive region of the 3D printing nozzle. A comparison between the models will be discussed in Sect. 3.5.

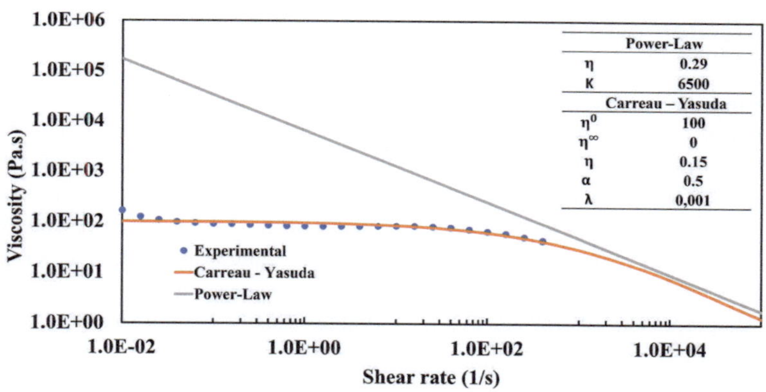

Fig. 5 Estimation of power law and Carreau–Yasuda coefficients

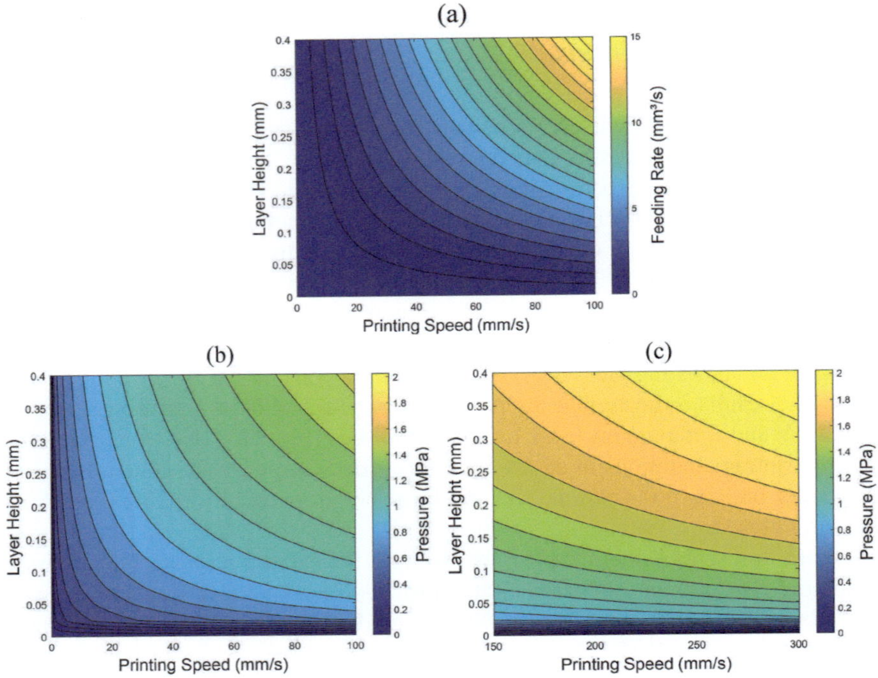

Fig. 6 Analytical model estimations: **a** Flow rate and extrusion speed. **b** Pressure at lower speeds. **c** Pressure at higher speeds

3.3 Mechanical Characterization of the Springs

Tensile tests were conducted to obtain springs stiffness constants. The commercial springs were made of AISI 302 stainless steel. Assuming Hooke's law and neglecting frictional forces, stiffness constants of 0.36 N/mm and 1.07 N/mm were determined from Fig. 8 for reference springs 599 and 565, respectively. Considering that the springs would be heated to a temperature close to that of the hot-end, 200 °C, the steel stiffness was reduced by 25% (Antoun 2004) for the evaluation of experimental pressure development.

The displacements estimated using the image analysis system showed a standard deviation of 5% for spring 565 and 10% for spring 599. Given the difference in stiffness between the two springs, spring 599 was used for tests at lower extrusion speeds (50–100 mm/min), while spring 565 was employed for tests at higher extrusion speeds (150–250 mm/min).

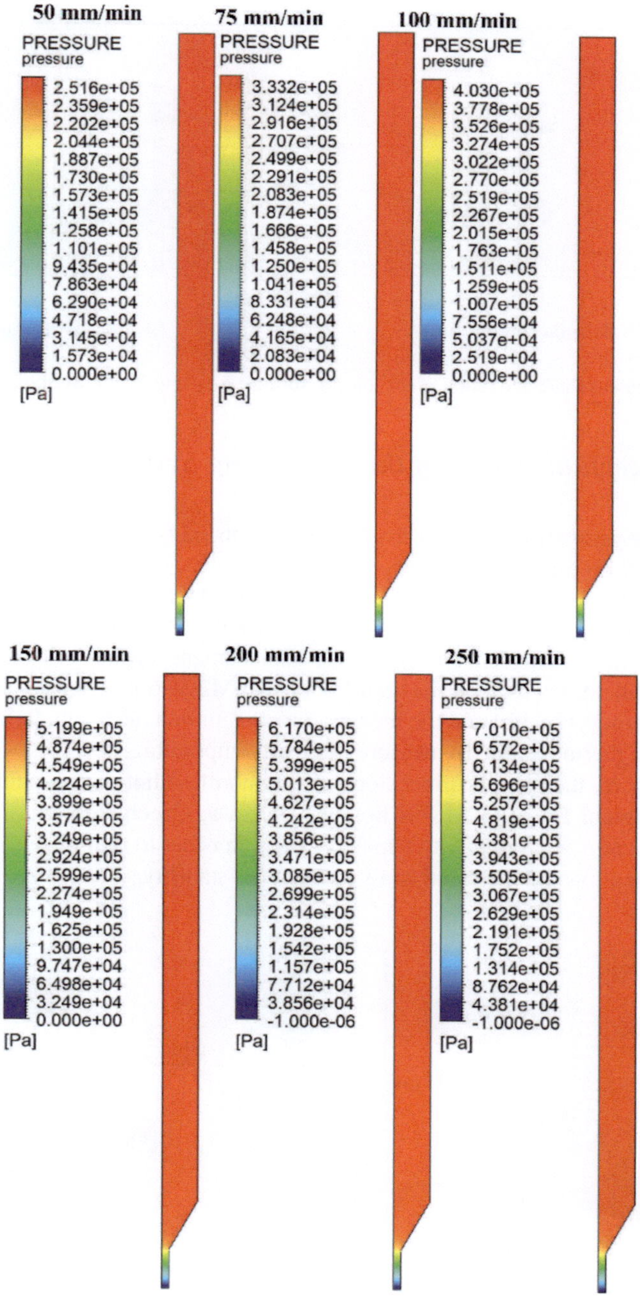

Fig. 7 Numerical pressure estimates for extrusion speeds from 50 mm/min to 250 mm/min

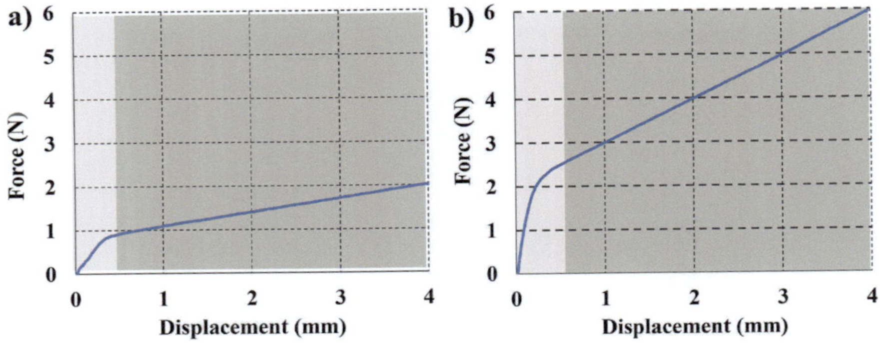

Fig. 8 Mechanical characterization of the tension springs: **a** Spring 599; **b** Spring 565

3.4 Conceptual Design of the Monitoring System

In Fig. 9, the geometric model of the FFF 3D printhead (specifically in the hot-end) is presented with implemented design changes.

A designated region for force transducer placement was incorporated into the hot-end's heating block, and a custom-designed nozzle was developed to enable both the mounting of the transducers and sufficient displacement for the springs.

Key modifications include the addition of two M2 × 6 mm screws (Fig. 9c) for spring mounting (Fig. 9b) and the inclusion of an alumina tube (Fig. 9a), which will later house a thermocouple to measure the melt temperature.

Furthermore, the overall dimensions of the modified hot-end remain consistent with the original Prusa i3 MK3S hot-end (Prusa®, Czech Republic), measuring 23 mm × 62 mm × 23 mm with a mass of 68 g, in order to minimize the impact of these modifications on the printhead's mechanical stability.

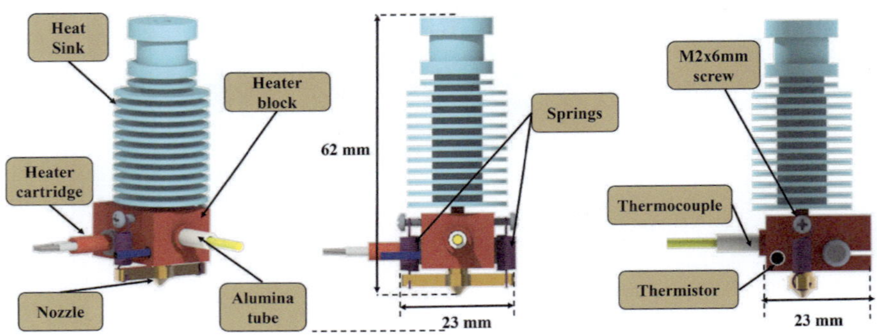

Fig. 9 Geometric modeling of the modified hot-end

3.5 Exploratory Prototype and Functional Testing

Figure 10 shows the prototype assembled for the functional tests carried out to support the conceptual validation of the developed monitoring system.

The heating block and the modified nozzle were manufactured using ASME C22000 copper alloy (Fig. 10a). The results of the functional tests, compared with the simulation outcomes, are presented in Fig. 11.

Experimental pressure values ranged from 1.27 MPa to 3.32 MPa for extrusion speeds between 50 and 250 mm/min, consistently exceeding the analytical and numerical predictions. Moreover, an exponential rise in pressure was noted at speeds above 200 mm/min, a trend also observed by Serdeczny et al. (2020) and Kattinger et al. (2023). According to Kattinger et al. (2023), the discrepancies at higher extrusion speeds mainly arise because thermal effects are not sufficiently captured by one-dimensional, isothermal models. Additionally, aside from thermomechanical factors, there are inherent characteristics of the FFF process that conventional pseudoplastic flow models fail to account for.

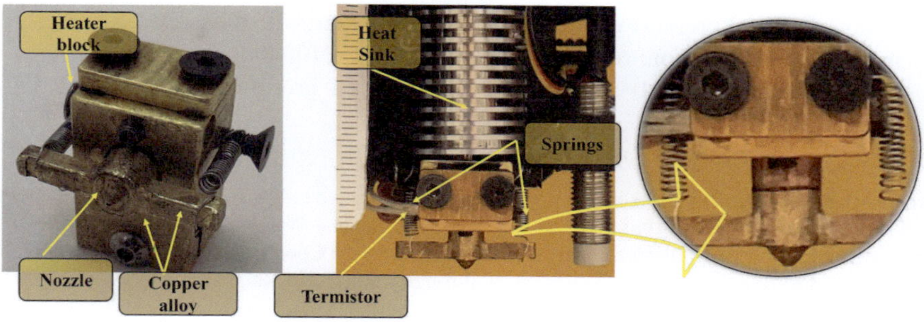

Fig. 10 Prototype built to validate the concept

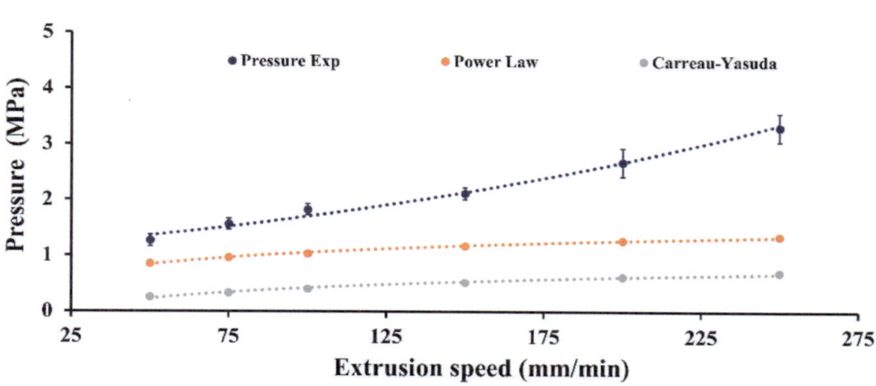

Fig. 11 Experimental pressure data for different extrusion speeds

4 Preliminary Conclusions

Based on the partial results presented, the following considerations can be made:

- A low-cost prototype (estimated cost of €80) for pressure monitoring at the nozzle of an additive manufacturing system was developed, and preliminary functional tests were conducted to validate the concept.
- Two analytical-experimental pseudoplasticity models (Power Law and Carreau–Yasuda) were implemented. The pressure simulation results from the Power Law model showed greater similarity with the experimental data; however, to better capture the material behavior at high shear rates, a capillary rheology study should be conducted.
- Experimental pressure ranged between 1.27 MPa and 3.32 MPa, exhibiting behavior consistent with values reported in the literature for commercial PLA filaments (Serdeczny et al., 2020; Kattinger et al., 2023);
- Springs with different stiffness values can be mounted to adjust the mechanical system's sensitivity.
- As a next step, the experimental results of this low-cost modular monitoring system for local pressure estimation will be compared with those from a dedicated modular pressure monitoring system employing a 4004A pressure microsensor (Kistler®, Switzerland) acquired by the research group (Lab3D-SEM-EESC-USP).

Acknowledgements This work was supported by the Associated Laboratory for Energy, Transport and Aeronautics (LAETA), University of Porto; the Mechanical Workshop of the Department of Mechanical Engineering (DEMec), University of Porto; and the Coordination for the Improvement of Higher Education Personnel (CAPES-Print), through scholarship No. 88887.916574/2023-00.

Responsibility for the Information The authors are fully responsible for the technical and scientific content of the information presented in this work.

This manuscript includes translated content generated with the assistance of ChatGPT (OpenAI, GPT-4, August 2025 version), an AI language model. The authors are responsible for verifying the accuracy of the final text.

References

Antoun BR (2004) Temperature effects on the mechanical properties of annealed and HERF 304L stainless steel. United States report. UNT Digital Library. https://digital.library.unt.edu/ark:/67531/metadc879331/0

Bellini A, Shor L, Guceri SI (2005) New developments in fused deposition modelling of ceramics. Rapid Prototyp J 11(4):214–220

Bose S et al (2018) Additive manufacturing of biomaterials. Prog Mater Sci 93:45–111

Bretas RES, D'ávila (2005) Reologia de polímeros fundidos. 2. ed. São Carlos: Edufscar

Coogan et al (2019) In-line rheological monitoring of fused deposition modelling. J Rheol 63:141–155

Chung PL et al (2024) Investigation of the melt and flow behaviour in the hot-end of a 3D printer using X-ray computed tomography. In: 13th Conference on industrial computed tomography (iCT 2024), Wels, Austria

Crump SS Apparatus and method for creating three-dimensional objects. US5121329. Filled 30 Oct. 1989. Granted 9 Jun 1992

Gardan J (2018) Smart materials in additive manufacturing: state of the art and trends. Virtual Phys Prototyp 14(1):1–18

Gibson I et al (2021) Additive manufacturing technologies, 3rd edn. Springer Nature, Switzerland

Go J, Schiffres SN, Stevens AG, Hart AJ (2017) Rate limits of additive manufacturing by fused filament fabrication and guidelines for high-throughput system design. Addit Manuf 16:1–11

Gonzalez Gutierrez J et al (2018) Additive manufacturing of metallic and ceramic components by material extrusion of highly filled polymers: a review and future perspectives. Materials 11(5):840

International Organization for Standardization (2011) ISO 376:2011: Metallic materials—Calibration of force-proving instruments used for the verification of uniaxial testing machines. ISO, Geneva, Switzerland

Kattinger J et al (2023) Analysis of melting and flow in the hot-end of a material extrusion 3D printer using X-ray computed tomography. Addit Manuf 76(25):103762

Khalid MY, Arif ZU, Ahmed W (2022) 4D printing: technological and manufacturing renaissance. Macromol Mater Eng 307(8):2200003

Michaeli W (2003) Extrusion dies for plastics and rubber, 3rd edn. Hanser Publications, Munich

Nzebuka CAC et al (2022) Numerical modeling of the effect of nozzle diameter and heat flux on the polymer flow in fused filament fabrication. J Manuf Process 82:585–600. https://doi.org/10.1016/j.jmapro.2022.08.029

Osswald TA et al (2018) Fused filament fabrication melting model. Addit Manuf 22:51–59

Phan DD et al (2018) Rheological and heat transfer effects in fused filament fabrication. J Rheol 62:1097–1107

Rauwendaal C (2014) Polymer extrusion, 5th edn. Carl Hanser Verlag, Munich

Read JR et al (2024) Online measurement for parameter discovery in fused filament fabrication. Integr Mater Manuf Innov 13(3):541–554

Roberson D et al (2014) Expanding the applicability of FDM-type technologies through materials development. Rapid Prototyp J 21(2):137–143

Serdeczny MP et al (2020) Numerical modeling of the polymer flow through the hot-end in filament-based material extrusion additive manufacturing. Addit Manuf 36:101454

Thompson MK et al (2016) Design for additive manufacturing: trends, opportunities, considerations and constraints. CIRP Annnals—Manufacturing Technol 65(2):737–760

Torrad, AR et al (2015) Characterizing the effect of additives to ABS on the mechanical property anisotropy of specimens fabricated by material extrusion 3d printing. Addit Manuf 6:16–29 ISSN 2214-8604

Application of Topological Optimization Procedure in Mechanical Soft Tooling Design

Marcos Vasques Pinho Cruz, Pedro Guilherme Carvalho de Souza Marconi, Geraldo José Belmonte dos Santos, Valter Estevão Beal, Cristiano Vasconcellos Ferreira, and Armando Sá Ribeiro Júnior

Abstract Plastic parts are widely used in industry due to their low weight, reduced cost and well-balanced mechanical behavior. In this context, soft tooling has emerged as an alternative, particularly for low-volume production. However, the lower thermal conductivity and mechanical strength of polymers used in 3d printers leads to challenges regarding their durability and long-term performance. To overcome the problems associated with their low mechanical strength, shell molds are usually filled with materials like epoxy resins or low-melting-point metal alloys. However, introducing these materials can cause deformations in the molds during the backfilling process, compromising the quality of the resulting plastic parts. Mold defects are often replicated in the final parts, posing a significant challenge to achieving high precision and quality. One solution to the problems caused by the backfilling process is to mechanically design the mold to withstand the pressures and forces typical of injection molding, incorporating reinforcing elements such as ribs. In this context, this study proposes an innovative approach that uses topological optimization for the application of ribs in polymeric molds, ensuring structural integrity without the need for filling materials. The aim is to achieve sufficient mechanical rigidity to eliminate dependence on filler materials, ensuring mold integrity and part quality. The developed method was compared with traditional techniques widely used in the automotive sector and also considering previous studies about the usage of additive manufacturing on soft tooling. The numerical application ensured the designed ribs

M. V. P. Cruz (✉) · P. G. C. de Souza Marconi · A. S. R. Júnior
Federal University of Bahia, Salvador, BA, Brazil
e-mail: vasquescruz@gmail.com

G. J. B. dos Santos
State University of Feira de Santana, Feira de Santana, BA, Brazil

V. E. Beal
SENAI CIMATEC University, Salvador, BA, Brazil

C. V. Ferreira
Federal University of Santa Catarina, Joinville, SC, Brazil

V. E. Beal and A. S. Ribeiro Junior (eds.), *Advances in Manufacturing Engineering Research*, Lecture Notes in Mechanical Engineering,
https://doi.org/10.1007/978-3-032-09858-0_2

met the established criteria, with precise control over parameters like height and volume. The study used polyamide (PA12) inserts manufactured via the Multi Jet Fusion process. The results indicated a reduction of up to 74% in the volume of the ribs compared to traditional techniques, eliminating the need for filler material. This approach improves the process efficiency and eliminates waste. Additionally, the method allows less experienced designers to achieve positive and optimized outcomes.

Keywords Shell molds · Soft tooling · Topology optimization · Multi jet fusion · Injection mold

1 Introduction

Plastic parts are essential in industries that require high production volumes. The increasing demand for them across various sectors has driven the need for increasingly efficient and cost-effective manufacturing methods. When it comes to high-volume plastic parts production, the injection molding process (IMP) is highly relevant, accounting for 32% of the total plastic product output (Shoemaker 2006). Traditionally in IMP, metal injection molds have been the primary option for manufacturing these parts. However, the cost of manufacturing a mold can be high, making many projects unfeasible, particularly those involving low-volume production. Polymer molds manufactured via 3D printing have emerged as an alternative for these cases, as their production can be carried out through different processes, and their cost is usually relatively low compared to steel or even aluminum molds (Fernandez et al. 2024; Ferreira and Mateus 2003). However, the development of polymer molds presents challenges related to maintaining their integrity in terms of mechanical properties and dimensional control due to the complexity of the loads associated with the process.

Polymer molds are typically manufactured in a shell form, followed by an internal filling stage with another material, which could be a low-melting-point metal alloy or epoxy resin, providing greater rigidity to the mold shell structure (Rahmati and Dickens 2007; Malloy 2010). However, the introduction of these materials can cause deformations in the molds during the filling process, compromising the quality of the produced plastic parts. Defects in the mold tend to reproduce in the final parts, which represents a significant obstacle to producing items with high precision and quality. One solution to mitigate the problems arising from the filling process is to design the mold mechanically in such a way that it can withstand the pressures and forces typical of the injection molding process, incorporating reinforcing elements like ribs. The application of ribs is a common technique to increase the rigidity and strength of plastic structures, but the traditional method of applying them relies entirely on the designer's experience, often leading to approaches that may be unsuitable for meeting the project's needs and criteria.

Regarding the use of structural reinforcements in molds, some studies explore different approaches and methods to optimize process performance. Li et al. (2017) proposed a method for designing the rib layout in shell-type structures based solely on the directions of principal stresses, improving mechanical behavior. Yang et al. (1996) investigated the effects of ribs in gas-assisted injection molding of plastic molds, highlighting the importance of rib width and the use of circular fillets at their base. Tanaka (2017) explored the effects of combining compression molding and injection molding on the mechanical properties of a ribbed shell-type plastic part. Regarding the use of topology optimization in manufacturing, Leal Pereira et al. (2017) optimized a mold made of 316L stainless steel to reduce weight and manufacturing time. The results showed a 41.1% weight reduction in the mold. Sinico et al. (2019) applied topology optimization to a metallic mold insert manufactured via Laser Powder Bed Fusion (LPBF), using an algorithm that accounts for the process's thermal aspects. The results showed a significant mass reduction of 50% and a 43% decrease in production time in the analyzed case study.

Considering this, the use of new technologies related to numerical simulations, such as topological optimization, emerges as a promising solution to address these challenges. Topological optimization is an advanced technique that allows designers to create structures with a more efficient material distribution, typically by reducing material usage. In this work, topological optimization will have an unconventional approach, providing directional information to the designer to help identify where ribs need to be applied and define their profile, thus enhancing the mechanical performance of the component.

Thus, this work proposes a method for designing ribs in polymer molds manufactured through 3D printing using topological optimization. By applying the proposed method, ribs can be created that ensure the necessary mechanical behavior without the need for internal filling. The importance of this research lies in the lack of literature on potential new uses of topological optimization aimed at increasing structural rigidity rather than just material reduction. Furthermore, the study will also allow a comparison of its effectiveness against traditional methods, contributing to the advancement of best design practices in the industry.

2 Materials and Methods

The methodology used in the development of this work initially involved conducting a literature review on polymer molds and optimization methods. Based on this study, a proposed method was developed and presented to guide the design of polymer molds manufactured by 3D printing, using topological optimization software. The proposed method was applied to the design of a mold that will be manufactured from polyamide (PA12) using the Multi Jet Fusion (MJF) process. The chosen geometry for the case study was an automotive fuel tank cap. Based on the results of the case study, the method was refined to provide better guidance in the mold design process.

To assess the impact of using the proposed method, two professionals with distinct experiences were asked to design the same mold using the traditional method. In a way to evaluate the method efficiency some comparison criteria were considered as time to develop each proposal, material savings, and process cost.

2.1 Proposed Procedure for the Optimized Molds Design

The steps of the proposed method in this work can be observed through the flowchart illustrated in Fig. 1, to demonstrate the evaluation criteria and possible procedure based on the results found. The description of each step is presented below.

The first step is to create a CAD model of the mold to be optimized, define the material with the main mechanical properties required by the software, and perform the calculation of the available volume for reinforcement. The available volume

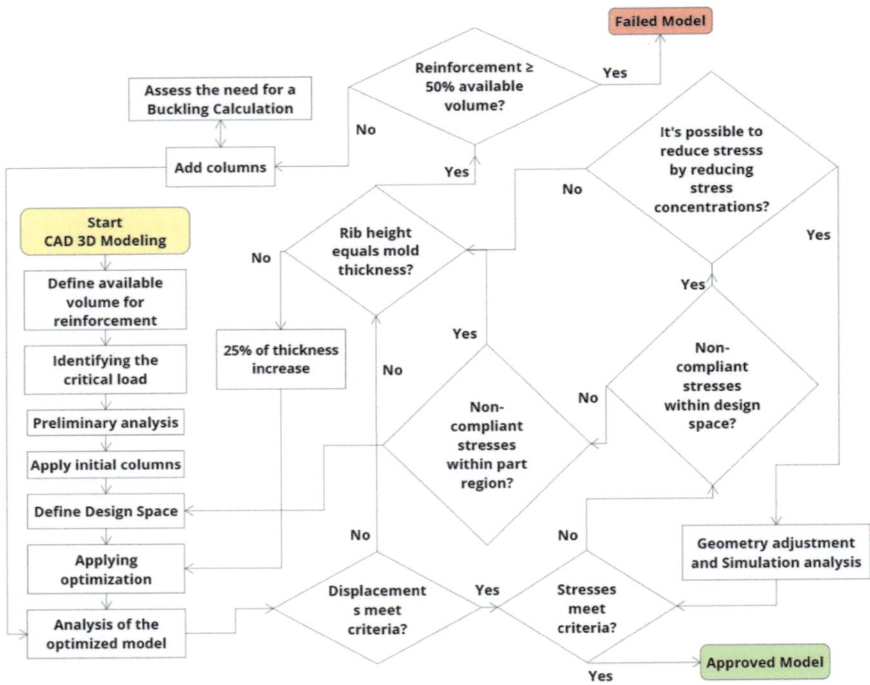

Fig. 1 Flowchart of the proposed method

(AV) should be the entire internal volume of the mold for a condition of maximum rib height, where the maximum height should be equal to the thickness of the mold shell.

Critical loads must be identified and defined, along with the appropriate regions for their application and the boundary conditions associated with the structure. Then, a preliminary analysis of the model should be performed, establishing the starting point to be used as a comparison with subsequent analyses. The values for maximum displacement and stresses should be recorded.

Once an initial analysis result of the structure in its rib-free condition is obtained, a column structure should be applied in regions where the greatest displacements are found. The upper face of these columns should make contact with the cavity holder, thus having movement restrictions associated with them. The column region that will be contained within the available volume should occupy between 5 and 10% of it. It is recommended that the column profile be circular, however the designer should evaluate and decide based on the complexity of the surface to which it will be applied.

2.2 *Optimization and Implementation of the Result*

The following parameters must be defined for the creation of ribs in the optimization objective configuration in the software: the percentage of the total volume to be preserved (between 40 and 45% of the design space), maximum and minimum element thickness, and symmetry conditions (if applicable, as determined by the designer).

In the first optimization step, an initial height for the optimizable volume corresponding to 25% of the mold shell thickness should be considered. Finally, topology optimization is performed. Once the optimization is completed, the resulting design must be implemented in the mold geometry using the chosen CAD modeling software.

The new optimized mold from this first iteration should be evaluated through a computational simulation to obtain the parameter values and compare them to the model approval criteria. Following the proposed methodology, the maximum displacement must be evaluated first; it must not exceed 15% of the injected part thickness or a project-specific criterion. If the displacement meets the approval criteria, the next evaluation should focus on the adopted stress criterion. If the maximum displacement does not meet the established criterion, an increase in the rib height within the optimizable area should be considered. The proposed method recommends incrementing the rib height by 25% in each cycle, with the height limit being equal to the shell thickness.

For each new height increment, the regions with the highest stresses must be evaluated. If these occur outside the design space, three possible actions may follow:

1. If the highest stresses occur in regions surrounding the design space, the design space should be expanded within its limitations to include these regions.
2. If the highest stresses are due to stress concentrators, these should be minimized with localized adjustments to the optimized geometry, followed by a new analysis.
3. If the highest stresses occur in the injected part region, the rib height limit should be increased.

When the rib height limit is reached, it must be assessed whether the rib volume has reached 50% of the available volume. If this limit has not yet been reached, information from the regions with the highest displacements identified in the previous analysis can be used to add new columns, aiming to minimize their number. This cycle should be repeated until all criteria are met or until the reinforcement volume reaches 50% of the available volume after each new column addition. If the criteria are not met and the volume limit is reached, the model is rejected.

3 Case Study

3.1 Mold Geometry, Dimensional Evaluation, and Boundary and Loading Conditions

As a starting point, this study used a simplified mold model provided by the DEMIBAV project—Demonstrator of Molds for Low-Volume Plastic Part Injection. The following geometries were included: fixed plate (female), movable plate (male), and ejector pins, illustrated in Fig. 2a, b and c, d, respectively.

The total dimensions of the simplified model assembly (fixed plate plus movable plate) are approximately 88 mm × 180.47 mm × 260 mm. The fixed and movable plates share the same width (180.47 mm) and length (260 mm) but differ in height, with the fixed plate being 53 mm high and the movable plate 34.5 mm high. The plastic part produced with this mold has a nominal thickness of 2 mm.

In the polymeric model, a shell thickness of approximately 5 mm was considered. Since both plates were to be fabricated as shells, it was necessary to extract the visible surfaces of the fixed and movable plates. Subsequently, a "thickening" process was performed on both plates to ensure that their cross-sectional area remained as constant as possible, as shown in Fig. 3a, b and c, d.

Polyamide 12 (PA12) was selected due to its high mechanical strength, thermal stability and compatibility with the Multi Jet Fusion additive manufacturing process, making it suitable to withstand the conditions of the injection process. PA12 is widely used in additive manufacturing for applications requiring good mechanical performance. The choice of this material is justified by the Multi Jet Fusion (MJF) manufacturing process adopted in the DEMIBAV project. Based on its thermal properties, it was determined that PA12 has a melting temperature of approximately 187 °C, making it suitable for withstanding the thermal and mechanical demands of the injection process. Some properties of PA12 are presented in Table 1.

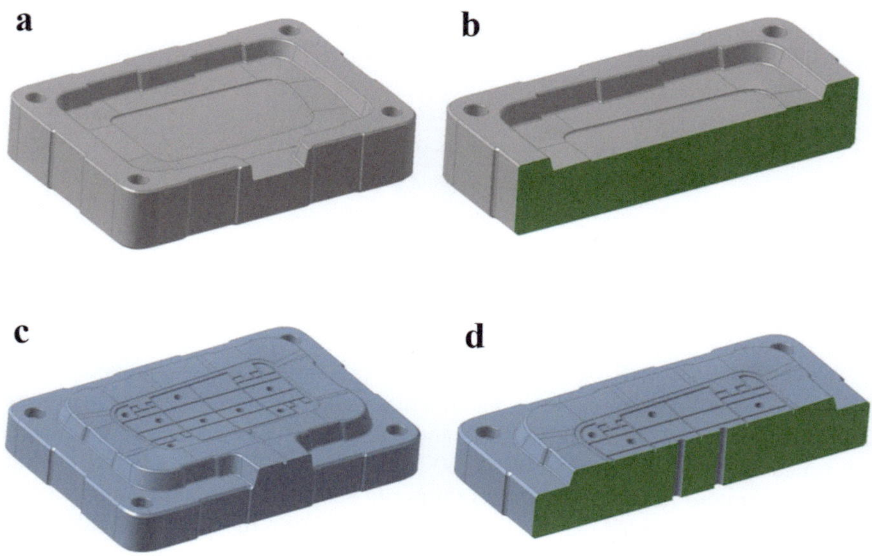

Fig. 2 Geometry of the fixed plate (female) and movable plate (male)

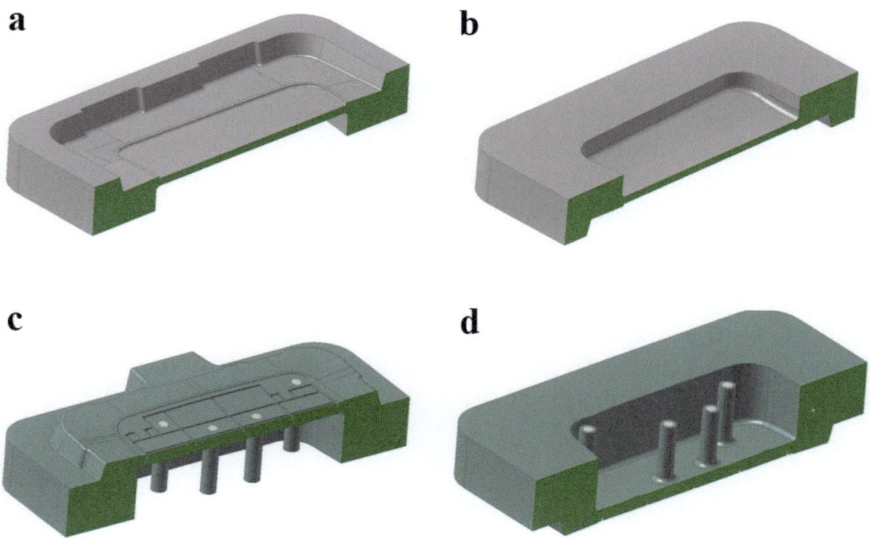

Fig. 3 Cross-section of the fixed plate (**a**, **b**), movable plate, and ejector pins (**c**, **d**)

Table 1 Properties of polyamide (PA12)

Properties	Values
Density (g/cm^3)	1.01
Tensile strength (MPa)—average	50
Tensile modulus (MPa)—average	1900
Elongation at yield (%)	10
Powder melting point (DSC) (°C)	187
Heat deflection temperature [@ 0.45 MPa, 66 psi] (°C)	175
Allowable stress (MPa)	16

Through the DEMIBAV project, data on polyamide were obtained, where polyamide (PA12) specimens were manufactured using the Multi Jet Fusion process with different layer orientations (25° and 90°). Tensile mechanical tests were performed at three temperature levels: 60 °C, 120 °C, and 150 °C, corresponding to the temperature range of the inserts during injection. Since numerical (rheological) simulations indicated that the maximum temperatures in the insert reach values below 120 °C, the data related to this test were analyzed to define the stress limits. The tests on specimens at 120 °C, with a layer orientation of 25°, indicated yield stresses of around 25 MPa. By adopting the project safety factor of 1.5, the allowable stress used for evaluating the inserts will be 16 MPa.

The loads considered in the analysis are associated with the injection molding process, and the data used were obtained through numerical rheological simulations using MoldFlow software. The model considered a static structural scenario, with the application of a mold clamping force of 52.6 kN, with the application region shown in Fig. 4a and c, and an internal pressure of 2.9 MPa on the internal walls of the mold cavity, with the application region shown in Fig. 4b and d.

For the boundary condition of the analysis, it is necessary to consider in the fixed plate the contact region between its structure and the cavity plate face, where the fixation between both restricts the relative degrees of freedom between them. Similarly, for the movable plate, in addition to the contact surface of the plate, there are also the ejector pins and their respective guides, which have restricted degrees of freedom on the faces that touch the cavity plate, as observed in Fig. 5.

3.2 Results of Topological Optimization

To enable comparison of the effects from the optimization method, an initial analysis was performed to check the stresses and deformations present in the current structure before the optimization. Table 2 shows the results obtained for the fixed plate and movable plate without ribs.

With this initial study, it was possible to infer some information about the results observed for the movable plate. Evaluating the stress results, the values are well above

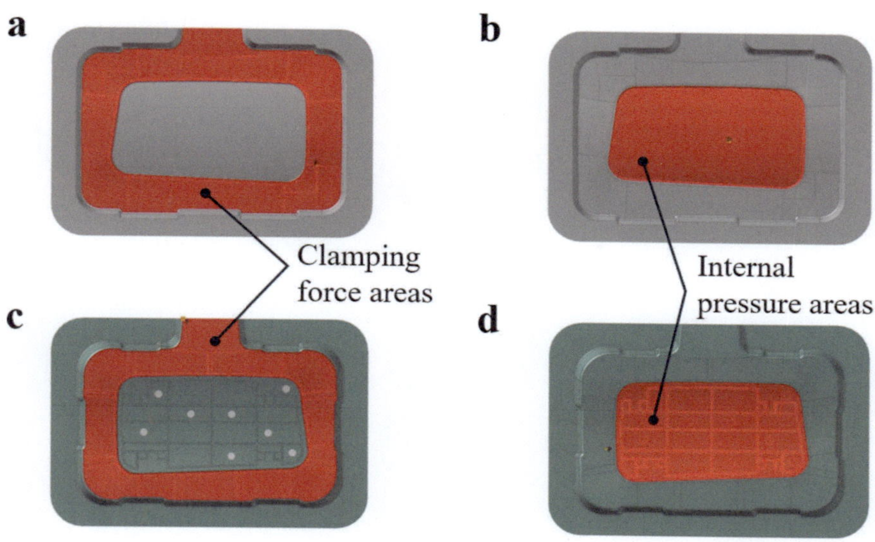

Fig. 4 Area of application of clamping force (**a** and **c**) and area of application of internal pressure (**b** and **d**)

Fig. 5 Support regions and degrees of freedom of the mold

Table 2 Results of the analyses in the initial reference model for the fixed plate

Geometry	Displacement (mm)	Maximum stress (MPa)
Fixed plate without ribs	9.2	(−) 32/183
Movable plate without ribs	1.82	(−) 38/82

the material's allowable stress (16 MPa), indicating mold failure. The displacement values obtained for both plates also exceeded the limit, considering this as 15% of the part's thickness (approximately 0.3 mm). It should be noted that the magnitudes of the values found in this first analysis were not considered, since the conditions to adopt the linear elastic analysis and small displacements were not met. However, the analysis clearly indicated the need for structural reinforcement.

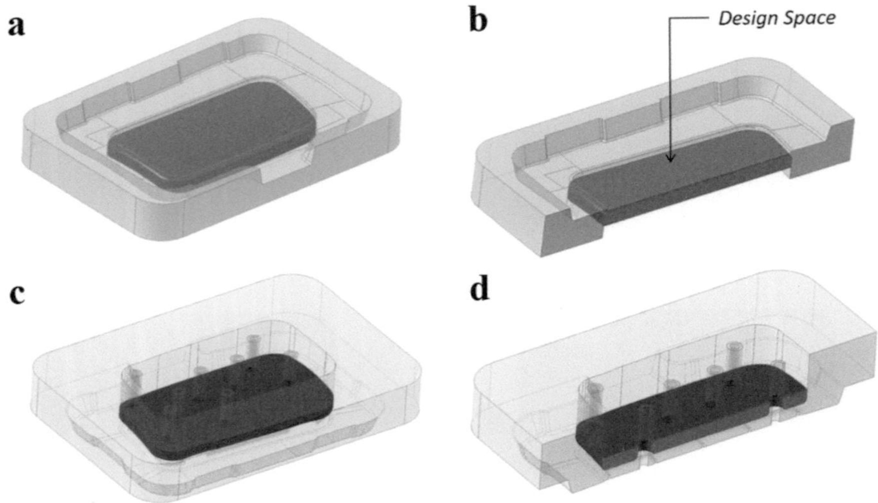

Fig. 6 Optimizable area of the fixed (**a**, **b**) and movable (**c**, **d**) plates

For the topology optimization, the first step was to define the optimizable area for both the fixed plate and the movable plate. In this case, when working with solids, a projected volume of the internal surface of the mold is created, as shown in Fig. 6.

To begin the topological optimization, the objective function must be established. In this case, the goal was to maximize rigidity in order to minimize deformations under the structure given the loads to be considered.

The rib height limit was defined as the thickness of the mold shell, 5 mm, and the maximum displacement criterion limit was set to 0.3 mm. Increments of 25% were performed with each increase in height of the design space. The first analysis was established at 1.25 mm in height, and successively until the height limit of 5 mm for the rib was reached.

Initially, the available volume (VD) was also identified. For a given maximum rib height of 5 mm, the VD found was 58,220 mm^3. The initial analysis also identified the regions of highest displacements to be considered in the subsequent step of adding columns.

The first columns were added as the starting point. The volume occupied by these columns was 3533.86 mm^3, corresponding to 6.06% of the available volume.

The optimization and analysis for the first cycle were performed using a height of 1.25 mm; however, it did not result in a maximum displacement value that could meet the established criterion. Therefore, the process was repeated for the subsequent four cycles until the height limit of 5 mm was reached. Table 3 lists the results of the analyses for the successive heights (1.25, 2.5, 3.75, and 5 mm) until the height limit was reached.

Table 3 Results from fixed plate

Cycle	Height	Displacement (mm)	Principal stress (MPa)
(1)	Without ribs	9.2	(−) 32/183
(2)	1.25 mm	1.25	(−) 16.9/37.3
(3)	2.5 mm	1.07	(−) 13/27.2
(4)	3.75 mm	0.93	(−) 10.9/21.7
(5)	5 mm	0.83	(−) 14/20.2

Upon reaching the maximum height limit for the rib, it was necessary to verify the reinforcement volume obtained and compare it with the total available volume for ribs, determining the capacity available for adding structural columns.

The geometry used to verify the volume was obtained in the fifth optimization cycle (Table 3). The additional columns followed the information obtained through the analysis of regions with the highest displacements in that same cycle.

For this configuration presented, the applied reinforcement volume was 22,279 mm^3, corresponding to 38.26% of the available volume. It is also noted that the volume of columns applied at the beginning of the study was 3533.86 mm^3, corresponding to 6.06% of the available volume, resulting in a total of 44.32% of the available volume. In this way, it was still possible to add columns in regions of the highest displacements. Figure 7 shows the optimization result with 5 mm rib height for the final layout.

After adding the new columns, totaling a volume of 29,101 mm^3 (49.9% of the available volume), both criteria were met in cycle 6 of the optimization for the fixed plate, where a maximum displacement of 0.304 mm was obtained, along with principal stresses of (−) 5.1 MPa and 8.59 MPa, as can be observed in Fig. 8a and b, respectively.

In a similar manner to the steps taken for the fixed plate, the available volume for the movable plate was also identified, where for a given maximum height of 5 mm, the VD found was 53,250 mm^3. The regions of maximum displacements were also identified. In this way, the first columns were added as the starting point. The volume

Fig. 7 Isometric view of the optimized mold (fixed plate and movable plate)

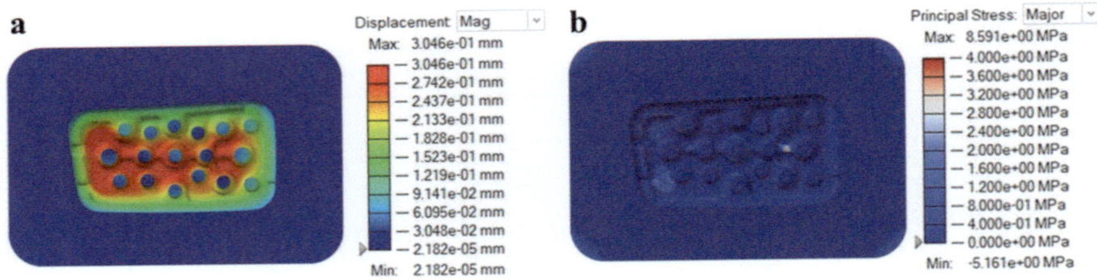

Fig. 8 Results of displacement and the principal stresses from optimization cycle 6 (fixed plate)

Table 4 Results from movable plate

Cycle	Height	Displacement (mm)	Principal stress (MPa)
(1)	Without ribs	1.82	(−) 38/82
(2)	1.25 mm	0.75	(−) 17.1/28.5
(3)	2.5 mm	0.36	(−) 22.9/75.7
(4)	3.75 mm	0.72	(−) 18.9/27.9
(5)	5 mm	0.57	(−) 17.1/19.8

occupied by these was 2641.41 mm^3, which corresponds to 4.9% of the available volume.

The values obtained from this first analysis did not meet both established criteria, and thus, subsequent cycles (for heights of 2.5, 3.75, and 5 mm) were carried out and evaluated based on the flowchart proposed by the method. Afterward, the height limit of 5 mm was reached, still not meeting the established criteria, as can be observed in Table 4.

After reaching the height limit, it was necessary to verify the reinforcement volume and compare it with the total available volume for ribs. The applied reinforcement volume was 19,320 mm^3, equivalent to 36.2% of the available volume, and the value for the initial columns was found to be 2641.41 mm^3, corresponding to 4.9% of the available volume. After summing both values, a total of 41.8% was found, which allowed the addition of new columns.

After adding the new columns, totaling a volume of 26,073 mm^3 (48.96% of the available volume), both criteria were met in round 6, where a maximum displacement of 0.308 mm and principal stresses of (−) 11.0 MPa and 10.0 MPa were obtained, as can be observed below in Fig. 9a and b, respectively.

3.3 Results of the Traditional Method

To perform a comparison of the proposed method with the traditional rib application method, the collaboration of two experienced designers in plastic part design for the automotive industry was obtained. Designer 1 has 21 years of experience and

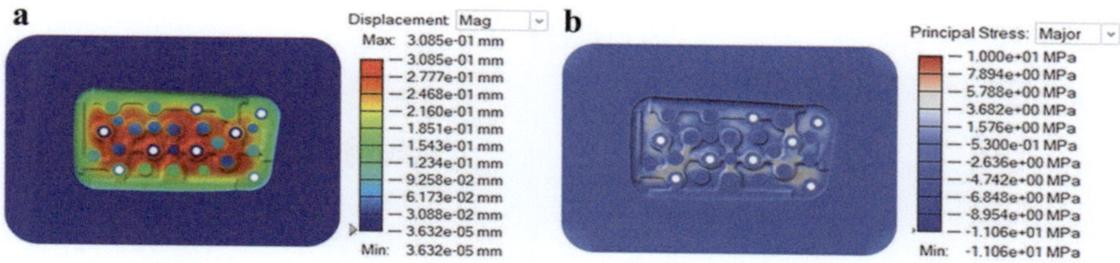

Fig. 9 Results of displacement and the principal stresses from the optimization cycle 6 (movable plate)

Designer 2 has 32 years of experience both with extensive expertise in plastic part design for the automotive sector.

The CAD geometry used as the starting point in the study of the method (without ribs) was provided, and each of the designers was asked to apply ribs according to their respective experiences. A material with details on fixations, boundary conditions, as well as manufacturing process information and the acting loads was sent. No information related to studies conducted through the proposed method in this work was provided. With this, modification and verification cycles began, where the geometry with ribs was sent for evaluation through numerical simulation, followed by a subsequent submission of the result report to the designer.

In this stage, the following parameters were obtained: Effective development time (thinking and applying the solution), total volume of ribs, and total correction cycles.

3.3.1 Results from Designer 1

The results for the evaluation of the fixed plate can be seen next in Table 5 and Fig. 10. In Table 5, the values of displacements and stresses obtained for each of the proposed models and corrections made with the designer can be verified. Similarly, the results for the movable plate can be found in Table 6.

Thus, after two attempts, sufficient values were obtained to meet the established criteria for both plates.

Table 5 Results from designer 1 for the fixed plate

Model	Displacement (mm)	Principal stress (MPa)	Volume of ribs (mm^3)
Without ribs	9.2	(−) 32/183.0	0
Model 1	6.5	(−) 152.6/199.3	19,820
Model 2	0.23	(−) 9.12/10.2	90,100

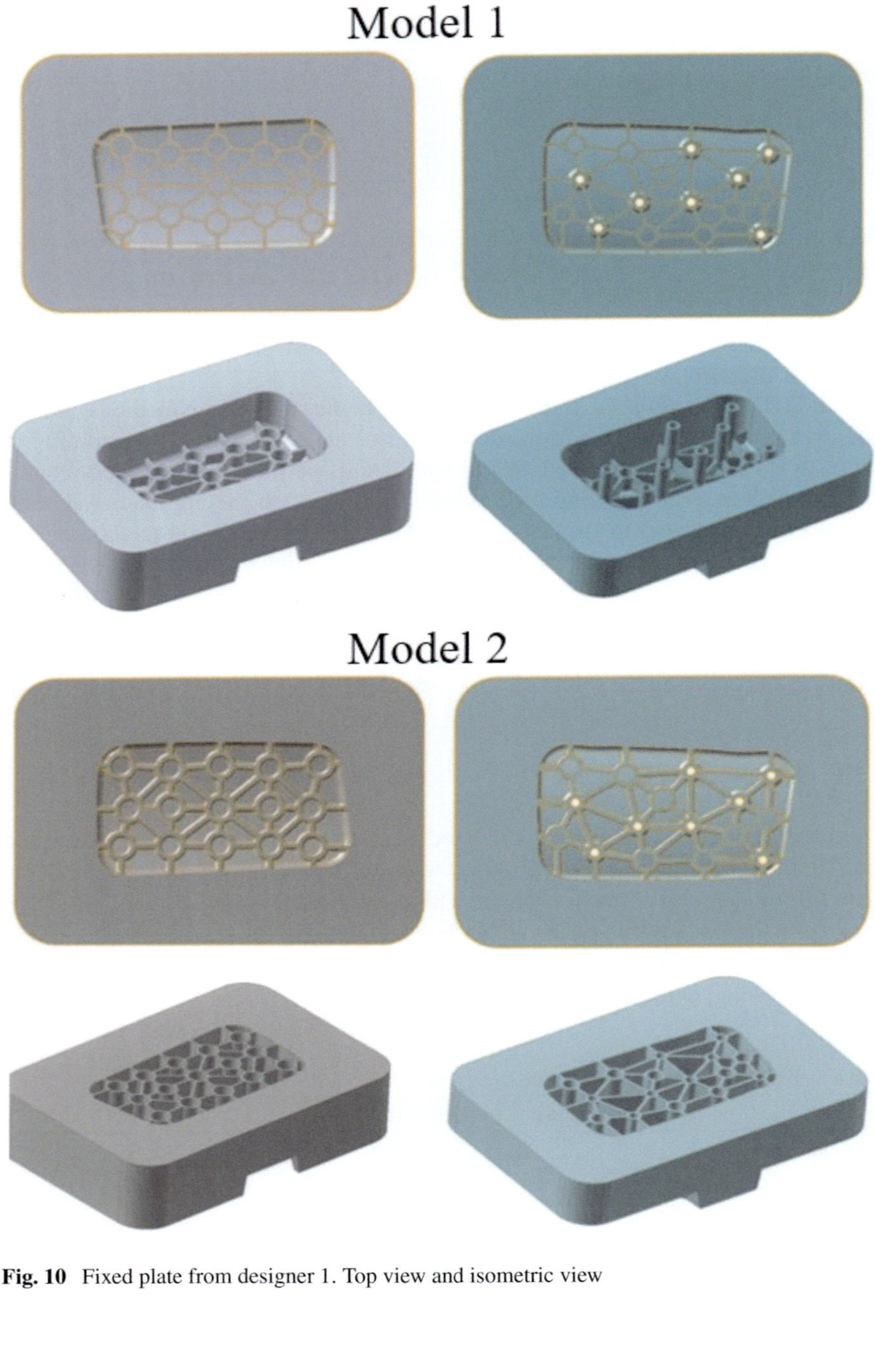

Fig. 10 Fixed plate from designer 1. Top view and isometric view

Table 6 Results from designer 1 for the movable plate

Model	Displacement (mm)	Principal stress (MPa)	Volume of ribs (mm^3)
Without ribs	1.82	(−) 38/82	0
Model 1	1.2	(−) 40.2/52.8	22,070
Model 2	0.25	(−) 5.1/14.1	84,450

3.3.2 Results from Designer 2

The results for the evaluation of the fixed plate can be seen in Table 7 and Fig. 11. Additionally, Table 8 presents the results for the movable plate. The tables results show the two attempts made by the Designer 2 on reducing the maximum displacement and principal maximum stress with the minimum material volume added using reinforcement ribs design approach.

The results regarding effective time, number of attempts, and the volume of reinforcements used for the three evaluated approaches are presented in Table 9.

During the evaluation step of the traditional process performed by Designer 1, it was observed that their starting point was more conservative in terms of rib height. Based on their assessment of potential regions of maximum displacements, honeycomb-type rib profiles were adopted. However, not meeting the established criteria led them to seek a maximum rib height condition, which resulted in a significant increase in material volume, thus ensuring compliance with the criterion.

The second evaluation of the traditional process was carried out by Designer 2, who had a different approach compared to Designer 1 in terms of their starting point. Rib profiles of maximum height were used as the initial proposal, but with large gaps between them. After the result from the first attempt, ribs were added to the most stressed regions and those with greater displacements, ensuring the established criteria were met.

It can be observed that the common point between the final geometries of both designers is the use of the maximum rib height, where the cavity plate region was used as the support point.

Table 7 Results of the analyses performed by Designer 2 for the fixed plate

Model	Displacement (mm)	Principal stress (MPa)	Volume of ribs (mm^3)
Without ribs	9.2	(−) 32/183.0	0
Model 1	0.43	(−) 26.7/13.4	78,790
Model 2	0.29	(−) 13.4/10	112,700

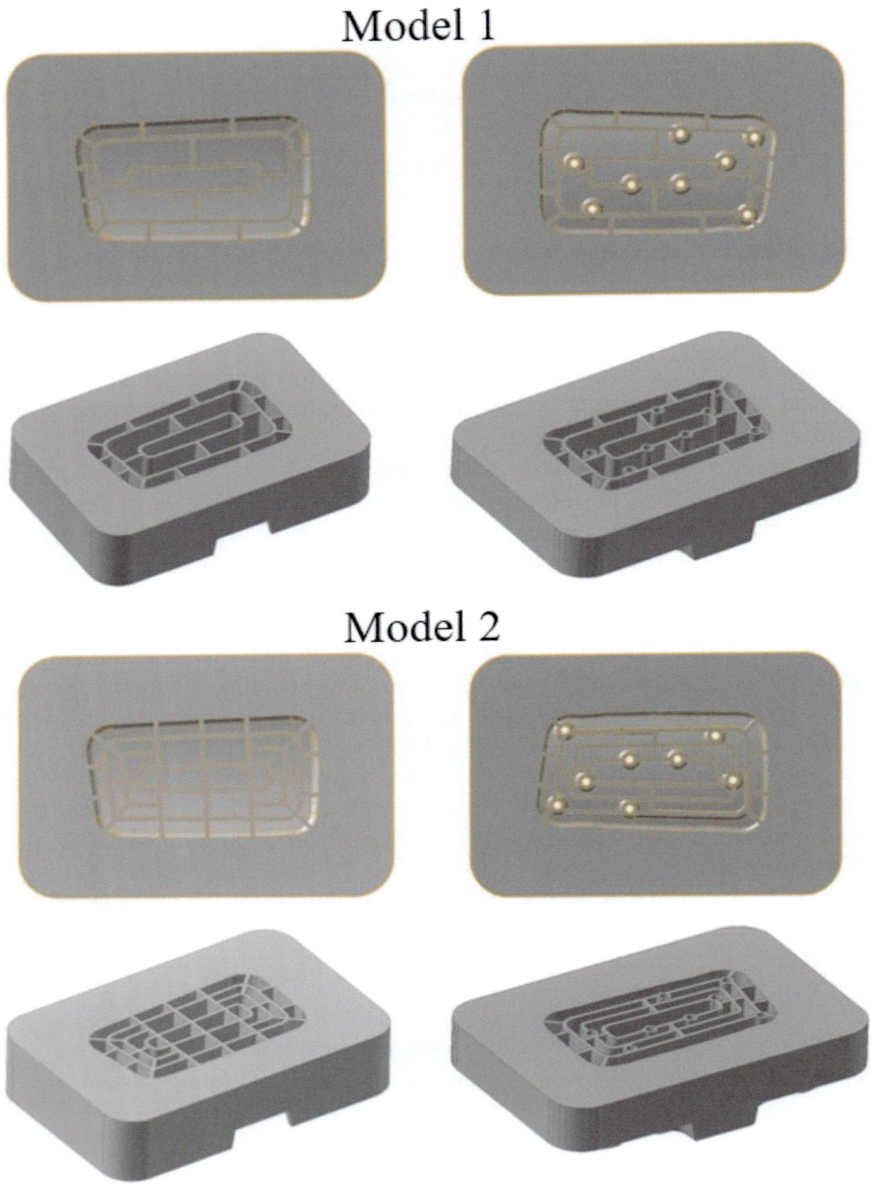

Fig. 11 Fixed plate from Designer 2

Table 8 Results of the analyses performed by designer 2 for the movable plate

Model	Displacement (mm)	Principal stress (MPa)	Volume of ribs (mm^3)
Without ribs	1.82	(−) 38/82	0
Model 1	0.36	(−) 16.8/15.1	78,790
Model 2	0.28	(−) 8.2/7.6	96,510

Table 9 Comparison of volume, cycles, and effective time

	Volumes of ribs (mm^3)		Cycle (attempt)	Effective time (h)
	Fixed plate	Movable plate		
Proposed method	29,101	26,073	6	10
Designer 1	90,100	84,450	2	3.5
Designer 2	112,700	96,510	2	1

4 Conclusions

The proposed method, evaluated in comparison to the traditional approach, presented an optimized approach both in the geometric aspect through topological optimization and in cost reduction. The evaluation steps were essential to avoid over dimensioning the structure, allowing for incremental adjustments and precise corrective actions, resulting in more targeted and accurate decisions for the designer.

The proposed method allows limiting the rib height relative to the shell thickness of the mold, ensuring effectiveness and better contact with the cavity plate in critical regions. Although it requires more cycles and a longer development time (10 h for both plates, compared to 3.5 and 1 h in traditional methods), its application is suited to the complexity of the study, providing more precise and optimized results.

The proposed method ensures proper structural sizing by controlling the total volume of ribs. For the fixed plate, the volume was 29,101 mm^3, while for the movable plate it was 26,073 mm^3. In comparison, the volumes generated by Designers 1 and 2 for the fixed plate were 209.6% and 287.2% larger, respectively, and for the movable plate, 223.9 and 270.1% larger. Larger volumes increase thermal zones in the additive manufacturing process, which can cause deformations. Therefore, the smaller volume obtained with the proposed method stands out positively compared to the traditional process, improving the final quality of the component.

The proposed method demonstrated effectiveness in the application of internal ribs in 3D-printed polymer molds, using topological optimization and eliminating the need for fill materials such as alloys or resins. Additionally, it stood out for volume reduction, cost savings, and improvement in the final component quality, being accessible even to less experienced designers.

Acknowledgements This work was carried out as part of the activities of the DEMIBAV project—Demonstrator of molds for the injection of low-volume plastic parts (27194 * 33), Line IV—More Competitive Brazilian Tooling, of the MOVER Program, formerly known as Rota 2030. The program is managed by FUNDEP and promoted by the Brazilian Federal Government, to whom the authors would like to thank. In addition, the authors would like to thank the Federal Universities of Bahia, Santa Catarina and SENAI CIMATEC, as well as FEESC, FEP and the other partners and ICTs of the DEMIBAV project.

Responsibility for the Information The authors are the only responsible for the printed material included in this paper. AI tools were used only to review written text. Tool: ChatGPT, Version: GPT-4o.

References

Fernandez E, Debrie S, Edeleva M, D'hooge DR, Cardon L (2024) Bridging experimental data and flow modelling to maximize the lifetime of additive manufactured molds. In: AIP conference proceedings, vol 3158, no 1. AIP Publishing

Ferreira JC, Mateus A (2003) Studies of rapid soft tooling with conformal cooling channels for plastic injection moulding. J Mater Process Technol 142(2):508–516

Leal Pereira P, Silva A, Gonçalves LA, Beal VE (2017) Simulation study to evaluate the use of fully 3D printed injection molds optimized by topology. In: Proceedings of the 24th ABCM international congress of mechanical engineering

Li W, Zheng A, You L, Yang X, Zhang J, Liu L (2017) Rib-reinforced shell structure. In: Computer graphics forum, vol 36, no. 7, pp 15–27

Malloy RA (2010) Plastic part design for injection molding: an introduction, 2nd edn. Hanser Publications, Munich, Germany

Rahmati S, Dickens P (2007) Rapid tooling analysis of Stereolithography injection mould tooling. Int J Mach Tools Manuf 47(5):740–747

Sinico M, Ranjan R, Moshiri M, Ayas C, Langelaar M, Witvrouw A, Van Keulen F, Dewulf W (2019) A mold insert case study on topology optimized design for additive manufacturing. In: International solid freeform fabrication symposium. University of Texas at Austin

Shoemaker J (2006) Moldflow design guide: a resource for plastics engineers, vol 10. Hanser Verlag

Tanaka K, Karasuno K, Ryuichi N, Katayama T (2017) Effects Of press pressure on the mechanical properties of rib root for glass fiber reinforced polypropylene composites, molded by press and injection hybrid molding. WIT Trans Eng Sci 116:317–326

Yang SY, Huang FZ, Liau WN (1996) A study of rib geometry for gas-assisted injection molding. Polym Eng Sci 36(23):2824–2831

Corrosion Performance of AISI 410L Martensitic Stainless Steel Parts Manufactured by Laser Directed Energy Deposition in Different Parametrizations

Jurandir Marcos Sá de Sousa, Milton Pereira, Bruno Borges Ramos, Cristiano Binder, and Juliana Ribeiro da Cruz

Abstract Martensitic stainless steel (MSS) has high hardness, mechanical strength, tribological performance, and corrosion resistance at milder temperatures. Additive manufacturing (AM) has brought industrial possibilities, such as customized geometries, on-demand production, and lead-time decrease. AM by laser directed energy deposition (L-DED) is intended for the building and repair of large parts. However, challenges arise due to the martensite brittleness and L-DED complex thermal history. Alternatively, lower C-content MSS can be used, as they offer a comparable property set with easier processability. In the L-DED, careful parameterization is required. There is still a dearth of literature regarding the L-DED processing of MSS for corrosion control. Within this context, the current study attempts to assess the corrosion resistance of a low C MSS (0.01 wt.%) multilayer samples built by AM L-DED AISI 410L in two parameterizations. Production (PRO), with a higher deposition rate, is meant for the building of bulker parts. Resolution (RES), with a lower deposition rate, is meant for more complex geometries. Two L-DED experiments were performed, one for PRO and the other for RES. Standardized multilayer samples were built in an argon-controlled atmosphere to mitigate the insertion of oxygen and impurities. Archimedes' density, microstructure, hardness, and electrochemical corrosion by potentiodynamic polarization in a 3.5% NaCl solution at room temperature analyses were performed. Samples of a commercial hot-rolled and annealed 410 MSS with a higher C-content (0.15 wt.%) were used as reference material (410-REF). Results showed compatible density levels (99.95%) in both L-DED optimized parametrizations. The L-DED microstructure obtained was biphasic with ferrite and refined martensite laths concentrated at the grain boundaries, owing to the laser

J. M. S. de Sousa (✉) · M. Pereira · B. B. Ramos · C. Binder
Federal University of Santa Catarina, Florianópolis, SC, Brazil
e-mail: jurandirmarcos37@gmail.com

J. M. S. de Sousa
Senai Innovation Institute for Manufacturing Systems and Laser, Joinville, SC, Brazil

J. R. da Cruz
Polytechnic School of Engineering, São Paulo, SP, Brazil

V. E. Beal and A. S. Ribeiro Junior (eds.), *Advances in Manufacturing Engineering Research*, Lecture Notes in Mechanical Engineering,
https://doi.org/10.1007/978-3-032-09858-0_3

fast cooling and AISI 410L lower C-content. The RES parameter produced a more refined microstructure, with a larger grain boundary area and 7% higher hardness than the PRO one, due to the faster solidification of the first condition. The 410-REF, in turn, showed a coarse ferrite plus tempered martensite and Cr-carbides, with a significantly lower hardness regarding both L-DED conditions. Electrochemical test data revealed a more positive corrosion potential Ecorr and a lower corrosion current Icorr for the RES parameter, which suggests a more effective passivation of this regarding the L-DED PRO and 410-REF. This indicates that lower C-content AISI 410L processed by L-DED at lower energy densities will be more resistant to corrosion in the presence of chlorides.

Keywords Laser directed energy deposition · AISI 410L · Refined microstructure · Effective passivation · Corrosion resistance

1 Introduction

Martensitic stainless steels (MSS) are used in industries with significant economic impact, such as oil and gas, cutlery, tooling, and mining. This is because of the balance between their cost and property set, which includes high hardness, mechanical strength, tribological performance, and corrosion resistance at milder temperatures. Components applied in these sectors are exposed to harsh conditions, such as temperature variations, humidity, and the presence of chemical elements. Therefore, they must be manufactured using processes and materials that provide effective protection. Currently, MSS parts are predominantly built by casting, hot rolling, and wrought. Additionally, post-heat treatments to homogenize are necessary to homogenize the microstructure and temper the martensite phase (Hosseini et al. 2020; Zhu et al. 2021).

To mitigate issues related to a fully martensitic microstructure, such as reduced toughness and tribological performance owing to embrittlement, as well as decreased corrosion resistance owing to C-enrichment and sensitization caused by the nucleation of Cr-rich carbides, low C-content MSS has been finding extensive applicability (Huang et al. 2019; Khodabakhshi et al. 2020; Zhai et al. 2023). Within this class, special attention is given to AISI 410L with a C-content of approximately 0.01 wt.%, which, besides showing simpler processability, provides a property set comparable to those of higher C-content MSS (Meng et al. 2022; Moura et al. 2023; Kang et al. 2024).

In the context of processing these components, additive manufacturing (AM) by laser directed energy deposition (L-DED) has added logistic benefits, such as the possibility of manufacturing out-of-scale parts, customized geometries, and processability of materials with poor machinability/weldability, reducing post-processing steps, and lead time. In L-DED, defined as a process in which focused thermal energy is used to melt materials as they are being deposited, besides logistical, there are also metallurgical advantages, such as the formation of a strength metallurgical bond with the substrate, narrow heat-affected zones, precise dilution control, and the achievement of refined microstructures (ASTM F3187 2023; Gradl et al. 2022; Hakami et al. 2024).

Thus, the manufacturing of low C-content MSS parts by AM L-DED combines the advantages of material and process. However, several challenges remain. Besides the brittle martensite nature, there is a possibility of Cr-carbide nucleation. These carbides can be beneficial for mechanical and tribological performance, but they may impair corrosion resistance by causing sensitization, which is the Cr-depletion in the regions surrounding the carbides. Furthermore, this can hinder the formation and stability of the Cr-oxide (Cr_2O_3) passive film, which is primarily responsible for keeping the corrosion resistance of SS. A reduction in thickness or breaks in the passive film expose the material to the environment, accelerating the corrosive action. Exposure of the surface to water (H_2O), oxygen (O_2), hydrogen sulfide (H2S), and chlorides (Cl^-) are among the main corrosion agents (Cai et al. 2018; Kiremit et al. 2023; Ramos et al. 2024; Cavilha et al. 2024). In the L-DED process, there is a cooling out of equilibrium with high solidification rates, which results in refined and heterogeneous microstructures (Lai et al. 2019; Gruber et al. 2020; Paes et al. 2022).

Thus, considering that both the MSS composition and manufacturing process characteristics affect the material's performance, it is essential to understand the corrosion resistance behavior of low C-content MSS processed by L-DED aiming at AM applications. In the literature, the number of studies on this research topic is still limited.

Fang et al. (2019) analyzed corrosion resistance by potentiodynamic polarization in 3.5% NaCl solution at room temperature of the Fe–Cr–Ni–Mn–Mo–Nb–Si MSS alloy processed by L-DED in the as-built state and of the reference material (FV520B MSS). It was observed that the corrosion potential (Ecorr) of L-DED was 0.001 V, while that of the reference material was −0.283 V. Corrosion current (Icorr) was also lower in L-DED. This result was attributed to the L-DED microstructure, which did not show appreciable segregation at the grain boundaries, as well as the non-existence of Cr-depletion regions. Wang et al. (2022) compared the corrosion resistance by potentiodynamic polarization in different solutions of 18Ni300 Maraging MSS manufactured by L-DED. The results obtained in the 3.5% NaCl solution showed that the L-DED Icorr was lower compared to AM by laser powder bed fusion (L-PBF). Besides, L-PBF break potential (Ebreak) surpassed that of L-DED, thus indicating better stability of the passive layer established in the former. L-PBF Ecorr was also less positive concerning L-DED, so its trend to corrode is greater. This result was attributed to the best L-DED chemical composition homogeneity, a result conferred to the lower number of L-DED thermal cycles, molten pools, and a slower cooling speed regarding L-PBF. Kochta et al. (2021) evaluated the corrosion resistance by potentiodynamic polarization at 3.5% and 0.035% NaCl solutions at room temperature of the Fe85Cr4Mo8V2C1 alloy manufactured by L-PBF, and in the wrought state. The Ecorr difference between the L-PBF and wrought was minimal, owing to the chemical composition similarity. However, the more refined microstructure and even distribution of the L-PBF phases delayed the corrosive activity. Besides, L-PBF showed a trend toward higher I_{corr} and lower E_{corr} concerning the wrought, pointing to a lower corrosion rate of the former. Yazdanpanah et al. (2022) investigated the correlations between different types of porosities (circular and irregular) on corrosion resistance by potentiodynamic polarization in 3.5% NaCl solution at room temperature of 316L SS built by L-PBF. Samples with higher density exhibited greater Ebreak and lower breakdown current (I_{break}). With the transition from rounded to irregular porosities, a substantial E_{break} reduction, and I_{break} increase were observed. Furthermore, I_{break} variation between the groups with the same porosity type can be attributed to the amount and dimensions of these. Corroded surfaces evaluated by scanning electron microscope (SEM) attested that samples with irregular porosities are more susceptible to pitting, crevice, and stress corrosion cracking mechanisms.

In this context, the present study investigates the corrosion resistance of low C-content MSS (0.01 wt.%) AISI 410L additively manufactured by L-DED, using two parameter sets, namely Production (PRO) and Resolution (RES). The first set has a higher deposition rate, intended for AM of larger components with simpler geometries. The second set has a lower deposition rate, aimed at AM of components with more complex geometries. The purpose is to understand the effects of these different L-DED parameters on microstructure, hardness, and electrochemical corrosion responses of corrosion potential (Ecorr), corrosion current (I_{corr}), polarization resistance (Rp), and aspects of corroded surfaces. A commercially wrought and annealed higher C-content 410 MSS (0.15 wt.%C) was used as a benchmark. Previous works by the authors have studied the parameterization of AISI 410L by L-DED (Sousa et al. 2021a, 2023b); the effects of build orientation (vertical and horizontal concerning the substrate) on mechanical performance (Sousa et al. 2024c); the effects of different post-heat treatment routes on mechanical performance (Sousa et al. 2023); and the effects of L-DED parameters and post-heat treatment on tribological performance (Sousa et al. 2024e).

2 Experimental Procedures

The as-received 410 MSS wrought and annealed square plates (244 × 244 × 50 mm) were used as the substrate for manufacturing L-DED multilayer samples. The feedstock was AISI 410L MSS powder (Höganäs Inc.), supplied in the gas-atomized state and with a particle size range of 53–150 μm. Table 1 exhibits the nominal chemical composition of the substrate and feedstock provided by the manufacturers.

RPMI 535 from RPM Innovations Inc. was the L-DED system used for multilayer sample manufacturing. The mentioned system is composed of the build platform, a continuous 100 μm fiber laser source doped with Ytterbium (Yb) with a maximum laser power of 3300 W (YLS-3000-CT) and 1064 nm wavelength from IPG Photonics Inc., and the auxiliary systems (chiller and powder suppliers). Laser beam parameter product (BPP = 10.0) and the beam quality parameter ($M^2 = 29.7$) were evaluated

Table 1 Nominal chemical composition of the substrate and the feedstock employed, as supplied by the manufacturers

Element (wt.%)	Cr	Mn	Si	Ni	C	Fe
410 substrate	13.5	1.0	1.0	0.7	0.15	Bal
AISI 410L powder	12.8	0.8	0.8	0.7	0.02	Bal

by the Focus Monitor FM + beam profilometer from PRIMES Inc. In the L-DED setup, a 25° discontinuous 4-coaxial nozzle was employed.

In the characterizations of the L-DED multilayer, it was used: Archimedes' density (METTLER TOLEDO ME204/A); optical stereoscope (OS, ZEISS V8); optical microscopy (OM, ZEISS AXIO M2M); scanning electron microscope (SEM, ZEISS SUPRA 55VP); X-ray diffractometer (XRD, RIGAKU MINIFLEX600); Vickers hardness tester (WILSON 402 MVD); electrochemical potentiostat (PGSTAT 302, METROHM).

Before L-DED multilayer samples were manufactured, 410 substrate plates shown in Fig. 1a were blasted and cleaned in ethyl alcohol. Figure 1b shows a representative sample of AISI 410L powder, that was embedded, sanded (#80–1200 mesh), polished (0.3 um diamond solid-suspension), and evaluated in cross-section by OM. No deleterious defects were perceived, but only a few porosities (yellow indications). SEM was employed in the powder morphology analysis, in which just some red highlighted satellite particles were observed. XRD was also performed, as will be discussed later.

The optimized L-DED parameters employed to manufacture multilayer samples were previously obtained. The method used in this process was based on the work of Thiesen (2021). These parameters have already been employed in other works by the group: design of experiments (Sousa et al. 2021a, 2023b); an investigation of the anisotropy effect on the mechanical properties as a function of the L-DED build orientation (vertical and horizontal) regarding the substrate (Sousa et al. 2024c); a

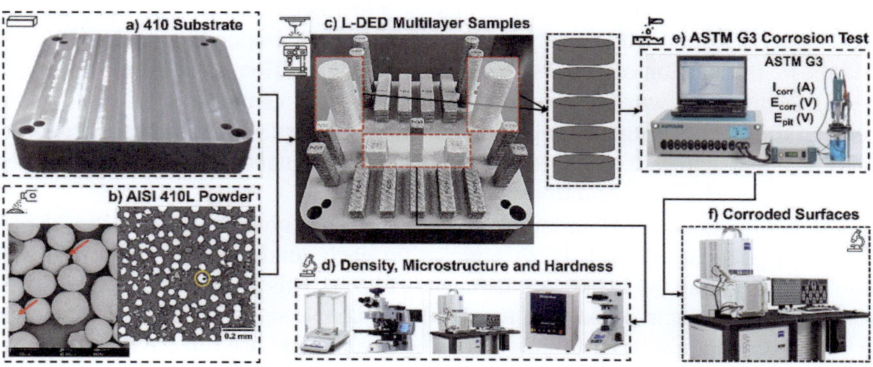

Fig. 1 Illustrative scheme of the experimental route

Table 2 L-DED optimized parameters employed in the multilayer samples manufacturing

L-DED parameters|condition	Production	Resolution
Spot size (Ø, mm)	2.54	1.78
Powder feedrate (F, g/min)	30	25
Laser power (P, W)	1800	1300
Travel speed (S, mm/min)	635	800
Laser power density (W/mm^2)	356	523
Hatch spacing (HS, mm)	1.93	1.03
Layer height (LH, mm)	1.22	0.84

study dedicated to finding an appropriate post-heat treatment route to improve the microstructure characteristics, hardness, and mechanical properties of AISI 410L MSS L-DED multilayer samples (Sousa et al. 2023); and the investigation of the different L-DED parametrizations (production and resolution) and post-heat treatment route on the tribological performance (Sousa et al. 2024e). From these, the optimized L-DED parameters for Production (named as PRO) and Resolution (named as RES) conditions are shown in Table 2. L-DED multilayer sample manufacturing was performed in an argon-controlled atmosphere, maintaining the O_2 level below 5 ppm. A 45° rotation angle between layers' slicing and the deposition of one layer at a time on each sample were the strategies adopted, aiming to reduce oxidation and excessive heat-input along with the L-DED manufacturing.

Two sets of samples were prepared to investigate the effect of L-DED parameters (PRO and RES), as illustrated in Fig. 1c, one for each condition. Aiming to assess the microstructure and hardness, cubes measuring 25 × 25 × 25 mm^3 were produced, as shown in Fig. 1d. Multilayer samples were built for the electrochemical corrosion test. The microstructure and hardness samples were cross-sectioned, metallographically prepared, and evaluated by OM to detect defects. These were also investigated concerning Archimedes' density; porosity and grain boundaries quantification by OM and Image J software; OM microstructure; SEM–EDS microstructure and chemical composition; and XRD phases' identification (Fig. 1d). The cautions were taken in Archimedes' analysis, concerning room temperature control (23 °C) and the non-formation of blisters. For porosity quantification, sample cross-sections were machined by wire electron discharge (W-EDM) as close as possible to the center. After metallographic preparation, a color contrast evaluation was performed in triplicate with the aid of ImageJ software, aiming to quantify the grain boundaries under different L-DED conditions.

Electrochemical corrosion samples were prepared in compliance with the ASTM G3 (2019) standard. Five L-DED samples for each condition were then W-EDM machined to 10 mm thickness, as illustrated in Fig. 1c. This procedure was completed 10 days before the start of the tests, to provide the sample surfaces with sufficient time to establish a natural passivating oxide. The potentiostat used in the electrochemical corrosion test, shown in Fig. 1e, was composed of three electrodes: a sample (WE, work electrode) with a 28.5 mm^2 exposed area, a saturated calomel electrode

(SCE, reference electrode), and a platinum electrode (CE, counter electrode). Tests were performed in a 3.5 wt.% NaCl aqueous solution. Cathodic polarization at −773 mV (vs. SCE) for 30 min was performed first. Before polarization, the open-circuit potential (OCP) was monitored for 60 min. Once the OCP evaluation was concluded, potentiodynamic polarization was performed at a 0.10 mV/s scan rate. NOVA 2.1 software was used to process the data. After the test, the corroded surfaces of representative samples of each L-DED condition were investigated by SEM, as can be seen in Fig. 1f.

For all the analyses described above, as-received 410 MSS commercial reference substrate samples (named SUB-AR) were machined by W-EDM and exposed to the same experimental route.

3 Results and Discussion

This section describes the results and discussion and is divided into three subsections. The first presents the main characteristics of microstructure and hardness. The second exhibits corrosion data, and the third demonstrates the aspects of the corroded surfaces. All these results were discussed in terms of the PRO and RES of the L-DED parameters.

3.1 Microstructure and Hardness of the Different L-DED Parametrizations

The optimized L-DED parameters under both conditions were adequate, as the analyzed multilayer geometries did not present significant geometric deviations regarding the 3D CAD model or harmful defects, such as cracks and lack of fusion (Fig. 2a). Only rounded porosities were found in the OM quantification analysis, as exemplified in Fig. 2b, possibly resulting from gas blister entrapment during the fast cooling imposed by the laser beam. The porosity size appeared to be larger for the PRO condition compared to RES (Reddy et al. 2018; Aversa et al. 2020). Both conditions exhibited an Archimedes density of 7.70 g/cm^3, representing 99.5% of the nominal material density, which is 7.74 g/cm^3.

Figure 3 shows the cross-section OM microstructure of both L-DED conditions. Besides confirming the absence of defects, molten pool boundaries' characteristic of AM was seen. The solidification mode was columnar, with refined grains. It was observed a ferritic matrix with refined martensite laths concentrated at the grain boundaries. These features can be attributed to the laser fast cooling and low C-content of AISI 410L. Thus, there was insufficient time for the grains to grow by diffusion and insufficient C-content to establish a fully martensitic microstructure. Comparing the L-DED conditions in Fig. 3a and Fig. 3b, RES exhibited an area

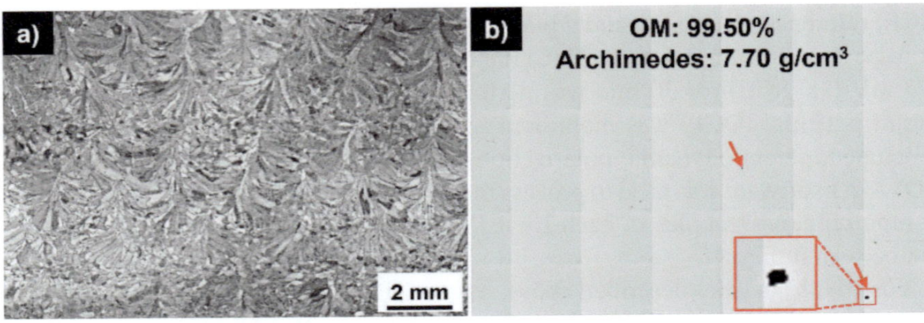

Fig. 2 Archimedes density and porosity quantification: **a** OM image and **b** OM map

occupied by grain boundaries approximately 7% larger than PRO, which attests to the formation of refined grains in the former. This response can be attributed to the faster cooling of RES. The layer dimensions for the optimized L-DED parameters also point in this direction. For PRO, the height was 1.44 mm versus 0.84 mm for RES. Despite occupying a larger area and requiring fewer layers to build the same multilayer geometry, PRO provides slower cooling and greater re-melting of previous layers, resulting in coarser grains and a more homogeneous microstructure concerning RES. Vickers hardness was another indication, as RES exhibited an average of 239 ± 12 $HV_{0.5}$, 6% higher than the 228 ± 11 $HV_{0.5}$ for PRO (Zhai et al. 2023; Shen et al. 2020; Tian et al. 2021). In Fig. 3c, SEM analysis confirmed these aspects, showing regions of the biphasic L-DED microstructure more clearly. In Fig. 3e, SEM–EDS analysis did not exhibit significant concentration differences in the AISI 410L nominal elements. However, the local hardness evaluated on the ferritic and martensitic regions revealed an average of 30% higher in the martensitic region. Some small, rounded dots were also observed, and they exhibited significantly higher contents of Si and O, indicating that they are likely Si-rich oxides. In Fig. 3d, SUB-AR exhibited a homogeneous microstructure composed of coarse tempered martensite grains, ferrite, and carbides. The average hardness value of 189 ± 5 $HV_{0.5}$ was 17% lower than both L-DED conditions. In Fig. 3f, SEM–EDS revealed a uniform chemical composition, with elevated contents of Fe and Cr on the precipitates. In Fig. 4, XRD diffractograms indicate the predominance of the BCC structure in the AISI 410L powder, and in both L-DED conditions, which corresponds to the ferrite and martensite phases. The main peaks were found at 44°, 64°, and 82°. A peak of tetragonal body-centered (BCT) martensite was found at 82°, partially overlapping with a ferrite/BCC-tempered martensite. This peak was more pronounced in the RES, which is consistent with its higher amount of martensite. SUB-AR, in turn, showed peaks of ferrite and BCC martensite. During SUB-AR manufacturing, part of the carbon-supersaturated martensite was transformed into Fe_3C cementite and $Cr_{23}C_6$-carbides (Zhu et al. 2021; Khodabakhshi et al. 2020; Kang et al. 2024; Lai et al. 2019; Lu et al. 2015; Nayebi et al. 2021).

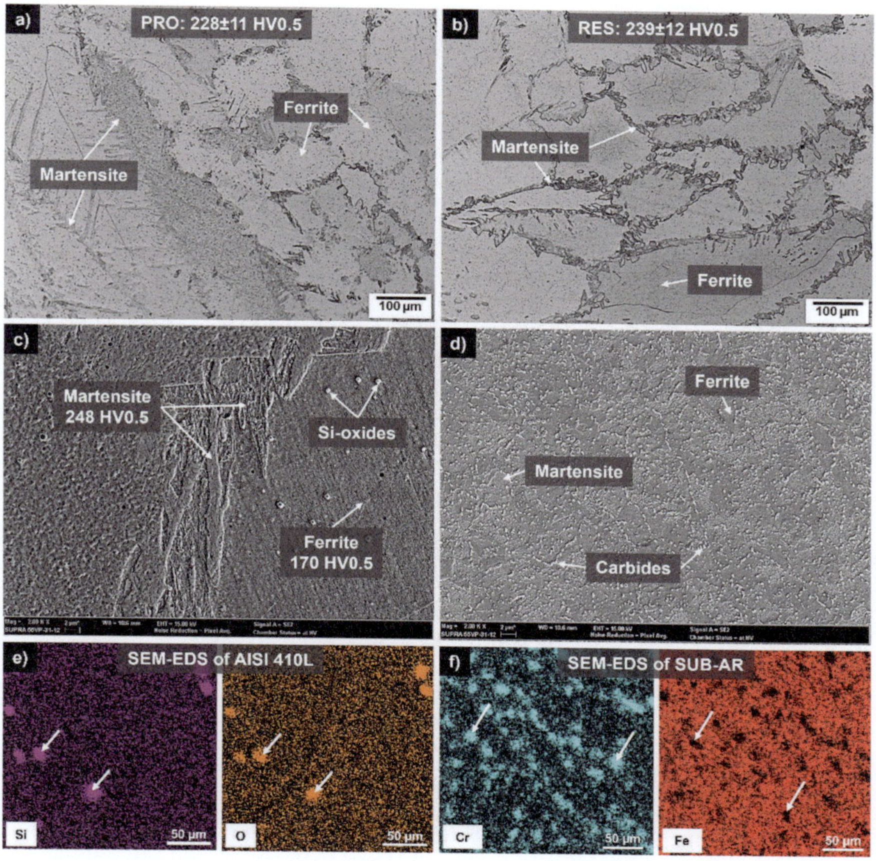

Fig. 3 OM of L-DED: **a** PRO and **b** RES; SEM of **c** L-DED and SUB-AR; SEM–EDS of **e** L-DED and **f** SUB-AR

3.2 Corrosion Data of the Different L-DED Parametrizations

The corrosion performance was evaluated in terms of corrosion potential (E_{corr}), corrosion current (I_{corr}), and polarization resistance (R_p). Using the Tafel extrapolation method, the average values of these corrosion parameters for different L-DED conditions and SUB-AR are listed in Table 3. Based on these results, some trends can be discussed.

Figure 5 shows the E_{corr} values of the different L-DED parameters and SUB-AR. As can be seen, the RES condition exhibited a significantly more positive E_{corr} than the PRO. On the other hand, SUB-AR showed the most negative value among the three conditions. These results indicate a trend of greater corrosion resistance and a more noble character in both L-DED conditions regarding SUB-AR. This indicates that even with a heterogeneous microstructure compared to the traditional SUB-AR,

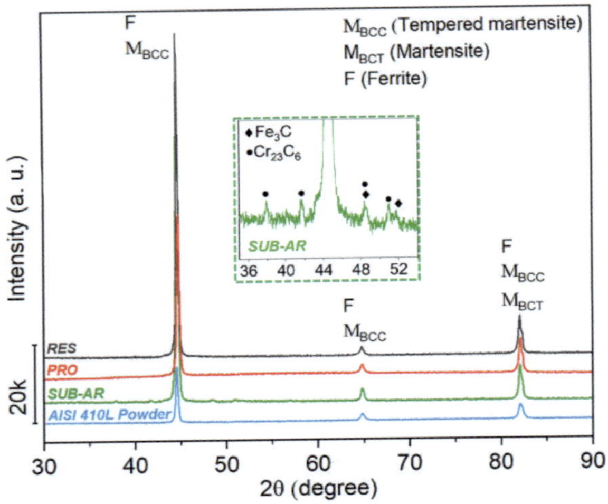

Fig. 4 XRD diffractogram of the different L-DED conditions and SUB-AR

Table 3 Tafel parameters of L-DED conditions and SUB-AR tested in a 3.5% NaCl solution

L-DED condition	PRO	RES	SUB-AR
E_{corr} (mV)	−258	−30	−504
I_{corr} (μA/cm^2)	0.142	0.004	1.185
R_p (Ωxcm^2)	9.54E + 04	2.72 $\pm$ 06	1.14 $\pm$ 04

AISI 410L exhibited better performance. Besides, the L-DED parameters used in their manufacture were robust (Kochta et al. 2021; Gollapudi 2012; Powell et al. 2023).

Figure 6 illustrates the corrosion current (I_{corr}) behavior in the same direction. The RES condition resulted in a lower average I_{corr} concerning PRO. On the other

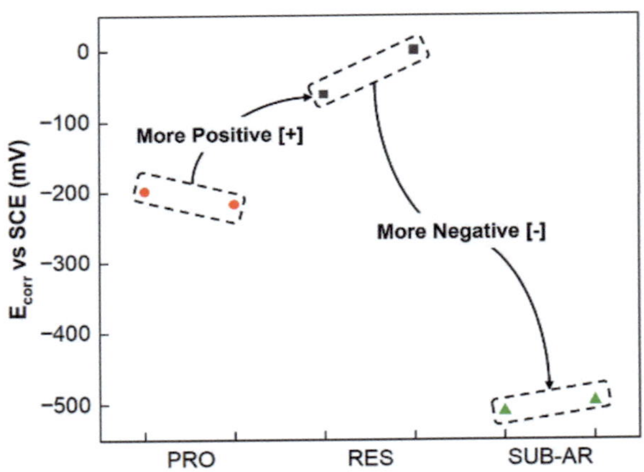

Fig. 5 Potential corrosion (E_{corr}) under different L-DED conditions and SUB-AR

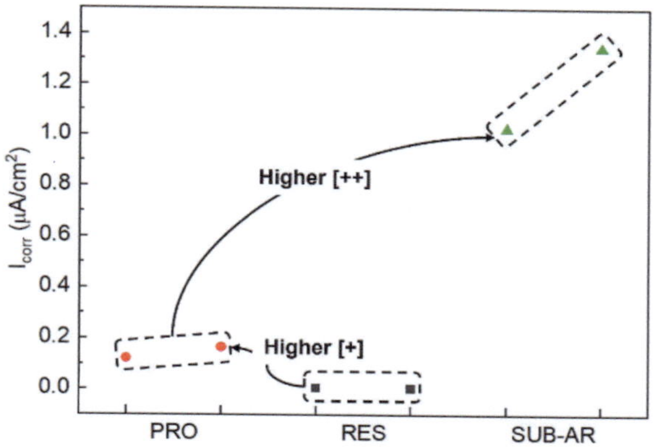

Fig. 6 Corrosion current (I_{corr}) of the different L-DED conditions and SUB-AR

hand, SUB-AR showed much higher values than both L-DED conditions. Combining a more positive E_{corr} and a lower I_{corr} for RES supports its greater corrosion resistance when exposed to a medium containing chlorides (Cl^-), such as the 3.5% NaCl solution used in this study. A higher E_{corr} indicates that the material resulting from this L-DED parameterization is more noble, whereas a lower I_{corr} suggests that it will degrade more slowly. Polarization resistance (R_p) was inversely proportional to I_{corr}, significantly higher in the RES condition regarding PRO. SUB-AR, on the other hand, exhibited values considerably lower than both L-DED parameters. The combination of more positive R_p and E_{corr} values suggests a lower pitting potential (E_{pit}), such that the presence of more localized corrosion mechanisms tends to be reduced. It is worth noting that the action of concentrated corrosion mechanisms, such as pitting, is the most common in industrial applications of AISI 410L, such as oil and gas (Ramos et al. 2024; Cavilha et al. 2024; Fang et al. 2019; Nath et al. 2019).

Figure 7 presents the polarization curves with E_{corr} and I_{corr}, reinforcing the discussions. Besides, observing the anodic passivation and metal dissolution regions was also possible after the breakdown potential (E_{break}) was reached. The latter, the inflection point between the cathodic and anodic curves, indicates the stability of the passive layer formed on the sample surface. In this work context, the RES condition exhibited a significantly higher E_{break} (0.17 V) concerning PRO (−0.30 V) and SUB-AR (−0.50 V). Besides, RES was the only condition that showed an anodic passivity plateau before passivity breakdown. These observations suggest that the passive layer formed on RES was more continuous, compact, and thicker than those formed under the other studied conditions (Ramos et al. 2024; Gollapudi 2012; Hufenbach et al. 2020; Ko et al. 2021).

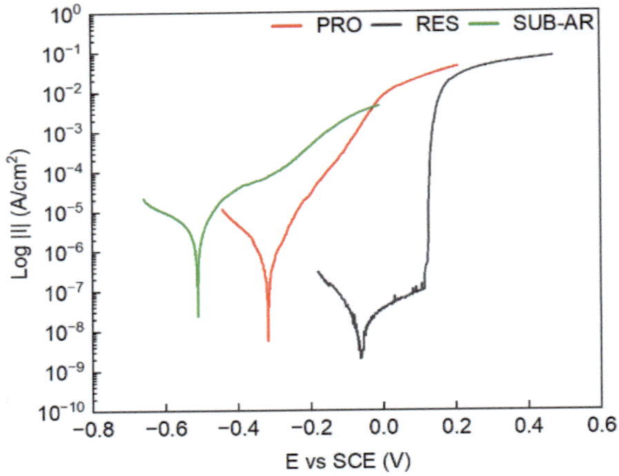

Fig. 7 Polarization curves of the L-DED AB conditions and SUB-AR

3.3 *Main Aspects of the Corroded Surfaces*

Several influence sources can be identified to justify the performance difference between the conditions studied. The main discussion points are the start oxidation level of L-DED samples, microstructure refinement, precipitates, and passive layer characteristics. Regarding the start oxidation level of the samples, it is believed that this did not exert a significant effect, as the same precautions were taken for all the samples tested. Powder feedstock was handled with minimal exposure to the ambient atmosphere, and all the multilayer manufacturing steps were performed in a controlled Ar atmosphere, with an internal O_2 level below 10 ppm. Furthermore, before polarization, all samples underwent the same passivation route. In this context, Fig. 8 presents the SEM–EDS analysis conducted on the surface of a representative sample from the PRO and RES conditions, as well as from the SUB-AR, to examine the type and level of oxidation. In the maps of both conditions, it was observed that almost all elements showed a homogeneous distribution. Cr and O can be attributed to the AISI 410L's nominal chemical composition and the Cr_2O_3 passive layer. Besides, Fe_2O_3 oxidation is likely to have formed during the test. In the SUB-AR, higher levels of C and Cr were detected in the Cr-carbides (Fang et al. 2019; Krell et al. 2017; Bonagani et al. 2019; Cruz and Bertazzoli 2020).

Corrosion behavior observed in the L-DED conditions can also be attributed to the different levels of microstructure refinement. The martensitic phase is less resistant than the ferritic, which impaired the RES performance, owing to its greater martensite concentration, because of its more refined microstructure. However, corrosion responses depend on the combined action of several factors. In this sense, the larger surface area occupied by grain boundaries also indicates higher surface energy. Therefore, the more RES refined microstructure offers greater resistance against corrosive agents. Martensite at the grain boundaries indicates a higher amount of C in solid solution in this phase. Therefore, Fe-matrix will be more depleted in this element, which,

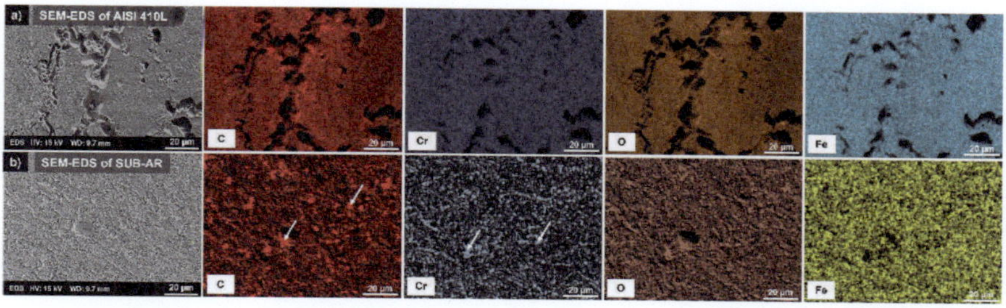

Fig. 8 SEM–EDS maps of the corroded surfaces of the different L-DED conditions and SUB-AR

in turn, possesses poor corrosion resistance. A higher C-content in solid solution in martensite also suggests a lower availability of this element to form precipitates, such as Cr-carbides. In this regard, the coarser PRO microstructure allowed greater availability of C and time, due to slower cooling, to form a larger amount of Cr-carbides in the Fe-matrix. The presence of these carbides can activate the galvanic corrosion with ferrite and sensitization by capturing Cr from Fe-matrix. As a result, the regions around these Cr-carbides are depleted in Cr, which is the main element responsible for the formation and maintenance of the passive layer. Another justification for the better RES performance may be the characteristics of its Cr_2O_3 passive layer. More refined microstructures favor this. In such cases, the passive layer tends to be more continuous, compact, and thicker. It is suggested that the higher volume fraction of grain boundaries makes it more difficult for corrosion mechanisms, such as pitting, to nucleate. On the other hand, coarser grains tend to exert the opposite effect, providing the formation of discontinuous and thinner passive layers, which do not provide sufficient protection against corrosive agents. Results suggest that PRO condition and SUB-AR conditions suffered from this effect (Kochta et al. 2021; Nath et al. 2019; Jangir et al. 2018; Cruz et al. 2024).

OM and SEM corroded surface analysis shows that the main mechanism detected in both L-DED conditions was intergranular corrosion, with the martensite present at the grain boundaries being preferentially corroded regarding the ferritic matrix. This mechanism was evident in Fig. 9a, c OM for PRO and RES, respectively. In these, martensitic grain boundaries appear dark, while the ferrite remains light. SEM evaluation revealed a greater number of pits in Fig. 9b PRO and larger pit sizes concerning Fig. 9d RES. As previously mentioned, PRO exhibited a trend for larger porosity dimensions. Therefore, the observed pits must have formed over porosities. Furthermore, assuming the presence of Cr-carbides, the larger Cr-depleted Fe-matrix area favored greater pit formation in PRO (Hufenbach et al. 2020; Cruz et al. 2024). In the SUB-AR, Fig. 9e OM showed a uniform corrosion mechanism. Figure 9f SEM revealed pits and generalized corrosion. These responses can be attributed to the homogeneous annealed microstructure of this condition. Additionally, several Cr-carbides were identified as being homogeneously distributed throughout the structure. Thus, SUB-AR Fe-matrix is Cr-depleted, facilitating the sensitization. The

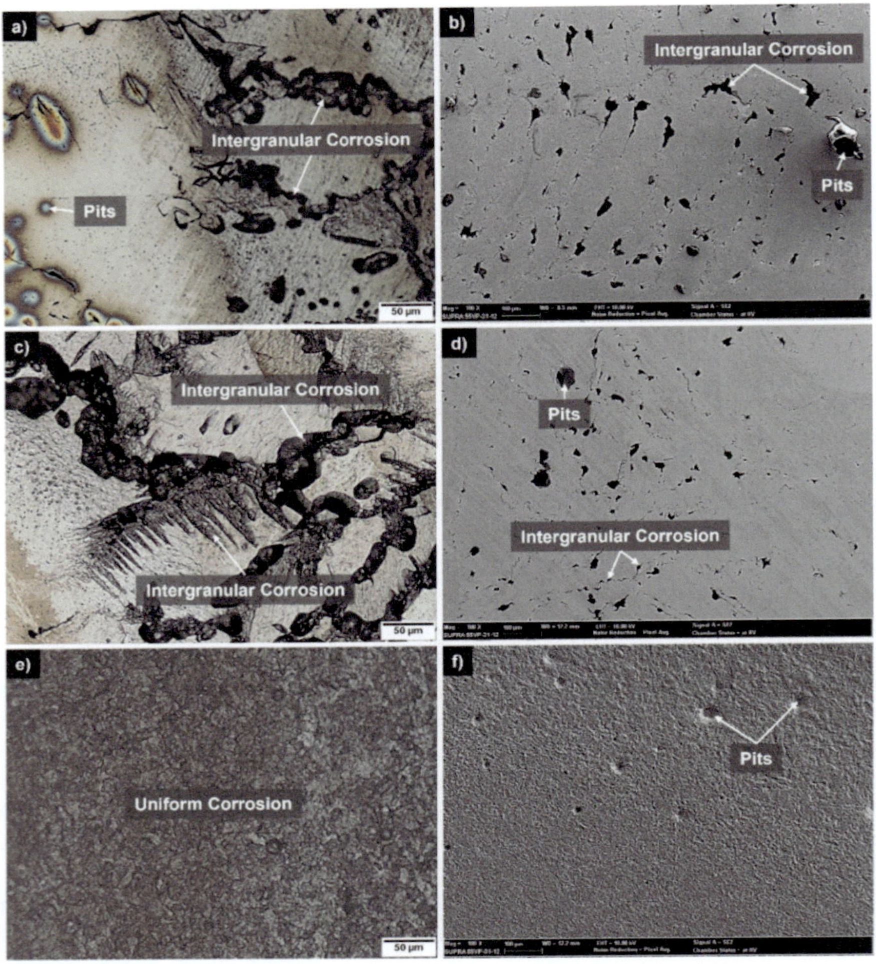

Fig. 9 Corroded surfaces of *PRO*: **a** OM and **b** SEM; *RES*: **c** OM and **d** SEM; SUB-AR **e** OM and **f** SEM

coarse microstructure also failed to support the formation and maintenance of the passive layer (Krell et al. 2017; Bonagani et al. 2019; Shahriari et al. 2021).

4 Conclusions

This study evaluated the electrochemical corrosion performance in a 3.5% NaCl solution at room temperature of multilayer samples of low C-content AISI 410L MSS manufactured by L-DED under different parameterization conditions, with

PRO showing a higher deposition rate, and RES having a lower deposition rate. Based on the results obtained, it is possible to infer:

- Both optimized L-DED parameter conditions showed no detrimental defects, and the density level was above 99.5%. The only difference was a trend for larger rounded porosities in the PRO condition.
- The microstructure of both conditions consisted of a ferritic matrix and refined martensite laths concentrated at the grain boundaries. This feature can be attributed to the low C-content of AISI 410L and the fast cooling imposed by the laser beam. Due to the lower energy density, RES condition cooled faster, resulting in a more refined microstructure and a larger area occupied by grain boundaries with refined martensite laths, which have higher hardness. No changes in chemical composition were observed. SUB-AR, on the other hand, exhibited a coarse microstructure with tempered martensite, Fe_3C cementite, and $Cr_{23}C_6$ Cr-carbides.
- The RES condition exhibited significantly superior corrosion performance concerning PRO and SUB-AR. The former's more refined microstructure provided higher surface energy and support for the Cr_2O_3 passive layer formation and maintenance. Martensite laths concentrated at the grain boundaries removed C from the Fe-matrix, reducing corrosion activity and the likelihood of sensitization phenomena. The coarser microstructure and larger porosities in PRO exerted the opposite effect, triggering pitting formation.
- In general, corrosion performance in a chloride environment of L-DED-processed low C-content AISI 410L, in both parameterizations (PRO and RES), was significantly superior to that of commercially available reference SUB-AR.

5 Responsibility for Information

The authors are solely responsible for the information included in this work.

6 Competing Interests

The authors have no conflicts of interest to declare that are relevant to the content of this chapter.

Acknowledgements We thank Precision Engineering Laboratory and Materials Laboratory from the Federal University of Santa Catarina and Senai Innovation Institute for Manufacturing Systems and Laser Processing for their technical-scientific support.

References

Hosseini SA, Gheisari K, Moshayedi H et al (2020) Basic alloy development of low-transformation-temperature fillers for AISI 410 martensitic stainless steel. Sci Technol Weld Join 25:243–250. https://doi.org/10.1080/13621718.2019.1681159

Zhu H, Ouyang M, Hu J et al (2021) Design and development of TiC-reinforced 410 martensitic stainless steel coatings fabricated by laser cladding. Ceram Int 47:12505–12513. https://doi.org/10.1016/j.ceramint.2021.01.108

Huang S, Li D, Zhang L et al (2019) Tailoring the mechanical properties of laser cladding-deposited ferrous alloys with a mixture of 410L alloy and Fe-Cr-B-Si-Mo alloy powders. Materials 12:1–12. https://doi.org/10.3390/ma12030410

Khodabakhshi F, Farshidianfar MH, Gerlich AP et al (2020) Effects of laser additive manufacturing on microstructure and crystallographic texture of austenitic and martensitic stainless steel. Addit Manuf 31:100915. https://doi.org/10.1016/j.addma.2019.100915

Zhai W, Wu N, Zhou W (2023) Laser metal deposition of low carbon 410L stainless steel and heat treatment. Mater Sci Eng A 872:144987. https://doi.org/10.1016/j.msea.2023.144987

Meng L, Li W, Shi Q et al (2022) Effect of partitioning treatment on the microstructure and properties of low-carbon ferritic stainless steel treated by a quenching and partitioning process. Mater Sci Eng A 851:143658. https://doi.org/10.1016/j.msea.2022.143658

Moura AN, Alcântra CM, Vieira EA et al (2023) Microstructure, crystallographic aspects and mechanical properties of AISI 420 martensitic stainless steel after different thermomechanical process routes. Mater Chem Phys 305:127723. https://doi.org/10.1016/j.matchemphys.2023.127723

Kang HK, Lee H, Oh C-S et al (2024) Microstructure and mechanical properties of laser direct energy deposited martensitic stainless steel 410. Micromachines 15:1–14. https://doi.org/10.3390/mi15070837

ASTM F3187 (2023) Standard guide for directed energy deposition. ASTM International 1–22. https://doi.org/10.1520/F3187-16R23

Gradl P, Tinker DC, Park A et al (2022) Robust metal additive manufacturing process selection and development for aerospace components. J Mater Eng Perform 31:6013–6044. https://doi.org/10.1007/s11665-022-06850-0

Hakami A, Ojo SA, Abere DV et al (2024) Advancements in metal additive manufacturing: opportunities, limitations, impact on properties, and potential solutions: a review. Prog Addit Manuf 9:1–49. https://doi.org/10.1007/s40964-024-00885-6

Cai Y, Zhao Y, Ma X et al (2018) Influence of environmental factors on atmospheric corrosion in dynamic environment. Corros Sci 137:163–175. https://doi.org/10.1016/j.corsci.2018.03.042

Kiremit S, Cremer J, Stallmeier A et al (2023) Development of an in-situ micro-corrosion cell for the investigation of pitting corrosion on austenitic and ferritic stainless steels. Corros Mater Degrad 4:104–119. https://doi.org/10.3390/cmd4010007

Ramos BB, Vicente FA, Hammes G et al (2024) Enhancing corrosion and tribology performance of stainless steel with DLC coatings: effects of doping and multilayered structures. Surf Coat Technol 477:130334. https://doi.org/10.1016/j.surfcoat.2023.130334

Cavilha F, Medeiros VK, Salinas-Barrera V et al (2024) Systhesis and characterization of Ti-13Ta-6Sn foams produced using mechanical alloying, the space holder method and plasma-assisted sintering. Materials 14:1–25. https://doi.org/10.3390/met14101145

Lai Q, Abrahams R, Yan W et al (2019) Influence of deposition materials, process parameters and heating conditions on material characteristics of laser-cladded hypereutectoid rails. J Mater Process Technol 263:1–20. https://doi.org/10.1016/j.jmatprotec.2018.07.035

Gruber S, Grunert C, Riede M et al (2020) Comparison of dimensional accuracy and tolerances of powder bed based and nozzle based additive manufacturing process. J Laser Appl 032016:1–13. https://doi.org/10.2351/7.0000115

Paes LES, Pereira M, Xavier FA et al (2022) Lack of fusion mitigation in directed energy deposition with laser (DED-L) additive manufacturing through laser remelting. J Mater Process 73:67–77. https://doi.org/10.1016/j.jmapro.2021.10.052

Fang JX, Dong SY, Li SB et al (2019) Direct laser deposition as repair technology for a low transformation temperature alloy: microstructure, residual stress, and properties. Mater Sci Eng A 748:119–127. https://doi.org/10.1016/j.msea.2019.01.072

Wang Q, Huang J, Guan L et al (2022) Electrochemical corrosion behavior of 18Ni 300 maraging steel obtained by laser cladding deposition and selective laser melting in corrosive mediums: a comparative study. J Mater Eng Perform 31:8389–8404. https://doi.org/10.1007/s11665-022-06849-7

Kochta F, Gebert A, Kühn U et al (2021) Improved corrosion behavior of a novel Fe85Cr4Mo8V2C1 tool steel processed by laser powder bed fusion. J Alloys Compd 867:158887. https://doi.org/10.1016/j.jallcom.2021.158887

Yazdanpanah A, Franceschi M, Rebesan P et al (2022) Correlation of lack of fusion pores with stress corrosion cracking susceptibility of L-PBF 316L: effect of surface residual stresses. Materials 15:1–16. https://doi.org/10.3390/ma15207151

Sousa JMS, Gutjahr J, Thiesen A et al (2021a) Laser directed energy deposition (L-DED) of AISI 410L: development of an operational map and the effect processing parameter on the bead geometry. In: Proceedings of the 26th International Congress of Mechanical Engineering—COBEM 2021 Florianópolis, Brazil

Sousa JMS, Pereira M, Ferreira HS et al (2023b) Laser directed energy deposition process optimization parameters for AISI 410L single layers. In: Proceedings of the 27th International Congress of Mechanical Engineering—COBEM 2023, Florianópolis, Brazil

Sousa JMS, Pereira M, Cruz JR et al (2024c) Mechanical performance of laser directed energy deposition produced AISI 410L martensitic stainless steel parts in different build orientations. In: Proceedings of the 12th National Congress of Mechanical Engineering—CONEM 2024, Natal, Brazil

Sousa JMS, Pereira M, Cruz JR et al (2023) Influence of post-processing heat-treatment on the mechanical performance of AISI 410L stainless steel manufactured by the L-DED process. J Laser Appl 035:1–9. https://doi.org/10.2351/7.0001135

Sousa JMS, Pereira M, Thiesen et al (2024e) Tribological performance of 410L martensitic stainless steel parts manufactured by laser directed energy deposition. J Eng Tribol 239:715–735. https://doi.org/10.1177/13506501241290383

Thiesen (2021) Selection of processing parameters for the laser directed energy deposition process applied to additive manufacturing: a methodological proposal. Master's thesis, Graduate Program in Engineering and Mechanical Sciences, Federal University of Santa Catarina, Joinville, Brazil

ASTM G3 (2019) Standard practice for conventions applicable to electrochemical measurements in corrosion testing. ASTM International 1–9. https://doi.org/10.1520/E1415-22

Reddy L, Preston SP, Shipway PH et al (2018) Process parameter optimization of laser clad iron based alloy: predictive models of deposition efficiency, porosity and dilution. Surf Coat Technol 349:198–207. https://doi.org/10.1016/j.surfcoat.2018.05.054

Aversa A, Saboori A, Librera E et al (2020) The role of directed energy deposition atmosphere mode on the Microstructure and mechanical properties of 316L samples. Addit Manuf 34:101274. https://doi.org/10.1016/j.addma.2020.101274

Shen LC, Yang XH, Ho JR et al (2020) Effects of build direction on the mechanical properties of a martensitic stainless steel fabricated by selective laser melting. Materials 13:1–18. https://doi.org/10.3390/ma13225142

Tian Y, Chadha K, Aranas C (2021) Laser powder bed fusion of ultra-high-strength 420 stainless steel: Microstructure characterization, texture evolution and mechanical properties. Mater Sci Eng A 805:140790. https://doi.org/10.1016/j.msea.2021.140790

Lu SY, Yao KF, Chen YB et al (2015) The effect of tempering temperature on the microstructure and electrochemical properties of a 13 wt.% Cr-type martensitic stainless steel. Electrochim Acta 165:45–55. https://doi.org/10.1016/j.electacta.2015.02.038

Nayebi B, Najafi H, Farnia A (2021) Microstructural, hardening and wear characteristics of surface re-melted AISI 410S stainless steel via fiber laser process. J Mech Sci Technol 35:4419–4426. https://doi.org/10.1007/s12206-021-0912-6

Gollapudi S (2012) Grain size distribution effects on the corrosion behavior of materials. Corros Sci 62:90–94. https://doi.org/10.1016/j.corsci.2012.04.040

Powell J, Koti D, Garmendia X et al (2023) Assessing the quality and productivity of laser cladding and direct energy deposition: guidelines for researches. J Laser Appl 35:012024. https://doi.org/10.2351/7.0000897

Nath SD, Irrinki H, Gupta G et al (2019) Microstructure-property relationships of 420 stainless steel fabricated by laser-powder bed fusion. Powder Technol 343:738–746. https://doi.org/10.1016/j.powtec.2018.11.075

Hufenbach J, Sander J, Kochta F et al (2020) Effect of selective laser melting on microstructure, mechanical, and corrosion properties of biodegradable FeMnCS for implant applications. Adv Eng Mater 22:1–8. https://doi.org/10.1002/adem.202000182

Ko G, Kim W, Kwon K et al (2021) The corrosion of stainless steel made by additive manufacturing: a review. Metals 11:1–21. https://doi.org/10.3390/met11030516

Krell PD, Li S, Cong H (2017) Synergistic effect of temperature and HCl concentration on the degradation of AISI 410 stainless steel. Corros Sci 122:41–52. https://doi.org/10.1016/j.corsci.2017.03.027

Bonagani SK, Vishwanadb B, Tenneti S et al (2019) Influence of tempering treatments on mechanical properties and hydrogen embrittlement of 13 wt% Cr martensitic stainless steel. Int J Press Ves Pip 176:103969. https://doi.org/10.1016/j.ijpvp.2019.103969

Cruz JR, Bertazzoli R (2020) Characterization of corrosion within friction stir weld zones of an API X-70 steel using a novel microcell setup. J Mater Eng Perform 29:98–108. https://doi.org/10.1007/s11665-019-04531-z

Jangir DK, Verma A, Sankar KM et al (2018) Influence of grain size on corrosion resistance and electrochemical behaviour of mild steel. Int J Res Pap Publ 6:287–288. https://doi.org/10.22214/ijraset.2018.4479

Cruz JR, Sousa JMS, Ferreira HS et al (2024) Impact of feedstock recycling on the corrosion resistance of inconel 625 processed by L-DED. Mater Res 27:1–16. https://doi.org/10.1590/1980-5373-MR-2023-0578

Shahriari A, Ghaffari M, Khaksar L et al (2021) Corrosion resistance of 13wt.% Cr martensitic stainless steels: additively manufactured CX versus wrought Ni-containing AISI 420. Corros Sci 184:109362. https://doi.org/10.1016/j.corsci.2021.109362

Development of Parameters for the Manufacture of Fe-3.5wt%Si by Laser Powder Bed Fusion (L-PBF)

Luane Maria de Brito Prata, Mariana Sizenando Lyrio, and Hugo Ricardo Zschommler Sandim

Abstract Additive manufacturing is a manufacturing method increasingly used in industrial processes. The laser powder bed fusion (L-PBF) manufacturing process allows the construction of complex geometries and mechanical properties equivalent to traditional manufacturing processes. However, the quality of the component depends on the manufacturing parameters used to process the material. Additive manufacturing of soft magnetic materials such as silicon steels is an alternative route to strengthening the crystallographic texture. However, the presence of pores and other defects in silicon steel components is detrimental to their magnetic properties. The objective of this work is to develop and validate parameters for manufacturing silicon steel with high density for powder layer thicknesses of 30 μm and 60 μm. The 60 μm thick layer was proposed to cover a processing window that allows for higher productivity, as it reduces scanning time by using twice the standard 30 μm thickness. For the first fabrication, laser power values of 200, 300, and 400 W, scanning speed values of 400, 600, 800, and 1000 mm/s and hatch distances of 80, 100, and 120 μm were selected for each layer thickness for fabrication. For the second fabrication, a regression was performed with the results obtained to construct a response surface for each processing window in order to determine the best set of parameters for validation. Through density analyses by the Archimedes method and porosity by image, it was possible to obtain density results higher than 99.5% for the validation parameters.

Keywords L-PBF · Silicon steels · Parameters development · Density

L. M. de Brito Prata (✉)
Federal University of Santa Catarina, Joinville, SC, Brazil
e-mail: luane.prata@grad.ufsc.br

M. Sizenando Lyrio · H. R. Z. Sandim
University of Sao Paulo, Lorena, SP, Brazil

L. M. de Brito Prata
SENAI Institute for Innovation in Laser Manufacturing and Processing Systems, Joinville, SC, Brazil

V. E. Beal and A. S. Ribeiro Junior (eds.), *Advances in Manufacturing Engineering Research*, Lecture Notes in Mechanical Engineering,
https://doi.org/10.1007/978-3-032-09858-0_4

1 Introduction

L-PBF technology enables the construction of complex geometries and mechanical properties equivalent to traditional manufacturing processes (ASTM 2015; Munhoz 2017). It also allows the development of topologies not achievable using other manufacturing techniques. (Yadroitsev and Smurov 2010) The quality of a component manufactured by L-PBF depends on the manufacturing parameters. The main manufacturing parameters are scan speed, laser power, layer thickness, and hatch distance, from which the volumetric energy density is calculated. The combination of these parameters improves process control. Volumetric energy density (VED) is a parameter that allows us to monitor the amount of defects and influences the consolidation of the powder feedstock (Gu 2012) and used to compare samples under different sets of deposition parameters. VED is a parameter used following the processing logic of a line (single track) to a layer and then to a block (Yap 2016). Ng et al. (2009) show that very high or low VED values cause greater manufacturing defects. The work of King et al. (King 2014) shows that when VED is too high, it generates the keyhole effect in the manufactured sample. Scipioni et al. (2017) showed that this effect occurs in ferrous materials, such as 316L steel, above 150 J/mm^3 regardless of the thickness of powder layer.

The main defects that can occur during manufacturing are lack of fusion and gas-entrapped porosity. Lack of fusion is primarily influenced by specific energy; this defect occurs when the energy is insufficient to fully fuse the powder material. Gas entrapment porosity is related to powder flow, weld pool geometry, and powder quality. (Ng 2009) Cracks are defects that can also be observed. Their causes vary, such as lack of power during manufacturing and the composition of the manufactured material. In silicon steels, for example, Feliciano (2023) reported that higher silicon contents can promote more cracks during manufacturing.

Silicon (or electrical) steels are characterized as soft magnetic materials and can contain varying percentages of silicon depending on the application. Soft magnetic materials are defined as being magnetized with low excitation, achieving high values of magnetic induction, having high magnetic permeability and leading to lower energy losses per magnetization cycle (Rodrigues-Vargas 2023). Silicon increases alloy resistivity and stabilizes ferrite, and because of its properties, silicon steels are a material used in electric machines (rotary motors and transformers), where they are responsible for transporting and amplifying magnetic flux density. (Giannotta 2023) Several studies have sought to understand the relationships between process conditions, microstructure development, alloy composition, heat treatment, component roughness, and magnetic properties. Garibaldi et al. (2018) manufactured Fe-6.9%Si alloy using L-PBF, varying the scanning speed and linear laser energy. They characterized the raw material, studied heat treatment to relieve residual stress, and analyzed the magnetic properties of the samples. They concluded that the material's porosity and crystallographic texture are the main factors influencing its magnetic properties. Lyrio and coauthors (Lyrio 2024) manufactured Fe-3.5wt%Si samples using L-PBF with different scanning strategies and applied thermomechanical processing to evaluate the microstructure and magnetic properties of thin sheets. They observed that the scanning strategy influences the sample texture and the intensity of (abnormal) grain growth after annealing.

This work reports the effects of laser scanning speed, laser power, and hatch distance parameters on density and microstructure of a Fe-3.5 wt%Si (wt-%) silicon steel fabricated by L-PBF. Two fabrication runs were performed to evaluate these parameters: an exploratory experiment and a validation experiment. The as-built samples were analyzed by Archimedes' apparent density tests, image analysis, and microhardness for all samples performed in the validation experiment.

2 Experimental

The samples were fabricated from a powder feedstock with nominal composition Fe-3.5%Si (wt-%) supplied by M4P material solutions GmbH (Austria) with a particle size range suitable for L-PBF. The machine used was an SLM 125HL from SLM Solutions, with a 125 × 125 × 125 mm^3 chamber, controlled argon atmosphere and heated plate (316L stainless steel) with a maximum temperature of 200 °C. The laser source used was an IPG Photonics YLR-400-SM, with a maximum power of 400 W, a continuous-wave laser with a focal diameter (spot size) of 70–100 μm. The arrangement of the test specimens on the construction platform was planned using Magics® software. The scanning strategy used for this study was 45° relative to the gas flow, with a 90° change for each layer. The parameters studied are layer thickness, laser power, scanning speed, and hatch distance. Layer thickness indicates the height of the powder layer deposited to fabricate the component's layers and is a key factor in determining manufacturing time. Laser power indicates the energy of the laser beam striking the material. Scan speed, expressed in mm/s, represents the speed at which the laser travels through the powder bed. Hatch distance gives the overlap distance between the lines the laser travels to fabricate the layer and depends on the laser spot size. In this work, two fabrications were performed: the first fabrication was for the exploratory experiment and the second for the validation experiment.

2.1 Exploratory Experiment Design

The objective of the exploratory experiment was to evaluate the material processing with two-layer thicknesses, each with 27 parameters. Exploratory tests were based on ISO/ASTM (2018). The powder layer thickness of 30 μm was selected according to the literature, since this allows the manufacture of components with a higher level of precision, and from this, a productivity evaluation was proposed with a powder layer thickness of 60 μm, since this allows a reduction in manufacturing time for parts that do not have a high level of complexity.

The power, scan speed, and hatch distance were selected to ensure a solid component with low porosity and other defects from the volumetric energy density (VED). The combination of the parameters involved is given in J/mm^3 and its expression is given by Eq. (1). The variables in the equation are the nominal laser power (P), the hatch distance (h), the scan speed (v), and the layer thickness (t).

$$\text{VED} = \frac{P}{v.h.t} \tag{1}$$

For this study, experimental runs with a scanning speed of 400 mm/s for the 30 μm layer, which generated very high VED calculations (above 250 J/mm^3), were therefore disregarded. For the 60 mμ layer, on the other hand, the speed of 1000 mm/s generated low VED (below 40 J/mm^3), and these experiments were also disregarded. In total, 54 samples were fabricated and analyzed, with VED ranging from 35 to 278 J/mm^3. The $10 \times 10 \times 15$ mm^3 samples were fabricated for analysis of the parameters presented in Table 1. All of these samples were fabricated on the same build platform. For each layer, the powder spreader deposited a 30 μm layer, so samples with a layer thickness of 60 μm were left unscanned for one layer, allowing components of different thicknesses to be manufactured on the same platform. Table 1 presents the main L-PBF processing parameters for the exploratory tests.

After fabrication, the samples were removed from the substrate by wire EDM on an Agie Charmilles CUT 20 P and subjected to the Archimedes density test. This was performed on a Mettler Toledo balance with ethanol at room temperature, with two replicates to evaluate the volumetric density of each sample.

The image analysis to quantify porosity was performed after sectioning, resin embedding, and metallographic surface preparation. To section the samples, a perpendicular cut was made to the substrate to assess the manufacturing direction. A Zeiss light optical microscope was used to capture images for determining porosity and for microstructure analysis, and ImageJ® software was used for image processing. Vickers microhardness was calculated from the average of five measurements taken from different regions of the sample using a Wilson 402MVD microhardness tester with a load of 0.5 kgf (4.9 N) and a dwell time of 10 s.

2.2 Validation Experiments

Following the results of the exploratory experiment, the set of parameters with the best results was determined for the validation experiment. To determine these results, a response surface was generated using Python code for the absolute density values of each hatch distance, and regression was performed to identify the set of parameters that would yield the best results. The validation experiment was conducted with eight results from the data processing, obtained through analysis of the results of the exploratory experiment, and three replicates of each.

Table 1 L-PBF processing parameters for the exploratory experiment

Sample	VED [J/mm^3]	Powder layer thickness [μm]	Laser power [W]	Laser scanning speed [mm/s]	Hatch distance [μm]
1	138.89	30	200	600	80
2	111.11	30	200	600	100
3	92.59	30	200	600	120
4	104.17	30	200	800	80
5	83.33	30	200	800	100
6	69.44	30	200	800	120
7	83.33	30	200	1000	80
8	66.67	30	200	1000	100
9	55.56	30	200	1000	120
10	208.33	30	300	600	80
11	166.67	30	300	600	100
12	138.89	30	300	600	120
13	156.25	30	300	800	80
14	125.00	30	300	800	100
15	104.17	30	300	800	120
16	125.00	30	300	1000	80
17	100.00	30	300	1000	100
18	83.33	30	300	1000	120
19	277.78	30	400	600	80
20	222.22	30	400	600	100
21	185.19	30	400	600	120
22	208.33	30	400	800	80
23	166.67	30	400	800	100
24	138.89	30	400	800	120
25	166.67	30	400	1000	80
26	133.33	30	400	1000	100
27	111.11	30	400	1000	120
28	104.17	60	200	400	80
29	83.33	60	200	400	100
30	69.44	60	200	400	120
31	69.44	60	200	600	80
32	55.56	60	200	600	100
33	46.30	60	200	600	120
34	52.08	60	200	800	80
35	41.67	60	200	800	100
36	34.72	60	200	800	120

(continued)

Table 1 (continued)

Sample	VED [J/mm^3]	Powder layer thickness [µm]	Laser power [W]	Laser scanning speed [mm/s]	Hatch distance [µm]
37	156.25	60	300	400	80
38	125.00	60	300	400	100
39	104.17	60	300	400	120
40	104.17	60	300	600	80
41	83.33	60	300	600	100
42	69.44	60	300	600	120
43	78.13	60	300	800	80
44	62.50	60	300	800	100
45	52.08	60	300	800	120
46	208.33	60	400	400	80
47	166.67	60	400	400	100
48	138.89	60	400	400	120
49	138.89	60	400	600	80
50	111.11	60	400	600	100
51	92.59	60	400	600	120
52	104.17	60	400	800	80
53	83.33	60	400	800	100
54	69.44	60	400	800	120

3 Results and Discussion

3.1 *Exploratory Experimental Results*

All 54 proposed samples were fabricated. Fabrication lasted about 11.5 h, with a maximum oxygen concentration during the process of 0.1%. Figure 1 shows the build platform after removal from the machine.

The results of density measurements using the Archimedes' method in relation to the VED of the samples can be seen in Fig. 2. The relative density of the samples was compared with the density of a Fe-3.4%Si (weight%) silicon steel reported by Smithells (1976) as 7.86 g/cm^3. In density measurements using the Archimedes method, it can be observed that all manufactured samples resulted in dense samples, with densities above 90%. The highest results are observed in the range of 80–140 J/mm^3, with relative density results above 98.50%.

The porosity obtained by image analysis as a function of VED showed maximum relative density results above 99.5% for layer thicknesses of 30 µm and 60 µm. The highest density value achieved was 99.89% for a sample with VED of 104.17 J/mm^3 processed with a layer thickness of 30 µm, power of 200 W, hatch distance of 80 µm and scanning speed of 800 mm/s. The lowest result was 88.60% for a sample with

Fig. 1 Build platform with 54 as-built samples from the exploratory experiment

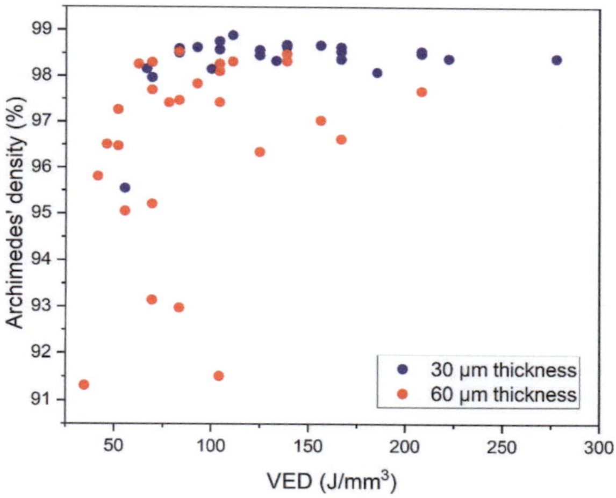

Fig. 2 Archimedes density for the samples manufactured in the exploratory tests as a function of the VED for layer thicknesses of 30 and 60 μm

VED of 83.33 J/mm^3 with a layer thickness of 60 μm, power of 200 W, hatch distance 100 μm, and speed of 400 mm/s.

Different types of defects were observed in the as-built samples: cracks, irregular porosities, and spherical (circular) porosities. Figure 3 shows each type of defect observed in different as-built samples. Among the defects observed, the presence of spherical pores is the one that occurs in smaller dimensions and in more isolated spaces, so samples containing only this defect obtain higher relative density values. A comprehensive review of the various types of defects in L-PBF-processed materials, their detection and control can be found in (Gordon 2020). Figure 4 presents two graphs showing the type of defect observed in the sample in relation to its relative

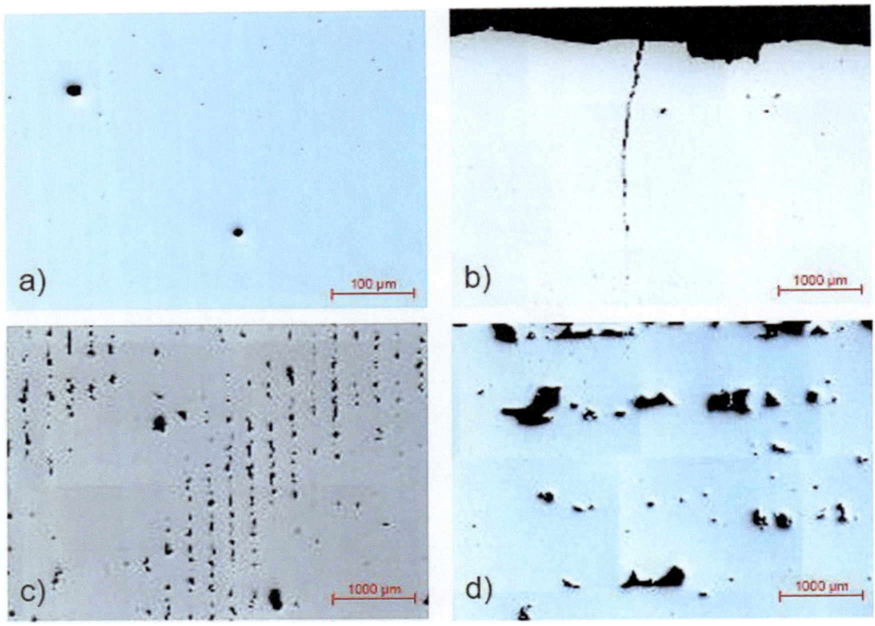

Fig. 3 Defects observed in non-etched as-built samples processed by L-PBF: **a** spherical (circular) pores; **b** cracks; **c** irregular pores due to insufficient overlap of the weld pool; and **d** irregular pores due to lack of fusion

density and the VED associated with the L-PBF processing parameters for each layer thickness.

Defects that cause irregular pores, such as lack of fusion and lack of pool overlap, occur primarily in samples with low VED values. The effect of irregular pores with a vertical pattern or lack of melt pool overlap, as shown in Fig. 3c, is associated with hatch distance values of 120 µm. This effect likely occurs because the hatch distance is greater than the laser focal diameter. This effect is mitigated in samples with higher power outputs; since higher power outputs direct more energy to the tracks. Cracks can be observed in both directions and occur mainly in samples manufactured with a thickness of 60 µm and a power of 400 W; however, this characteristic is not directly associated with the VED value.

The samples that obtained the highest Archimedes density, did not present cracks or porosity due to lack of fusion and with Vickers microhardness comparable to the wrought materials, were selected as the best parameter sets for each layer thickness. The samples manufactured with a 60 µm layer thickness obtained results with high dispersion in all tests performed, indicating that this provides a narrower processing window. Despite this, it was possible to obtain satisfactory results from a sample manufactured with a layer thickness of 60 µm. The sample with the best set of parameters was the one with a VED of 104.17 J/mm^3, manufactured with a power of 300 W, hatch distance of 80 µm and laser scanning speed of 600 mm/s. The

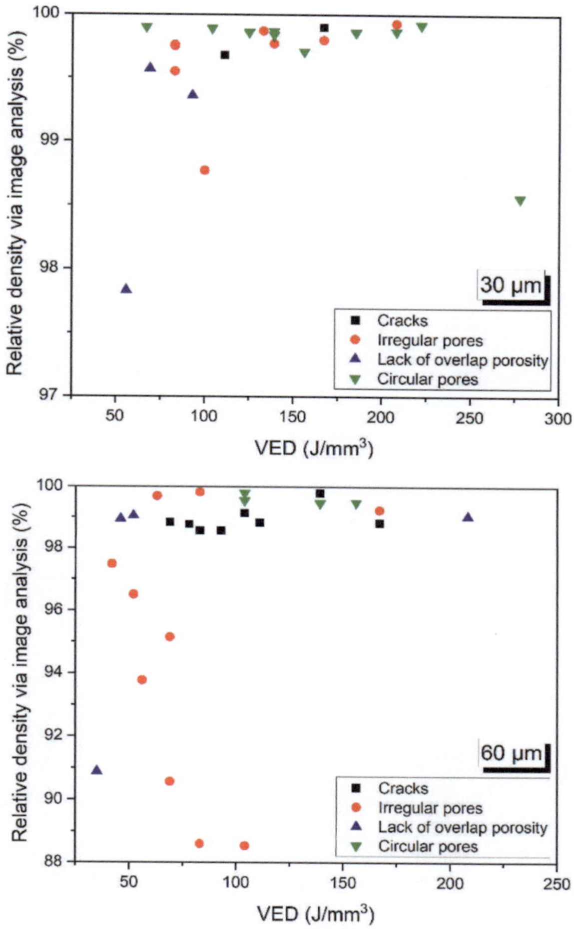

Fig. 4 Dispersion of the main types of defects observed in the sample in relation to the VED for as-built samples with layer thicknesses of: **a** 30 μm; **b** 60 μm

resulting Archimedes density was 7.585 g/cm^3 (98.76% relative density) or 99.89% when determined by image analysis. The only observed defects were spherical pores, which are less detrimental to mechanical properties.

The objective of the exploratory experiment was to find a set of parameters that would produce a dense sample with minimal defects. For the validation experiment, the results of the previous fabrication were used to perform a second fabrication using only parameters that presented samples with high densities to evaluate density, microstructure, and microhardness. The parameters were determined based on the analysis of the results of the exploratory experiment. The values of each layer thickness and each hatch distance were separated and a three-dimensional surface graph was plotted, with their respective laser power, scanning speed, and Archimedes density results. Figures 5 and 6 show the response surfaces for each set of parameters for samples manufactured with thicknesses of 30 μm and 60 μm, respectively.

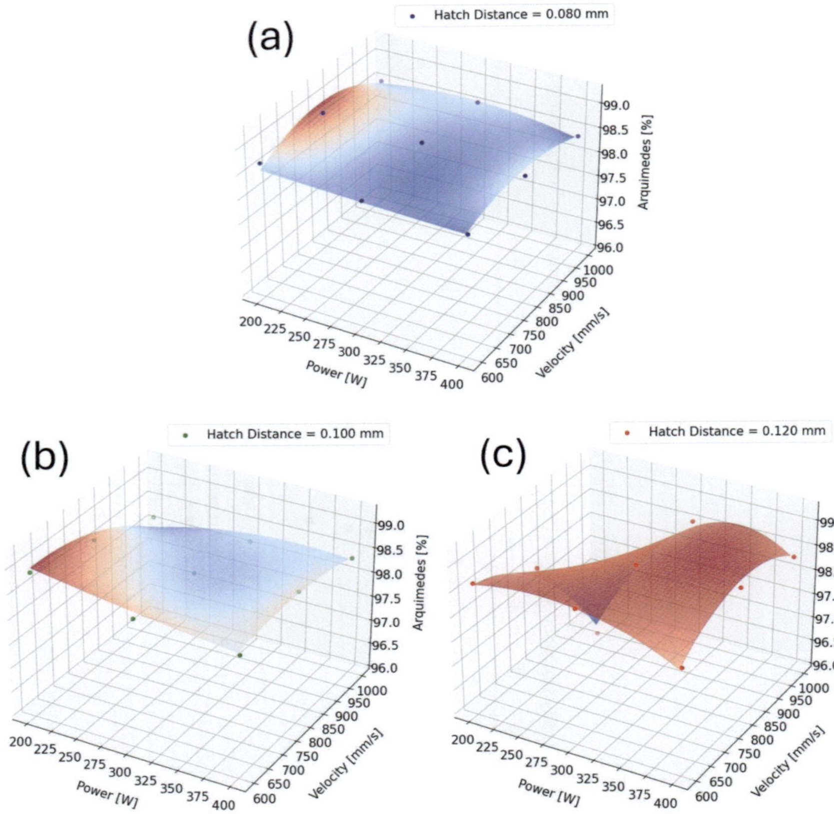

Fig. 5 Graphical representation of the response surface of the samples manufactured with a thickness of 30 µm for a hatch distance of: **a** 80 µm, **b** 100 µm; **c** 120 µm

After determining the response surface, the set of parameters that would theoretically yield the highest density value was calculated. These results are the parameters selected for the validation process. A set of parameters was then proposed that extrapolate this response surface to four dimensions, where the hatch distance can vary and is no longer predetermined. Two additional sets of parameters were obtained, one for each layer thickness.

3.2 *Validation Experiment*

Following the results of the first stage (exploratory tests), the set of parameters with the best results for the validation experiment was determined. To determine these results, a response surface was generated for the absolute density values of each hatch

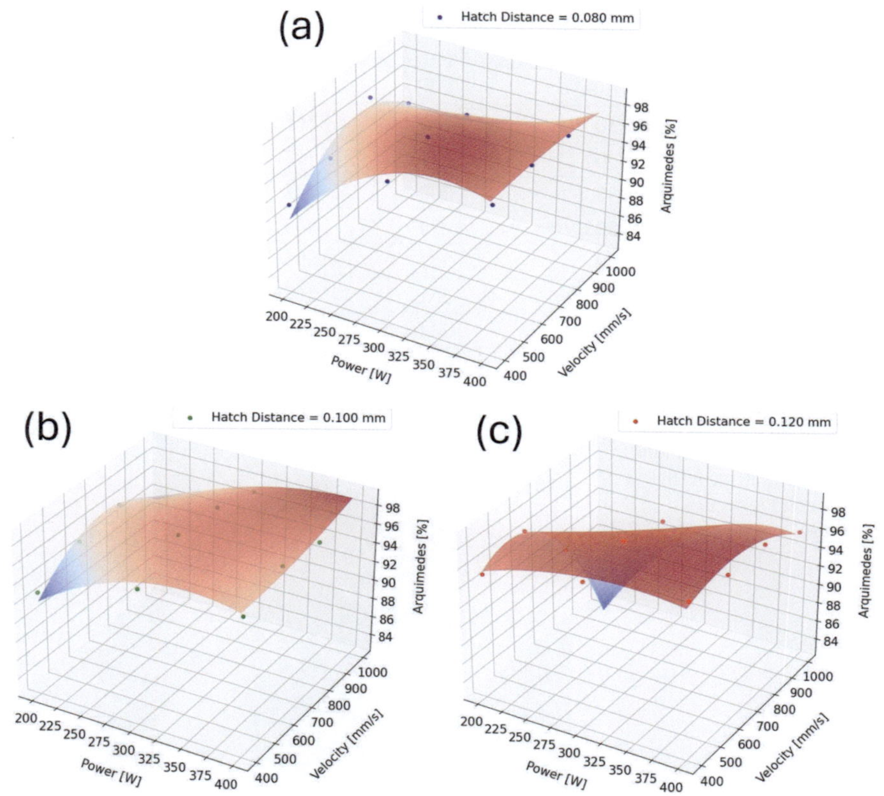

Fig. 6 Graphical representation of the response surface of the samples manufactured with a thickness of 60 μm for a hatch distance of: **a** 80 μm, **b** 100 μm; **c** 120 μm

distance, and regression was performed to identify the set of parameters that would yield the best results. The validation experiment was conducted with eight results from the data processing, obtained through analysis of the results of the exploratory experiment, and three replicates of each. Table 2 presents the parameters of the manufactured samples for the validation tests.

Each parameter set was fabricated with three replicates, totaling 24 successfully fabricated samples. Manufacturing of this validation batch also took approximately 11.5 h at a maximum oxygen concentration of 0.1%. Figure 7 shows the platform after removal from the machine. The results of the density measurement obtained by the Archimedes method, the image density calculated from the sample's sectioned surface area in ImageJ software, and the microhardness are presented in Table 3.

The validation experiment samples achieved excellent results, with Archimedes density values greater than 7.5 g/cm^3 and image density values greater than 99%. The computational evaluation revealed satisfactory results, equivalent to those estimated using the Archimedes method. The Vickers microhardness results for the as-built

Table 2 Calculated L-PBF processing parameters used in the validation experiments

Condition	VED [J/mm^3]	Layer thickness [μm]	Laser power [W]	Scanning speed [mm/s]	Hatch distance [μm]
1	102.46	30	200	765	90
2	105.07	30	200	793	80
3	106.23	30	200	627	100
4	117.43	30	358	848	120
5	122.69	60	351	596	80
6	129.57	60	351	565	80
7	83.33	60	344	689	100
8	132.05	60	400	420	120

Fig. 7 Build platform with 24 as-built samples from the validation experiment (8 distinct conditions in triplicate)

Table 3 Results of the validation experiment

Condition	VED [J/mm^3]	Density from Archimedes principle [g/cm^3]	Relative density from image analysis [%]	Vickers micro hardness [HV]
1	102.46	7.564 ± 0.006	99.36 ± 0.12	244 ± 5
2	105.07	7.558 ± 0.009	98.88 ± 0.05	240 ± 4
3	106.23	7.568 ± 0.012	99.53 ± 0.19	241 ± 2
4	117.43	7.559 ± 0.007	98.75 ± 0.05	249 ± 3
5	122.69	7.556 ± 0.005	98.70 ± 0.06	244 ± 4
6	129.57	7.567 ± 0.005	98.91 ± 0.09	250 ± 3
7	83.33	7.576 ± 0.006	99.09 ± 0.03	243 ± 3
8	132.05	7.527 ± 0.017	98.24 ± 0.11	247 ± 2

samples ranged from 240 to 250 HV-0.5, very close to the values reported in the literature by (Lyrio 2024, 2025) for different powder feedstock suppliers.

4 Conclusions

The aim of this work was to study how scanning speed, laser power, and hatching distance influence the processing of Fe-3.5%Si (wt%) fabricated by L-PBF. Archimedes density and image analysis tests were used to determine the best manufacturing parameters for producing dense parts. An exploratory study was proposed, manufacturing 54 samples varying these parameters. The main results of this study are as follows:

- Samples with a layer thickness of 30 μm show higher density results than those with a layer thickness of 60 μm. The difficulty in processing the latter can be observed because the fabrication of the 60 μm layer requires a greater amount of energy to melt the layer.
- High hatch distance values cause defects such as irregular pores due to insufficient melt pool overlap. The effect of irregular pores with a vertical pattern or lack of overlap melt pools is associated with hatch distance of 120 μm and low VED. High laser powers for a layer thickness of 60 μm can induce crack formation.
- Samples with the same VED presented different results in terms of density and types of defects. The VED alone is not representative, so it is necessary to analyze the entire set of parameters to determine the most appropriate processing window.
- The response surface selection method for manufacturing validation proved to be suitable for parameter selection processing, with relative density results exceeding 98%, reaching values of 99.5%.

Acknowledgements LMBP thanks the Institute for Innovation in Laser Manufacturing and Processing Systems, to Dr. H. R. Oliveira e M. Bento for their guidance. HRZS and MSL are also thankful to FAPESP for the financial support (Grant 2020/10174-0). HRZS is CNPq Fellow (National Council for Scientific and Technological Development, Brazil) under Grant 308.296/2021-4.

Competing Interests The authors have no conflicts of interest to declare that are relevant to the content of this chapter.

Responsibility for Information The authors are solely responsible for the information included in this work. No AI systems were used in the translation or writing of the text.

References

ASTM Standard 52900 (2015) Standard terminology for additive manufacturing—general principles—terminology. ASTM International. West Conshohocken, PA

Feliciano FR (2023) Development and characterisation of soft magnetic iron-silicon alloys for laser beam powder bed fusion additive manufacturing. Dissertation. Universidade de Lisboa

Garibaldi M, Ashcroft I, Hilier N, Harmon SAC, Hague R (2018) Relationship between laser energy input, microstructures and magnetic properties of selective laser melted Fe-6.9%wt Si soft magnets. Mater Charact 143:144–151

Gianotta N, Sala G, Bianchini C, Torreggiani A (2023) A review of additive manufacturing of soft magnetic materials in electrical machines. Machines 11:1–23

Gordon JV, Narra SP, Cunningham RW, Liu H, Chen H, Suter RM, Beuth JL, Rollett AD (2020) Defect structure process maps for laser powder bed fusion additive manufacturing. Addit Manuf 36:101552

Gu D, Hagedorn Y, Meiners W, Meng G, Batista RJS, Wissenbach K, Poprawe R (2012) Densification behavior, microstructure evolution, and wear performance of selective laser melting processed commercially pure titanium. Acta Mater 60(9):3849–3860

ISO/ASTM DIS 52902 (2018) Additive manufacturing—test artefacts—standard guideline for geometric capability assessment of additive manufacturing systems. ASTM International. West Conshohocken, PA

King WE, Barth HD, Castillo VM, Gallegos GF, Gibbs JW, Hahn DE, Kamath C, Rubenchik AM (2014) Observation of keyhole-mode laser melting in laser powder-bed fusion additive manufacturing. J Mater Process Technol 214:2915–2925

Lyrio MS, Aota LS, Sandim MJR, Sandim HRZ (2024) Additive manufacturing of Fe-3.5 wt.-%Si electrical steel via laser powder bed fusion and subsequent thermomechanical processing. J Mater Sci 59(9):4019–4038

Lyrio MS, Oliveira HR, Sandim MJR, Devulapalli V, Sandim HRZ (2025) Effect of the scanning strategy on texture of grain-oriented electrical steel (Fe-4wt%Si) processed via laser powder-bed fusion and subsequent thermomechanical processing. Mater Charact 221:114789

Munhoz ALJ, Costa CA, Ahrens CH, Carvalho J, Santos JRL, Silva JVL, Foggiatto JA, Lima MSF, Volpato N (2017) Manufatura Aditiva: Tecnologias e Aplicações da Impressão 3D. São Paulo: Edgard Blücher

Ng GKL, Jarfors AEW, Bi G, Zheng HY (2009) Porosity formation and gas bubble retention in laser metal deposition. Appl Phys A-Mater 97(3):641–649

Rodrigues-Vargas BR, Stornelli G, Folgarait P, Ridolfi MR, Miranda Pérez AF, Di Schino A (2023) Recent advances in additive manufacturing of soft magnetic materials: a review. Materials 16(16):5610

Scipioni U, Wolfer AJ, Mathews MJ, Delplanque JR, Schoenug JM (2017) On the limitations of volumetric energy density as a design parameter for selective laser melting. Mater des 113:331–340

Smithells CJ (1976) Metals reference book, 5th edn. Butterworths, London and Boston, p 1976

Yadroitsev I, Smurov I (2010) Selective laser melting technology: from the single laser melted track stability to 3D parts of complex shape. Phys Proced 5:551–560

Yap CY, Chua CK, Dong ZL (2016) An effective analytical model of selective laser melting. Virtual Phys Protot 11(1):20–26

ALTPRINT: Algorithm for Additive Planning with Local Variation of Filling of the Extruded Filaments

Ricardo Dias, Daniel Couto, Henrique Idogava, and Zilda Castro Silveira

Abstract This article presents an update to the AltPrint program by adapted to use two filament feeding mechanisms: direct drive and Bowden tube and also allows the generation of continuous trajectories, with the incorporation of modifications to extend the program to 3D printers with "Bowden" type filament feed and allow the generation of continuous trajectories. The AltPrint program is an algorithm developed to integrate into additive planning extruded material flow control, to generate parts with flexible regions. The computational implementation was designed from Open Source concept using the Python language, integrated with an open additive planning program. The complex multifactorial relationship between material flow control at the extrusion nozzle and indexed movement of the build platform as well as process parameters and environmental conditions, considering desktop 3D printers, directly affects the dimensional quality of the parts. These conditions may affect final quality of the mesostructure (such as, for example, lack or excess of interface between the deposited filaments, generating internal material discontinuities) and in the final part. These failures contribute to the reduction in the mechanical strength of the part, reduced surface and dimensional quality with pronounced ghosting by extrusion. On the other hand, control strategies on parameter "E" can contribute to the generation of preselected flexible regions, which is part of the proposal of this work. This numerical strategy allows the generation of geometries and parts with flexible regions through the generation of beams, similar to the behavior of mechanical elastic elements. However, different architectures of desktop 3D printing equipment are also sources of variability in the control of parameter "E." In this way, as result of these improvement proposals, experimental geometries demonstrating visual quality in the flexible regions for both types of filament feeding mechanisms as well continuous trajectory generation demonstrating the program's potential to generate functional parts with elastic regions.

R. Dias (✉) · D. Couto · Z. C. Silveira
University of São Paulo, São Carlos, SP, Brazil
e-mail: ricardo.dias78@hotmail.com

H. Idogava
Federal University of Vale do São Francisco, Petrolina, BA, Brazil

V. E. Beal and A. S. Ribeiro Junior (eds.), *Advances in Manufacturing Engineering Research*, Lecture Notes in Mechanical Engineering,
https://doi.org/10.1007/978-3-032-09858-0_5

Keywords Additive manufacturing · FDM · Functional parameterization · Process calibration · Flexible parts

1 Introduction

Over the course of nearly four decades, most of the processes that comprise additive manufacturing (AM) technology have significantly expanded their capabilities to produce prototypes in various forms—from low-fidelity to high-fidelity—including experimental and evolutionary prototypes, as well as the direct manufacturing of individual parts and, in some applications, functional technical subassemblies. From the perspective of the Design for Additive Manufacturing (DfAM) concept, aimed at the customized production of parts, the orientation by design attributes and engineering materials, in conjunction with the selected additive process, enables enhanced functionalities (Loh et al. 2018; Ngo et al. 2021).

Recent research has addressed new configurations of multi-axis AM systems (Chalvin et al. 2020; Roach et al. 2019; Schuh et al. 2020; Duarte et al. 2022), new strategies for toolpath optimization (Hu et al. 2022), adaptive digital slicing (Gokhale and Kala 2022; Chen et al. 2022), and advanced DfAM methodologies (Leary 2020)). The integration of new computational geometric design strategies, as well as the development of multiscale constitutive models with cellular and lattice-based design approaches, enables novel part functionalities (Paolini et al. 2019), such as the use of metamaterials (Mohsenizadeh et al. 2018; Askari et al. 2020; Santos et al. 2024), topology optimization (Gardan, and Schneider, 2015), assembly-free components that require no post-AM assembly (Wei et al. 2016; Hallmann et al. 2019), auxetic structures (Alomarah et al. 2020; Zhang et al. 2020), and functional multi-jointed mechanisms (Chen 2022; Schaechtl et al. 2020).

Despite some parametric variations depending on the additive technique employed, additive planning incorporates features from conventional manufacturing, particularly regarding the use of G-code programming for executing the additive workflow. However, the advantage of greater freedom in exploring different Fused Filament Fabrication (FFF) techniques using low-cost equipment, combined with the use of open-source digital slicing software and embedded electronics, can partially translate into the drawback of frequently requiring a calibration process for each equipment architecture and even for changes in raw material. These conditions are reflected in the generated mesostructure (e.g., insufficient or excessive interface between deposited beads, causing internal material discontinuities) and in the final part, resulting in reduced mechanical strength and poor surface finish, with pronounced undulations (Santana et al. 2017).

In this context, Moetazedian et al. (2021) argue that addressing the flow behavior of the extruded material begins with the design of the digital geometric model and the treatment of its features, adjusted in conjunction with digital slicing parameters and equipment precision (Majeed et al. 2021; Ryan et al. 2021).

Therefore, when using desktop 3D printing equipment, especially with open architecture, the calibration process is a fundamental step to achieve higher-quality parts. The raw material in solid (filament) form, typically a specific thermoplastic preselected from a single manufacturer, can exhibit variations in composition, such as the use of plasticizers and additives, which affect not only the extrusion temperature but also the entire additive manufacturing process strategy. Frontier knowledge in areas such as materials science/engineering materials, additive manufacturing/digital slicing software, and post-processing can lead to the development of more robust calibration methods, which ensure printability conditions related to filament stiffness and velocity during deposition, geometric characteristics of the calibrated nozzle, and filament behavior according to thermoplastic rheological parameters, including volumetric flow at the deposition nozzle, shear rate, and chemical composition.

In the context of this work, the extrusion-based additive technique using Fused Deposition Modeling (FDM) was selected due to its wide variety of thermoplastic filament materials (Balani 2019; Thompson et al. 2016), low operational and maintenance costs, open-hardware concept, and because it still presents research gaps as previously mentioned—particularly in the generation of parts with localized flexibility and the possibility of fabricating functional multi-jointed mechanisms. These mechanisms can serve applications ranging from consumer goods, mechanical connectors, electrical components, artificial baits, masks (Bonenberger 2016), to assistive technology, especially devices for activities of daily living and orthoses Silveira (2022).

Extrusion-based additive techniques offer numerous application opportunities that extend beyond purely technological fields, increasingly reinforced by developments in different research areas: (a) advanced materials: broader variety of engineering materials responsive to external *stimuli* (4D printing), biocompatibility and osteointegration (tissue engineering), composites enabling applications requiring higher mechanical strength or specific load activation (e.g., conductivity, bactericidal properties), customized pharmaceuticals, fabrication of orthoses and prostheses, and large-scale additive manufacturing (BAAM—Big Area Additive Manufacturing) (Roberson, et al., 2015; Ullah et al., 2023; Gardan, 2018; Gutiérrez, 2018; Silveira 2022; Khalid, et al., 2022); (b) new equipment configurations (Justino Netto, et al., 2021); (c) additive planning (Idogava et al. 2023); (d) process monitoring Dvorak (2019).

The program named AltPrint, presented in Idogava et al. (2023), is characterized as a method within additive planning, enabling controlled flexibility in functional regions through adaptations of the “E” parameter, related to the geometric characteristics of the filament/deposition nozzle and the flow rate of the molten material. The method involves varying the unidirectional width of extruded beads, integrated with digital slicing software, and has been made available on the open GitHub platform (Couto et al. 2021), applied to low-cost 3D printing equipment with direct-drive feed architecture. This article presents a refinement of the AltPrint program, with reprogrammed routines to incorporate Bowden-type 3D printer architectures and generate continuous printing toolpaths.

1.1 Considerations About FDM Technique

Regardless of the additive technique, additive planning begins with the import of a three-dimensional geometric model, which can be generated using Computer-Aided Design (CAD) software, micro-computed tomography images, scanners, among others. The model is then exported to a file in STL (ASCII-encoded) or AMF format. During this transition, the 3D model loses resolution, as its representation is generated through a triangular mesh to create the geometric surface of the object, with each line of code storing details about each triangle, such as the coordinates of each unique vertex.

The STL format has no provisions for representing colors, textures, materials, mesh structures, units, or other properties of the finished printed part. The format used to describe 3D models in Additive Manufacturing File format (AMF) is based on Extensible Markup Language (XML). AMF describes color, material, texture, and mesostructures. An AMF file is essentially a simplified set of triangles combined with a set of scripts defining edge curves and shape details. The advantages of the AMF format compared to STL are related to the use of these curved triangles, which improve accuracy while reducing the amount of data required to represent a model at the same fidelity as an STL file.

The simplified working principle of the FDM technique, schematized in Fig. 1a, uses solid feedstock in the form of a thermoplastic filament. The first part of the extrusion head contains a pair of drive rollers, responsible for pulling the filament—stored on a filament spool—and pushing it along a feed channel into the heating zone. In this region, the material transitions from a solid to a semi-solid state. The filament itself, when driven forward, acts as a plunger at the inlet of the print head, generating the pressure required to force the heated material through a calibrated nozzle.

While depositing the material onto the build plate or onto an already fabricated layer, the extruder head moves along the *X* and *Y* axes according to the object's design and infill parameters (Fig. 1b). After the construction of a layer, the build plate moves downward along the *Z*-axis in increments corresponding to the layer thickness (Crump 1992; Kretschek 2012; Santana 2015;).

The volumetric flow rate is controlled by the relationship between the deposition speed (s_d) specified by the user, as well as the expected bead width (w) and bead height (h). According to Chaunier et al. (2018), practical values show that an ideal layer height (h) should be equal to or less than 75% of the calibrated nozzle diameter ($h \leq 0.75$ Do). The slicing software adjusts the filament feed length and feed speed

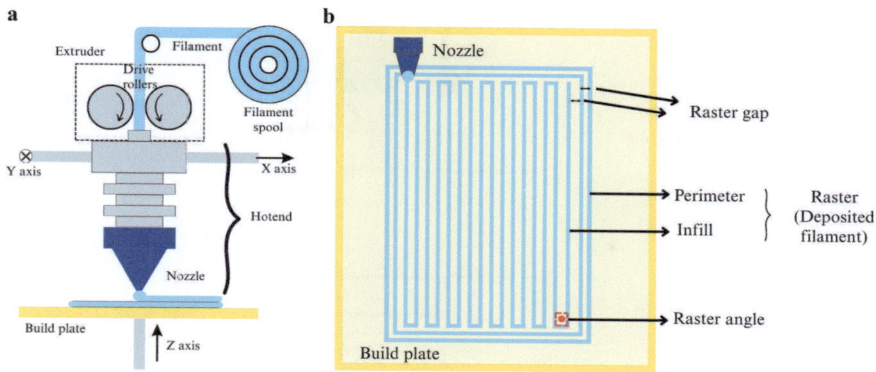

Fig. 1 FDM system: **a** schematic of the FDM/FFF process; **b** main elements and parameters associated with the FDM technique

(E and s_i, respectively) based on the print head speed (s_h), assuming a simplified rectangular cross-section for the deposited bead (section. The linear relationship between these variables is given by Eq. 1.

$$s_d Wh = s_i E \tag{1}$$

The deposition path, also referred to as the toolpath, consists of a series of coordinate locations, as the software used shares the same G-Code standard applied in Computer Numerical Control (CNC) equipment. These coordinates define the contour and infill pattern for a given layer of the part under construction. The path is generated according to the slicing parameters specified by the user, typically following a contour-and-raster strategy to preserve the external features of the part. Multiple toolpaths generated for a common rectilinear infill pattern are illustrated in Fig. 2.

More specifically, the deposition path corresponds to the trajectory traced by the center of the calibrated nozzle's circular cross-section. The target area to be filled is discretized according to the nozzle diameter (D_o), which often results in unfilled voids between deposited beads, thereby compromising the mechanical strength of the printed parts. The dimensions of these voids can be reduced by allowing adjacent beads to overlap; however, this approach entails a loss of dimensional accuracy in the overlap regions. Although the deposition direction may be kept constant across all layers, it is generally preferable to adopt weaving patterns—similar to those used in composite manufacturing—in order to achieve improved mechanical properties (Chaunier et al. 2018). The weaving pattern is defined by the raster angle, which is set individually for each layer.

In addition to bead geometry (which determines raster width and layer height) and raster angle, other fundamental printing parameters involved in toolpath generation include contour thickness, layer height, and raster gap (Fig. 1b). Once toolpaths are generated for each layer, they are converted into G-code, which can be considered the

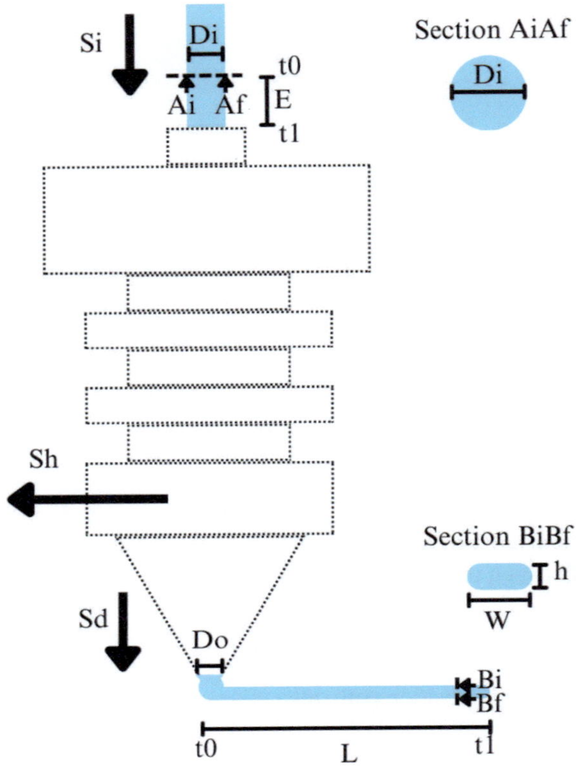

Fig. 2 Typical deposition parameters for indexed motion in extrusion-based equipment

standard CNC programming language used in fused filament fabrication (FFF). The G-code therefore encodes information about the amount of material to be deposited, the infill pattern, movement speeds (printing and travel moves), and the temperatures of both the extrusion system and the build platform. From a hardware perspective, 3D deposition is enabled by a positioning system integrated into the structural frame of the machine.

When the thermoplastic reaches its melting temperature within the reservoir, it is pressurized against the deposition nozzle by the incoming solid filament. The filament feed rate is a key parameter for ensuring consistent material distribution. The filament diameter (*Di*) is typically 1.75 mm and should remain as constant as possible so that the drive rollers can effectively advance it. The feeding mechanism can be mounted directly on the heated reservoir—referred to as a direct drive configuration—or fixed to the 3D printer's frame, in which case a polytetrafluoroethylene (PTFE) tube is used to guide the filament, a configuration known as a Bowden extruder. The pinch-roller/pulley system is responsible for advancing the filament into the heated reservoir (hot-end), as illustrated in Fig. 1a.

Accordingly, the system must deliver a constant volumetric flow of material to avoid under- or over-extrusion during deposition. The drive roller is pressed against the idler roller, generating a normal force, which is further amplified by a spring-loaded screw mechanism that increases the contact pressure. The heating power available is generally around 40 W, which imposes limitations on both the maximum feed rate and the achievable extrusion temperature (Chaunier et al. 2018). Insufficient heating results in feeding issues due to the excessively high viscosity of the molten material.

A fundamental criterion for 3D printing capability relates to the constitutive and rheological properties of the feedstock material, under the condition that the extrusion pressure must remain below the critical buckling stress. Another critical aspect concerns the control of material flow. In FDM printheads, the volumetric extrusion rate (V) is determined by the relationship between the feed rate (F_R) and the filament diameter (Di), as given by Eq. 2.

$$V = \frac{F_R \pi Di^2}{4} \tag{2}$$

In practice, the extruder is treated as an independent subsystem, and for a given filament diameter, the volumetric flow rate is calculated by the slicing software, taking into account the linear speed of the drive rollers (s_i) and the filament feed length (E) at the inlet of the print head (Chaunier et al. 2018). This relationship is expressed by Eq. 3:

$$V = s_i E \tag{3}$$

The feedstock can be conveyed using two types of filament transport systems to the extrusion nozzle: the feeding mechanism (extruder) positioned directly on the heated reservoir (Direct-drive) or the feeding mechanism mounted on the 3D printer frame, with the filament guided to the reservoir through a PTFE tube (referred to as a Bowden tube extruder). Figure 3a and b illustrate these two configurations.

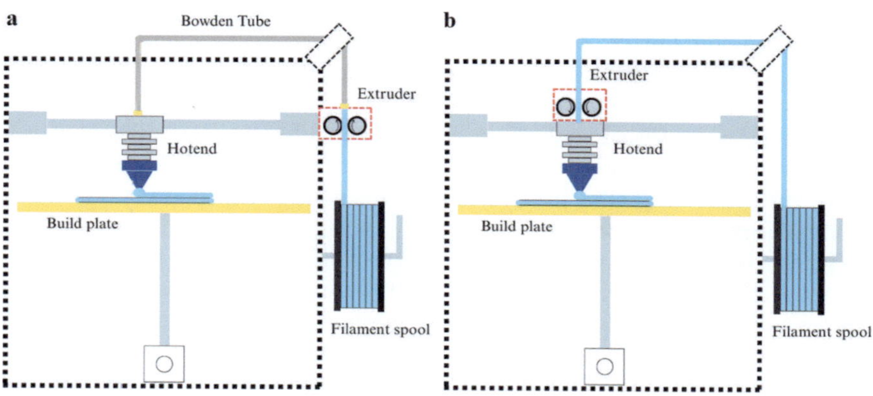

Fig. 3 Desktop 3D printer architecture: **a** Bowden tube; **b** direct-drive

2 Methodology

2.1 Materials

The thermoplastic copolymer Acrylonitrile Butadiene Styrene (ABS) was selected due to its widespread industrial application, offering moderate flexibility, impact absorption capacity, and availability at Lab3D (SEM-EESC-USP).

The filament-form ABS produced by GTMax 3D (Brazil) exhibits an amorphous molecular structure, with a glass transition temperature of 100 °C, melting temperature ranging from 210 °C to 230 °C, elastic modulus of 0.23 MPa, and tensile strength of 0.0044 MPa.

The low-cost 3D printing equipment used included the Sethi 3D AiP A3 (direct-drive) and the GTMax3D Core A1 Dual (Bowden tube), both Brazilian-made. The Sethi 3D AiP A3 has the following technical specifications: print volume of 9,240 cm^3 in an open chamber and layer resolution between 0.1 mm and 0.3 mm. The GTMax3D Core A1 Dual features a print volume of 15,000 cm^3 in a closed chamber, layer resolution between 0.05 mm and 0.32 mm, and maximum print and travel speeds of 120 mm/s and 200 mm/s, respectively. Regarding temperatures, the printhead reaches up to 295 °C, while the build platform can be heated up to 110 °C.

The software used for 3D modeling of the parts and test specimens in this work was Fusion 360, version 2.0.20508, developed by Autodesk with an educational license.

For the implementation of the algorithm, the Python programming language (version 3) was chosen. Python is an interpreted, interactive, multi-paradigm, open-source language that, combined with certain libraries, allows for an efficient and agile implementation of the algorithm (Venners 2003). Among the libraries employed in the project, the Shapely package (Gillies et al. 2020) provides tools for manipulation and analysis of planar geometric objects, which are very useful for toolpath generation; the Trimesh package (Dawson-Haggerty 2020) enables import and manipulation of STL files; the NumPy library (2019) provides numerical processing functions; and the Matplotlib library (Hunter (Hunter 2007) allows the creation of graphs and diagrams, useful for trajectory visualization. These libraries are hosted on GitHub and maintained under open-source licenses.

Additionally, the Visual Studio Code editor (version 1.96 for Windows 10), developed by Microsoft under an open license, was used to facilitate the development of the AltPrint program, integrating the Python interpreter and its dependencies, including the libraries mentioned above.

Finally, the software tools Repetier-Host (version 2.3.2), Prusa G-code Viewer (version 2.8.0), and PrusaSlicer (version 2.8.0) were employed. The first two were used for graphical visualization of the G-code generated by AltPrint, facilitating error correction and optimization of the printing trajectories, while the latter, a commercial slicing software, was used to compare the trajectory generation with that produced by the experimental slicing software under development.

2.2 Methods

2.2.1 Development of the AltPrint Program

The assumptions considered for the development of the AltPrint program (Alt = Alternative; Print = Printing) were: open-source code and language, and the use of publicly available libraries, including Shapely, NumPy, and Matplotlib. For the preliminary version of the program, the two main alternative functions within extrusion-based additive manufacturing were: (a) slicing geometries into pre-defined regions within the same layer and (b) enabling control of material deposition through variation of print speed and material flow parameters, thereby generating variable bead width, considering a direct-drive architecture for FDM 3D printheads. A later version of the program focused on adapting the slicer software for the Bowden tube architecture.

According to Costa and Covas (2014), two types of defects may arise in FFF processes due to thermomechanical interactions, mainly caused by excessive deformation of the extruded filament during cooling, which can compromise part dimensions, and by insufficient inter- and intra-layer adhesion of deposited beads. These conditions can negatively affect the mechanical strength of the final part. For polymer-based materials, adhesion depends on the rheological properties of the material and process parameters, with a direct relationship to the slicing software and deposition strategy.

Beyond the configuration of slicing coordinates for printed parts, other parameters can be varied through the toolpath program, such as material flow. The "E" parameter is responsible for controlling the linear quantity of filament from the spool that will be used during printing.

For extrusion-based additive manufacturing using solid filaments, the raster width is strongly related to a set of parameters: material feed rate in the heating subsystem (hot-end), nozzle diameter, extrusion temperature, layer height, and indexed movement determined by the print speed (Jap et al. 2019; McLouth et al. 2017). However, both commercial and open-source programs present limitations regarding the variability of extruded raster width, control of material deposition in specific regions, and variation of deposition properties within a single layer.

The "E" parameter (Idogava et al. 2023) depends on multiple factors, such as nozzle calibration, extruded material temperature, and system cooling level. When monitored, these variables allow control over the deposited raster shape through material flow management. Controlling the process via the E parameter (Idogava et al. 2023) results in thinner raster sections, which are more susceptible to bending and can be used for the fabrication of advanced functional structures without compromising physical integrity.

The study and calibration of these parameters were explored by Couto et al. (2021) and Idogava et al. (2023), producing structures with flexible regions through localized material variation in printed segments (as illustrated in Fig. 4). This approach

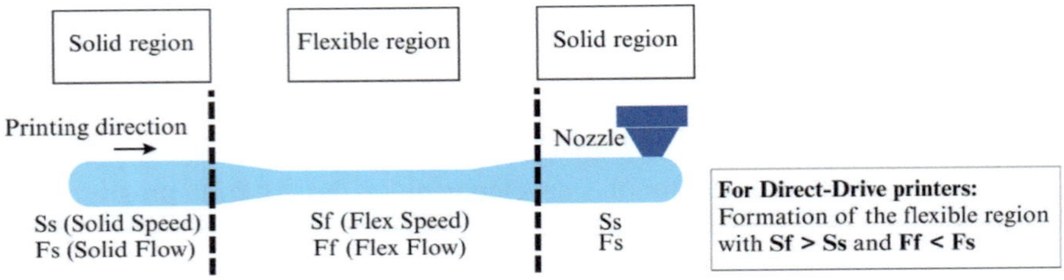

Fig. 4 Concept for the development of the AltPrint program: Deposition sequence to achieve a region with flexibility

demonstrates potential for the fabrication and mechanical evaluation of functional parts, such as snap-fit joints and mechanisms requiring flexibility.

In the updated version of the AltPrint program, the objective was to adapt the algorithm for Bowden tube-type printers, where the main difference compared to the previous direct-drive approach is that material deposition control occurs solely through variation of the print speed parameter, since the extruder is mounted on the printer frame. This design, illustrated in Fig. 5, results in a lighter hot-end, allowing higher print speeds with reduced quality loss compared to the direct-drive system, due to the lower inertia of the printhead during nozzle direction changes.

However, the material deposition parameter is extremely sensitive in this system, since the motion transmission for pushing or retracting the filament from the extruder to the nozzle is slower. This requires a more precise retraction value to prevent the filament from retracting so much that the molten material in the hot zone is pulled into the cold zone, solidifying and clogging the nozzle (Fig. 5—high retraction), or retracting too little, causing filament oozing during a travel move where no deposition is intended (Fig. 5—low retraction). These issues can lead to the formation of

Fig. 5 Filament behavior inside the hot-end due to improper retraction calibration: low retraction and high retraction

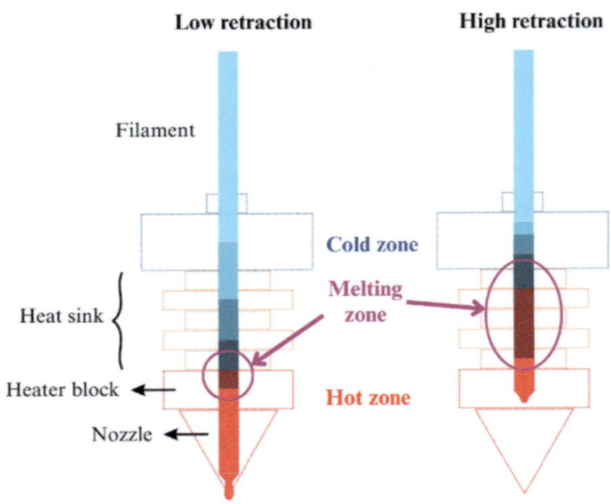

deformed bead bundles in the printed layers. In this context, it was observed that to generate the flexible region in 3D printers with a Bowden tube, the printing speed in the flexible region should be higher than the printing speed in the solid region, and the material flow for both regions should be the same value, that is, $S_f > S_s$ and $F_f = F_s$.

According to Dávila et al. (2016), extrusion processes commonly experience interruptions in material deposition between layers, particularly between the end of one layer and the start of the next, causing material accumulation points and compromising print quality. Therefore, a continuous printing toolpath offers considerable advantages. This can be achieved by defining strategies that minimize the distance between perimeters within the same layer, between perimeter and infill within the same layer, and between the infill of the current layer and the perimeter of the next layer, significantly reducing interruptions in material deposition. The fewer the Bowden tube system relies on filament retractions, the lower the risk of these issues occurring.

2.2.2 Algorithm Implementation with Extension for Bowden Tube Feed and Continuous Printing Toolpaths

The flowchart shown in Fig. 6 illustrates the logic implemented in the AltPrint program. Dashed boxes represent the improvements and adaptations introduced with the second version of the program: red boxes indicate adjustments of additive process parameters, integrated with strategies and algorithms for generating continuous toolpaths between layers, considering slicing direction and infill. Blue boxes represent the algorithms implementing the strategy for generating continuous printing toolpaths, following the continuity concept presented earlier.

This procedure was integrated into the AltPrint program (Idogava et al. 2023), enabling targeted flexible regions in combination with continuous printing toolpaths. The first and second versions of the AltPrint program are available at the following repositories: https://github.com/couto0/altprint and https://github.com/GustavoKanaiama/continuous_altprint.git.

3 Results

A visual comparison was performed between a part printed with the first version of the code (without continuous toolpaths) and the same part printed with the second version of the code (with continuous toolpaths). It was observed that the part without optimization exhibited the formation of a transverse bead bundle across the flexible region of the part, highlighted in green in Fig. 7b. The stacking of this bead bundle along the layers qualitatively resulted in a reduction of flexibility in this region. In contrast, for the part fabricated using the more optimized approach (Fig. 7c),

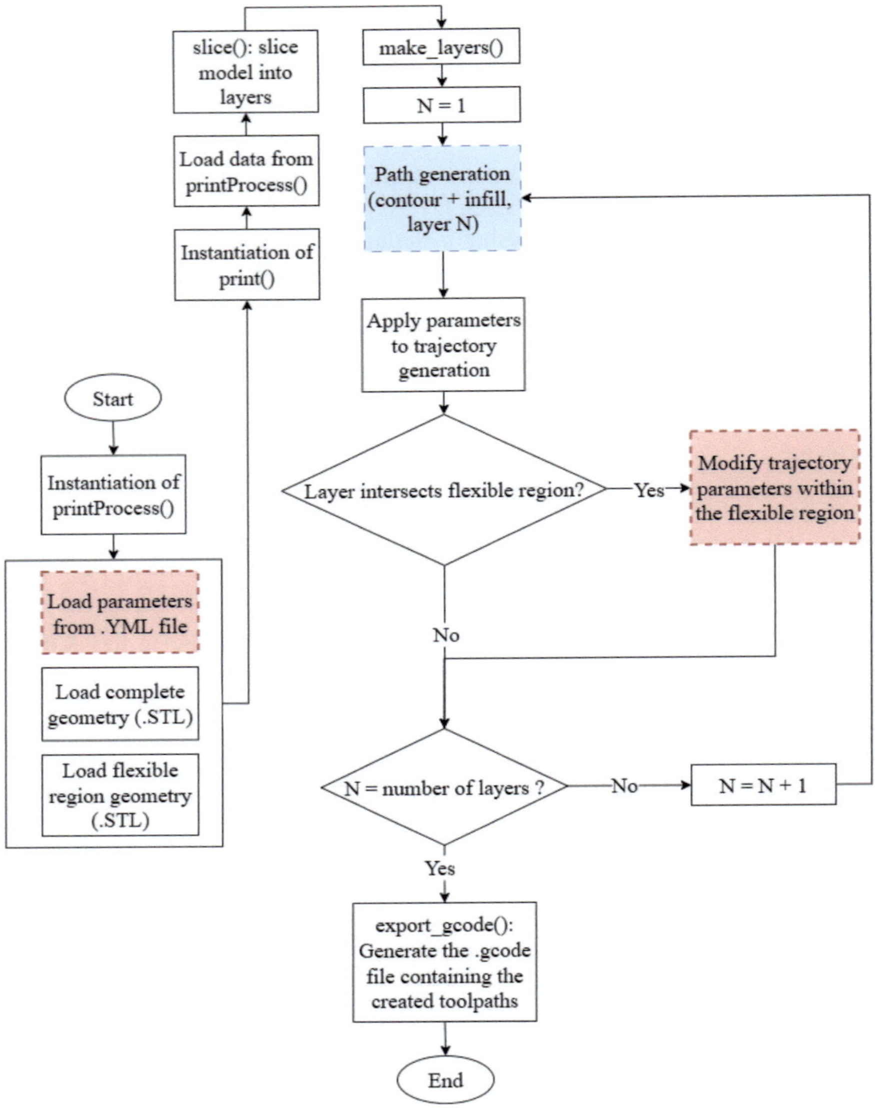

Fig. 6 Flowchart implemented for the AltPrint program

no irregular bead formation was observed, leading to greater flexibility in the pre-selected region compared to the previous approach.

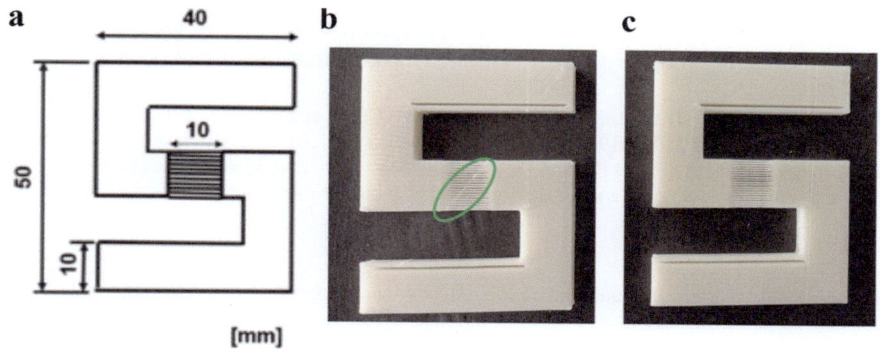

Fig. 7 "S"-shaped geometry used for comparison of implementations with ABS material: **a** geometry design with dimensions; **b** specimen printed without continuous toolpath optimization; **c** specimen printed with continuous toolpath optimization

4 Conclusions

The partial results presented in this work aim to demonstrate the development and experimental evaluation of the proposed algorithm integrated into additive planning for the FFF/FDM technique. The approach is based on variation of the extrusion rate combined with printing speed and retraction parameter control, as well as filament and nozzle geometric characteristics (correlated with the so-called "E" parameter).

This systematized approach, guided by the material extrusion rate and layer-by-layer fabrication, starting from additive programming and specifically digital slicing, allowed the generation of geometries with user-defined flexible regions.

It is important to differentiate between phenomena occurring before and after extrusion. The amount of material supplied at the nozzle outlet remains constant (disregarding polymer swelling effects). The deposition of variable-width segments results from the shear rate of the deposited material occurring after extrusion. In conjunction with the speed condition, it is necessary to ensure that material is present in the nozzle; otherwise, raster breakage may occur due to shear stresses.

The current version of AltPrint allows the generation of parts with layer-specific flexibility in user-selected regions, offering potential replacement of flexible mechanical elements, such as mechanical springs or snap-fits (as exemplified in Idogava et al. 2023). However, this version was only tested on direct-drive 3D printers using PLA thermoplastic material for cycling specimens and functional parts. The current version of the program has been expanded to Bowden tube 3D printer architectures and, importantly, generates continuous printing toolpaths, enabling more complex geometric variations and applicability to the most common desktop 3D printer architectures.

Future research will include additional cycling tests of "S"-shaped specimens to evaluate the three approaches: direct drive, Bowden tube, and continuous toolpath generation, as well as assessments through dynamic mechanical analysis (DMA), confocal imaging for morphology, and the fabrication of functional parts.

Acknowledgements The authors would like to thank CNPq and FAPESP, under grant numbers 2023/11404-7 and 118105/2019-1, respectively, as well as Prof. Dr. Marcelo Aparecido Chinellato and technician Ricardo Gomes Pereira (Department of Materials Engineering, EESC-USP) for their kind support with equipment and assistance in carrying out the experimental tests.

Declarations

Responsibility Notice The authors are the only responsible for the printed material included in this paper and the translations were performed using OpenAI's ChatGPT (GPT-5 model).

References

Alomarah A, Masood SH, Ruan D (2020) Out-of-plane and in-plane compression of additively manufactured auxetic structures. Aerosp Sci Technol 106:106107

Askari M et al (2020) Additive manufacturing of metamaterials: a review. Addit Manuf 36:101562

Balani SB et al (2019) Influence of printing parameters on the stability of deposited beads in fused filament fabrication of poly(lactic) acid. Addit Manuf 25:112–121

Bonenberger PR (2016) The first snap-fit handbook (3rd ed., pp. I–XXII). Hanser

Chalvin M et al (2020) Layer-by-layer generation of optimized joint trajectory for multi-axis robotized additive manufacturing parts revolution. Robot Comput Integr Manuf 65:101960

Chaunier L et al (2018) Material extrusion of plant biopolymers: opportunities & challenges for 3D printing. Addit Manuf 21:220–233

Chen Q et al (2022) Volumetric adaptive slicing of manifold mesh for rapid prototyping based on relative volume error. Rapid Prototyping J 28:602–626

Costa SF, Duarte FM, Covas JA (2014) Thermal conditions affecting heat transfer in FDM/FFE: a contribution towards the numerical modelling of the process. Virtual Phys Prototyping 10(1):35–46

Couto DMS, Idogava HT, Silveira ZC (2021) AltPrint: alternative 3D printing. Retrieved from https://github.com/couto0/altprint

Crump SS (1992) Apparatus and method for creating three-dimensional objects. US Patent 5121329

Dávila J et al (2016) Software to generate 3-D continuous printing paths for the fabrication of tissue engineering scaffolds. Int J Adv Manuf Technol 84:1671–1677

Dawson-Haggerty M (2020) Trimesh documentation. Retrieved from https://trimsh.org/index.html

Duarte J et al (2022) Curved layer path planning on a 5-axis 3D printer. Rapid Prototyping J 28:629–636

Gardan N, Schneider A (2015) Topological optimization of internal patterns and support in additive manufacturing. J Manufact Syst, 37:417–425. https://doi.org/10.1016/j.jmsy.2014.07.003

Dvorak C (2019) Measurement and analysis of pressure profile within big area additive manufacturing single screw extruder. In: solid freeform fabrication symposium proceedings.

Gardan J (2018) Smart materials in additive manufacturing: state of the art and trends. Virtual Phys Prototyp, 14(1):1–18. https://doi.org/10.1080/17452759.2018.1518016

Gillies S et al (2019) Shapely 1.7a2 documentation. Retrieved from https://shapely.readthedocs.io/en/latest/project.html

Gokhale NP, Kala P (2022) Development of a deposition framework for implementation of a region-based adaptive slicing strategy in arc-based metal additive manufacturing. Rapid Prototyping J 28:466–478

Gutiérrez TJ (2018) 3D printing of biopolymers: trends and opportunities for medical applications. In: Handbook of biopolymers. Jenny Stanford Publishing. 45–74

Hallmann M et al (2019) Optimization of build time and support material quantity for the additive manufacturing of non-assembly mechanisms. Proced CIRP 84:271–276

Hu Y, Jiang X, Huo G, Su S, Li H, Zheng Z (2022) A novel adaptive slicing algorithm based on ameliorative area ratio and accurate cusp height for 3D printing. Rapid Prototyp J, 28:453–465

Hunter J, Dale D, Firing E, Droettboom M (2012) Matplotlib: Version 3.0.3 [Computer software]. https://matplotlib.org/

Idogava HT, Couto DS, Santana L, Alves FJL, Silveira ZC (2023) AltPrint: new filling and slicing planning based on deposited material with geometry variation. Rapid Prototyping J

Jap NS et al (2019) The effect of raster orientation on the static and fatigue properties of filament deposited ABS polymer. Int J Fatigue 124:328–337

Justino Netto JM, Idogava HT, Frezzatto Santos LE, Silveira ZDC, Romio P, Alves JL (2021) Screw-assisted 3D printing with granulated materials: a systematic review. Int J Adv Manuf Technol, 115(9):2711–2727

Khalid A, Wei Y, Saleem MR, Lughmani WA (2022) Meso scale component manufacturing: a comparative analysis of non-lithography and lithography-based processes. J Micromech Microeng, 32(6):063002

Kretschek D (2012) Desenvolvimento de um cabeçote de extrusão por êmbolo para polipropileno granulado visando a manufatura aditiva.

Loh GH, Pei E, Harrison D, Monzón MD (2018) An overview of functionally graded additive manufacturing. Addit Manuf, 23:34–44. https://doi.org/10.1016/j.addma.2018.06.023

Leary M (2020) Design for additive manufacturing. New York: Elsevier

Majeed A et al (2021) A big data-driven framework for sustainable and smart additive manufacturing. Robot Comput Integr Manuf 67:102026

McLouth TD et al (2017) The impact of print orientation and raster pattern on fracture toughness in additively manufactured ABS. Addit Manuf 18:103–109

Moetzzedian A (2021) Convex (continuously varied extrusion): a new scale of design for additive manufacturing. Addit Manuf 37:101576

Mohsenizadeh A (2018) Additively manufactured lightweight metamaterials for energy absorption. Mater des 139:521–530

Ngo TD et al (2018) Additive manufacturing (3D printing): a review of materials, methods, applications and challenges. Compos B 143:172–196

NumPy (2019) NumFocus: open code, better science. Retrieved from https://numpy.org

Paloni A, Kollmannsberger S, Rank E (2019) Additive manufacturing in construction: A review on processes, applications, and digital planning methods. Addit Manuf, 30:100894

Roach DJ et al (2019) The m4 3D printer: a multi-material multi-method additive manufacturing platform for future 3D printed structures. Addit Manuf 29:100819

Roberson DA, Torrado Pérez A, Shemelya CM, Rivera A, MacDonald E, Wicker RB (2015) Comparison of stress concentrator fabrication for 3D printed polymeric Izod impact test specimens. Addit Manuf, 7:1–11. https://doi.org/10.1016/j.addma.2015.05.002

Ryan K (2021) Future of additive manufacturing: overview of 4D and 3D printed smart and advanced materials and their applications. Chem Eng J 403:126162

Santana L et al (2017) A study of parametric calibration for low-cost 3D printing: seeking improvement in dimensional quality. Mater des 135:159–172

Santana L (2015) Avaliação de uma impressora 3D baseada em projeto de código aberto na fabricação de peças em PLA (Master's thesis, Universidade Federal de Santa Catarina, Florianópolis)

Santos WF et al (2024) Exploring novel mechanics of metamaterial: unravelling deformation mode coupling and size effects through second-order computational homogenization. Int J Solids Struct 292:112724

Schaechtk P et al (2020) Tolerance analysis of additively manufactured non-assembly mechanisms considering joint clearance. Proced CIRP CAT 92:27–32

Schuh G et al (2020) Feature-based print method for multi-axis material extrusion in additive manufacturing. Proced CIRP 93:85–89

Silveira Z, de C (2022) Test bed problem: case study—design for customization of an assistive device for daily activities. In: VK.Manupati GD, Putnik & MLR Varela (eds.) Smart and sustainable manufacturing systems for industry 4.0, 1st edn. CRC Press, pp. 593-660.

Thompson MK et al (2016) Design for additive manufacturing: trends, opportunities, considerations, and constraints. CIRP Ann 65(2):737–760

Ullah M, Wahab A, Khan SU, Naeem M, ur Rehman K, Ali H, Ullah A, Khan A, Khan NR, Rizg WYAlkhalidi HM (2023) 3D printing technology: a new approach for the fabrication of personalized and customized pharmaceuticals. Eur Polym J, 195:112240

Venners B (2003) The making of Python: a conversation with Guido van Rossum—Part I. Retrieved from www.artima.com/intv/pythonP.html

Wei X et al (2016) A study on revolute joints in 3D printed non-assembly mechanisms. Rapid Prototyping J 22:901–933

Zhang J, Lu G, You Z (2020) Large deformation and energy absorption of additively manufactured auxetic materials and structures: a review. Compos B Eng 201:108340

Development of Customized Cooling Channels for Molds Using Numerical Simulations and Additive Manufacturing

Rafael Oliveira, João Pedro Cheloni, Gabriel Ribeiro, Rodrigo Silveira de Santiago, and Valter Estevão Beal

Abstract Among the various polymer manufacturing methods, the injection molding process stands out, being commonly used for producing highly complex parts in large production batches. However, due to the high production repetition rate and the frequent temperature variations inherent to the process, wear of the mechanical components becomes an intrinsic challenge. Consequently, metallic materials are typically employed in the mechanical assemblies of injection systems. The current demand for parts manufactured through injection molding requires high geometric precision in complex components, combined with the lowest possible cost. Therefore, it is essential to minimize waste, both in the mechanical components of the injection process and in the produced parts. Considering these requirements, two technological tools that contribute to the success of projects in this field are the use of additive manufacturing (AM) in combination with computational numerical simulations. The high thermal gradients involved in the plastic injection molding process, in which specific regions reach higher temperatures than others, often result in distortions in the manufactured parts. Therefore, it is essential to have a cooling system within the molds to promote a more uniform heat exchange and prevent geometric distortions in the produced components. An effective alternative for manufacturing customized cooling channels (conformal cooling) in complex metallic geometries, while ensuring efficient thermal exchange, is the Direct Energy Deposition (DED) process, in which a metallic wire is melted by a laser beam, forming a small molten pool and continuously depositing material layer by layer. In this study, different cooling channel designs are analyzed for a mold component used in the production of automobile cup holders. To enhance cooling efficiency, the channels incorporate complex architected structures known as Triply Periodic Minimal Surfaces (TPMS). The analyses are supported by numerical simulations based on the Computational Fluid Dynamics (CFD) method, aiming to estimate the pressure profile on the mold surface, pressure drop, flow velocity fields, and regions with higher and lower heat

R. Oliveira (✉) · J. Pedro Cheloni · G. Ribeiro · R. S. de Santiago · V. E. Beal
SENAI CIMATEC University, Salvador, BA, Brazil
e-mail: rafaelemos01@gmail.com

V. E. Beal and A. S. Ribeiro Junior (eds.), *Advances in Manufacturing Engineering Research*, Lecture Notes in Mechanical Engineering,
https://doi.org/10.1007/978-3-032-09858-0_6

transfer rates. The results obtained contribute to a more accurate definition of operational requirements, supporting the selection of the most suitable pump and the optimal process parameters for injection molding.

Keywords Additive manufacturing · Direct energy deposition · Conformal cooling · Injection mold · Computational fluid dynamics

1 Introduction

In the field of thermoplastic injection molding, efficient mold cooling is crucial to ensure both quality and productivity. According to Marques et al. (2015), the mold cooling time accounts for two-thirds of the total cycle time, making it parameter with the greatest impact on overall production time. Typically, mold cooling systems are constructed with linear, simple-shaped channels, and the geometric parameters for cooling channel spacing in steel molds are based on practical experience (Soares et al. 2023). However, this approach may be insufficient for molds with complex geometries, where optimized cooling is essential.

The application of conformal cooling involves the use of internal cooling channels within the mold, providing more uniform and efficient cooling. According to Wang and Lee (2023), the use of conformal channels can reduce cooling time by 30% compared to conventional methods, while also resulting in a more uniform temperature distribution throughout the part. Additive manufacturing, specifically the Direct Energy Deposition (DED) process, has emerged as a promising technology for producing these conformal cooling channels, as it enables the creation of complex and customized geometries that would be unfeasible or cost-prohibitive with traditional methods. The DED process, which uses metallic wires melted layer by layer by a laser beam, allows the fabrication of highly complex geometries that cannot be reproduced using conventional methods and, when possible, would incur high manufacturing costs.

In this context, the use of architected structures, known as Triply Periodic Minimal Surfaces (TPMS), in combination with cooling channels has shown great potential to enhance heat transfer in heat exchangers (Dutkowski et al. 2022; Kaur and Singh 2021). Moreover, studies have demonstrated, beyond experimental approaches, the design and optimization of these structures for heat exchangers (Mahmoud et al. 2023; Reynolds et al. 2023). Consequently, TPMS offer significant advantages, primarily due to weight reduction and increased thermal efficiency in heat exchange. However, architected structures also present challenges, such as manufacturing complexity and control of mechanical and thermal properties in critical regions of the manufacturing process, requiring a more in-depth analysis of their manufacturability and performance for the intended application (Cheloni et al. 2025).

To ensure the effectiveness of the different profile types selected in this study, an analysis using Computational Fluid Dynamics (CFD) was conducted. Through numerical evaluation of fluid flow behavior and heat transfer within the cooling channels, it is possible to assess the thermal performance of the mold, adjust inlet and outlet parameters, and refine the topology of the cooling channels. Consequently, CFD analysis reduces the need to manufacture a large number of prototypes and provides a detailed study of the operational conditions to which the mold may be subjected. This study aims to analyze the behavior of a mold with and without TPMS structures in the cooling channels, focusing on the use of conformal cooling channels manufactured via the DED process and evaluated through CFD. The objective is to deepen the understanding of how additive manufacturing, particularly DED, can be applied to enhance the efficiency and quality of industrial products and projects, specifically through the integration of cooling channels that conform to the mold geometry.

2 Methodology

Two types of cooling channels were designed for a metallic injection mold for plastic parts, which will be manufactured using the DED process. The mold in question is a male insert of a cupholder, as shown in Fig. 1, developed using SolidWorks software. The cooling channel features two inlets and two outlets for the cooling fluid (water). Figure 1 illustrates, through a cross-sectional view, the arrangement of the cooling channel within the mold.

The design strategy was based on maximizing the contact area between the mold and the fluid to enhance heat transfer, while also positioning a thermocouple near the center of the mold. This approach was adopted due to the geometric freedom provided by the additive manufacturing (DED) process. The mold features a minimum wall thickness of 2 mm, in accordance with the constraints imposed by the machine (Meltio 2023).

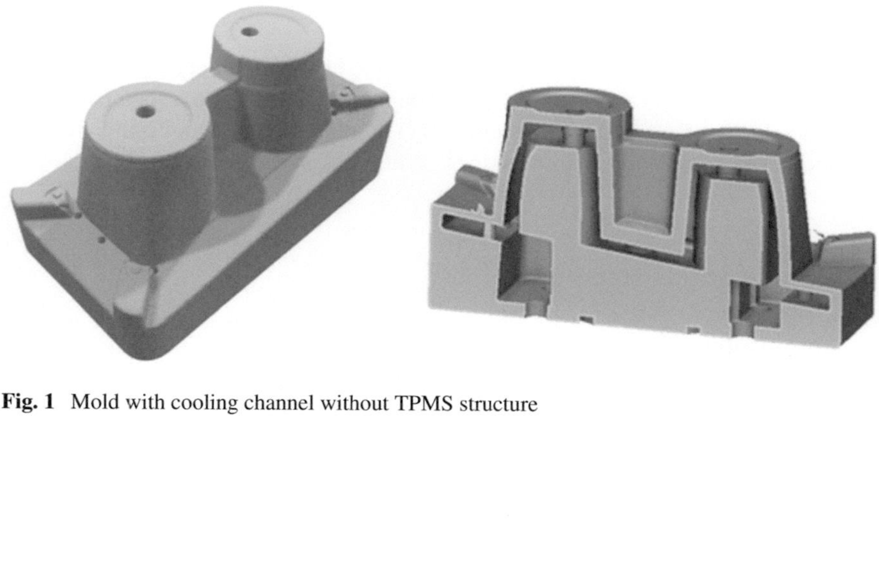

Fig. 1 Mold with cooling channel without TPMS structure

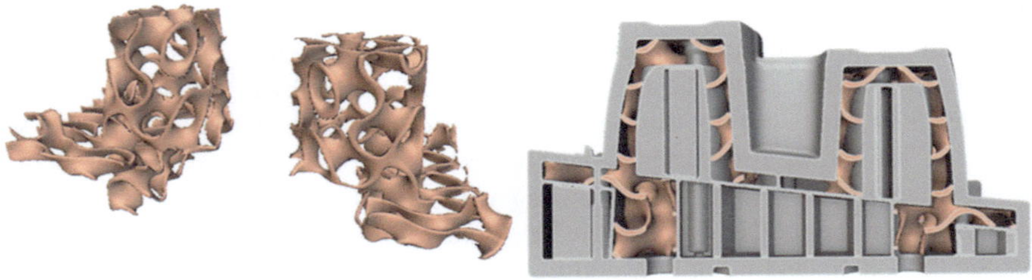

Fig. 2 TPMS structure used in the cooling channel

In order to enhance heat transfer efficiency during the injection molding process, an architected lattice structure of the Triply Periodic Minimal Surface (TPMS) type was incorporated into the cooling channel using nTopology software, which provides tools to handle complex geometries implicitly. The lattice cells used measured 30 × 40 × 40 mm, with a thickness of 4 mm. Figure 2 shows the TPMS structure integrated into the cooling channel.

Computational numerical simulations were conducted on the two types of cooling channels developed in this study to investigate the thermal behavior of the mold. To optimize the performance of the cooling channels, it is important to estimate the pressure profile on the mold surface, pressure drop, flow velocity fields, regions of higher and lower heat transfer rates, and wall shear stress.

For this purpose, a numerical model based on Computational Fluid Dynamics (CFD) was developed using AcuSolve software through the SimLab interface by Altair. Three different inlet flow rates were analyzed in this study: 0.01 kg/s, 0.2 kg/s, and 1 kg/s, where the lowest flow corresponds to laminar flow and the two higher flows correspond to turbulent regimes. The inlet diameter was 14 mm, with water properties of dynamic viscosity 0.001 Pa·s, density 1000 kg/m3, specific heat 4183 J/kg·K, and structural steel properties for the mold. A steady-state thermal simulation was performed, assuming an ambient temperature of 24 °C, a top surface temperature of 90 °C, and an outlet pressure of 0 Pa.

An average element size of 1.5 mm was assigned for both the mold and the fluid, with 10 elements in the boundary layer of 0.06 mm thickness. The turbulence model adopted was the Shear Stress Transport (SST) model. Figure 3a and b illustrate the simulation setup and boundary conditions, while Fig. 3c shows the mesh employed.

3 Results and Discussions

Preliminary results were obtained from CFD simulations for the mold structure with cooling channels, both without and with the presence of the TPMS structure. Figure 4 presents an isometric view, a cut along the YZ section, and a cut along the XZ section. In the latter, a narrow region housing the thermocouple can be observed, along with

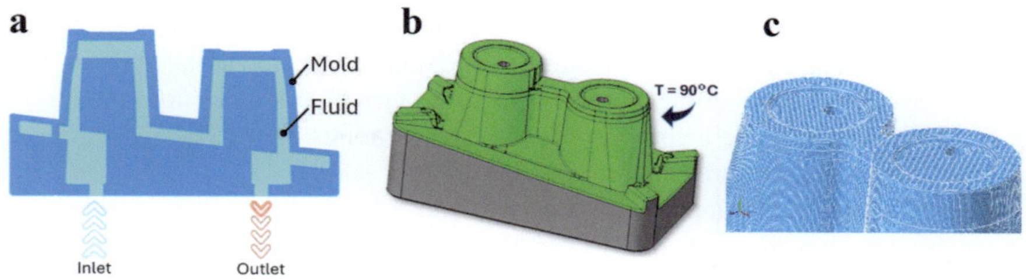

Fig. 3 **a** Scheme used in the numerical simulation **b** surface where a temperature of 90 °C was applied **c** mesh used in the model

temperature variations in the mold without TPMS, considering a surface temperature of 90 °C and a cooling fluid inlet temperature of 24 °C, for three different flow rates: 0.01, 0.2, and 1 kg/s.

Figure 4a, corresponding to a flow rate of 1 kg/s, shows a more uniform temperature distribution throughout the structure, maintaining the inlet and outlet temperatures, as well as the zone around the central hole, close to ambient temperature. In contrast, Fig. 4b, corresponding to a flow rate of 0.2 kg/s, exhibits significant differences compared to Fig. 4a, especially in the central region of the mold, where the fluid flow maintains a temperature of 24 °C, while surrounding areas reach slightly higher temperatures, around 60 °C. Similar behavior is observed in the difficult-to-access region near the central hole. Additionally, the outlet fluid temperature is moderately elevated, around 34 °C.

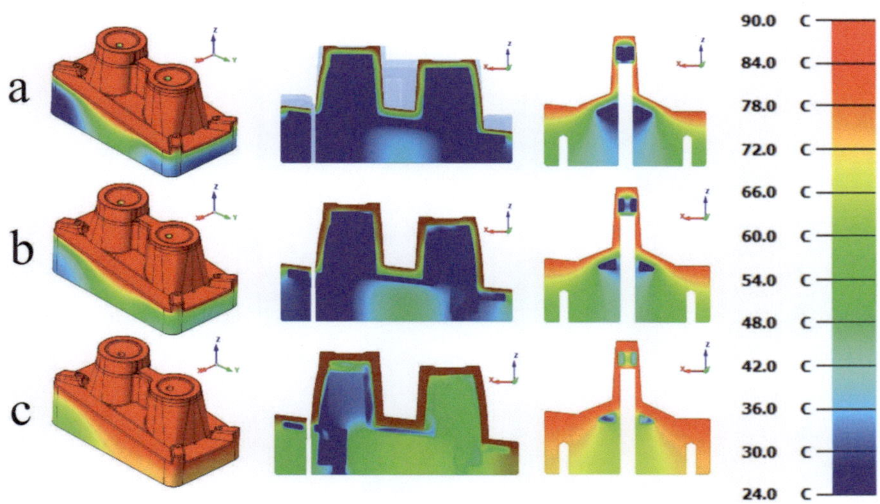

Fig. 4 Study of mold temperature variation without TPMS for flow rates of **a** 1 kg/s, **b** 0.2 kg/s and **c** 0.01 kg/s

Figure 4c, representing a flow rate of 0.01 kg/s, displays a more variable temperature profile with an overall higher average temperature compared to the previous two cases. The central flow region shows temperatures around 60 °C, as does the region of the central hole, while nearby areas reach temperatures above 70 °C. Moreover, the fluid outlet temperature is approximately 54 °C.

In Fig. 5, the mold temperature variation with the TPMS structure is presented under the same boundary conditions previously applied. The second column shows the YZ section cut, while the third column illustrates the XZ section cut. In Fig. 5a, the temperature variation for a flow rate of 1 kg/s is shown where the fluid enters and exits the system at ambient temperature. A balanced temperature distribution is observed throughout most regions of the mold through which the fluid flows, particularly in the difficult-to-access area near the central hole. Figure 5b, on the other hand, presents the temperature variation for a flow rate of 0.2 kg/s, using the same views. The temperature distribution is less uniform compared to Fig. 5a, with several regions of the mold reaching approximately 60 °C, especially in the central area of the structure, while the fluid outlet temperature is around 36 °C. Figure 5c represents a flow rate of 0.01 kg/s, which exhibited an undesired temperature variation. In this case, the fluid enters at ambient temperature but exits the structure above 60 °C. Higher temperatures were observed near the surface temperature (90 °C), with a minimum of 68 °C in the central region of the mold. Additionally, the central area exhibited a higher average temperature compared to the other two flow conditions.

Figures 6a and 7a illustrate the velocity profile variation in the mold without TPMS, while Figs. 6b and 7b correspond to the mold with TPMS structures. The constricted regions of the channels exhibit higher flow velocities, which directly influence the heat transfer rates. Considering that the base geometry channel contains

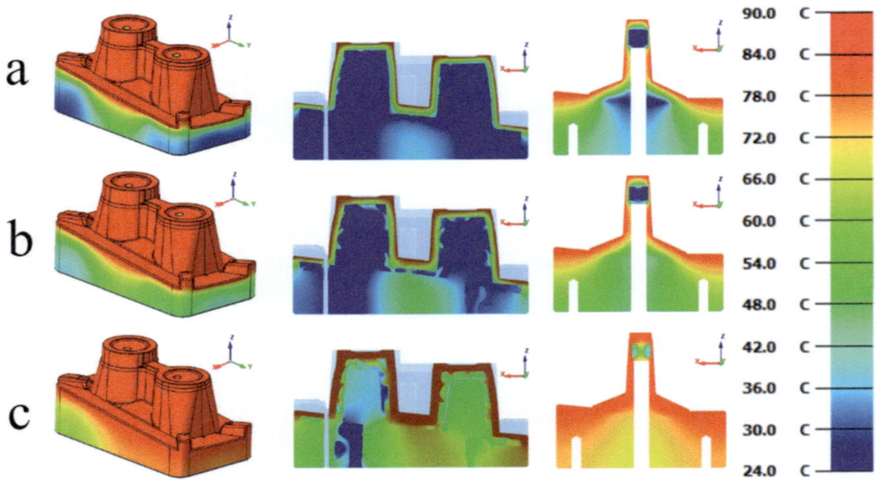

Fig. 5 Study of mold temperature variation with TPMS for flow rates of **a** 1 kg/s, **b** 0.2 kg/s, and **c** 0.01 kg/s

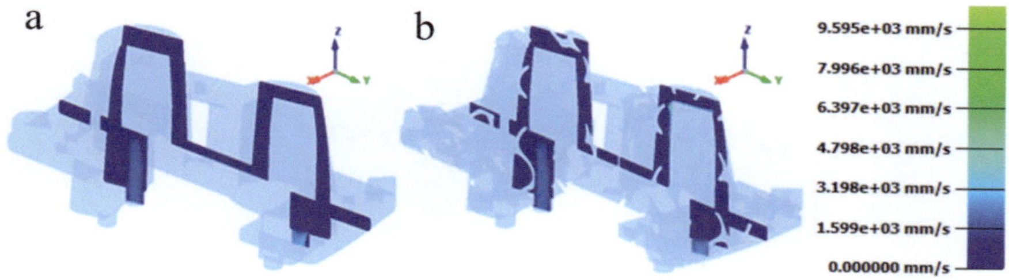

Fig. 6 Study of velocity variation in molds: **a** non-structured and **b** with TPMS structures for a flow rate of 1 kg/s

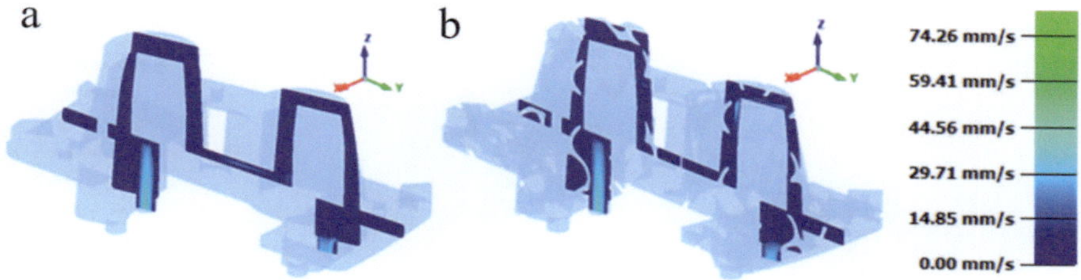

Fig. 7 Study of velocity variation in molds: **a** non-structured and **b** with TPMS structures for a flow rate of 0.01 kg/s

several regions with large hydraulic diameters and constricted sections near the transition between the two bodies, the base geometry shows the highest velocity gradients when evaluating the velocity field along the mold. In contrast, the TPMS geometry, with channels maintaining a uniform hydraulic diameter along their entire length, does not display this behavior. This directly impacts the pressure drop, due to greater channel constriction and flow development limitations, which in turn contributes to an increased heat transfer rate.

Figures 8a and 9a also show, through a cross-sectional view, the pressure drop profile analyzed along the mold without TPMS. Figures 8b and 9b, on the other hand, represent the pressure drop behavior of the mold with TPMS structures. The mold with TPMS structures exhibited a higher pressure drop than the non-structured mold due to its smaller hydraulic diameter. Consequently, operating predominantly in a turbulent flow regime, higher pressure drops lead to increased heat transfer rates, highlighting another advantage of using TPMS structures in mold fabrication.

Figure 10 presents the behavior of temperature variation, pressure drop, heat transfer rate as a function of flow rate, and the variation of the Nusselt number as a function of the Reynolds number, for three flow rates 0.01, 0.2, and 1 kg/s in both the mold without TPMS and the mold with TPMS. Figure 10a illustrates the temperature variation, which decreases as the flow rate increases, indicating a cooling process. This behavior was very similar in both geometries throughout the analysis; however, the mold with TPMS exhibited larger temperature variations. Figure 10b shows an

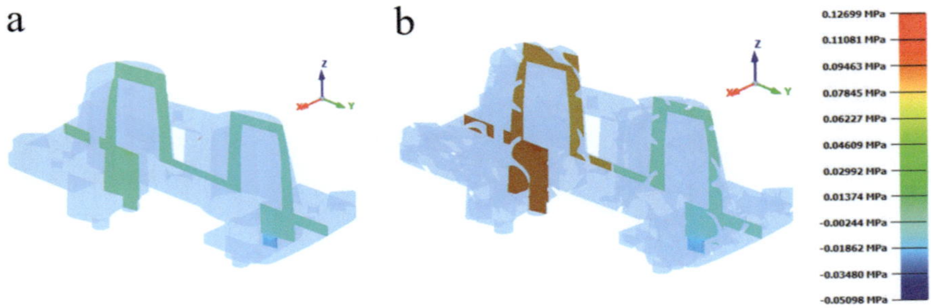

Fig. 8 Study of relative pressure variation in molds: **a** non-structured and **b** with TPMS structures for a flow rate of 1 kg/s

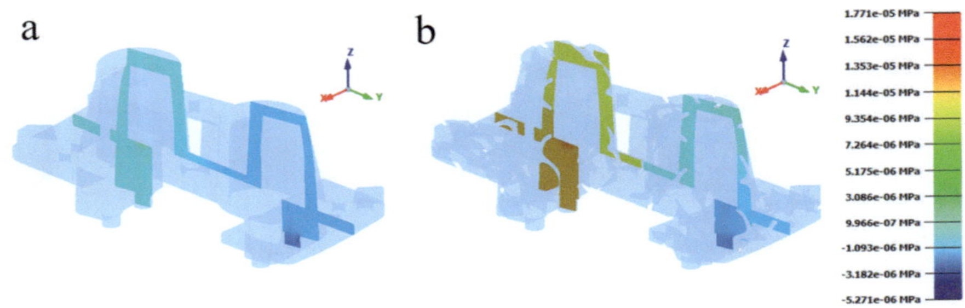

Fig. 9 Study of relative pressure variation in molds: **a** non-structured and **b** with TPMS structures for a flow rate of 0.01 kg/s

almost linear relationship between pressure drop and flow rate for the base mold, which is typical in internal turbulent flows. In contrast, the TPMS geometry exhibited a much steeper pressure drop due to the high surface curvature and greater capability to hinder flow development, a behavior consistent with literature findings. Figure 10c presents a considerable increase in the heat transfer rate with increasing flow rate, influenced both by the greater heat transfer area and by the vorticity generated in the flow due to the geometry. This indicates a higher cooling efficiency for the TPMS mold compared to the base mold at the same flow rate. Finally, Fig. 10d shows the relationship between the Nusselt and Reynolds numbers for the same system. Higher Nusselt numbers indicate higher convective heat transfer coefficients, resulting in greater heat transfer rates. Throughout the analysis, the TPMS mold displayed slightly higher Nusselt numbers compared to the mold without TPMS.

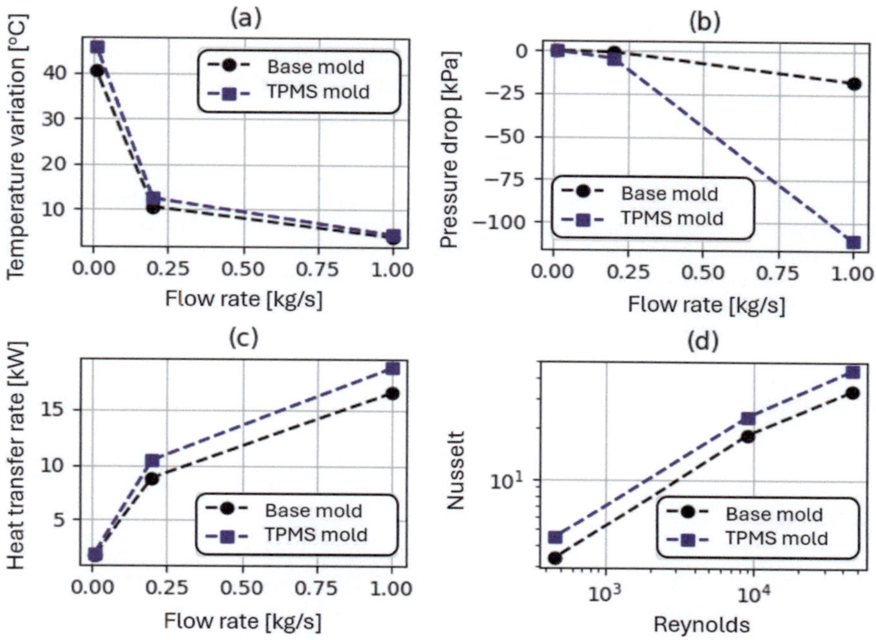

Fig. 10 Comparative analysis of the results obtained

4 Conclusion

This study presented a comparative investigation of the thermal performance of a cupholder mold, evaluating a conventional non-structured configuration and an architected mold incorporating a Triply Periodic Minimal Surface (TPMS) structure. The analysis focused on temperature distribution, heat transfer rate, velocity field, and pressure drop, with the objective of validating the enhanced cooling performance of the TPMS-based design within the mold's fluid flow cavity.

The results confirmed that the TPMS mold exhibited superior thermal efficiency under all evaluated flow rates. The customized TPMS cooling channel consistently achieved higher Nusselt numbers across the tested range, indicating improved convective heat transfer coefficients. Additionally, the TPMS configuration displayed higher pressure drops, which, in predominantly turbulent flow regimes, directly contributed to increased heat transfer rates.

Notably, the TPMS mold provided more effective cooling in regions of limited fluid accessibility, particularly in the narrow section where the thermocouple was positioned, compared to the non-structured mold.

Nevertheless, opportunities for further optimization remain. Future work should include a comprehensive parametric investigation considering a wider range of flow rates, different heated surface temperatures, and variations in the dimensions of the architected structure. Moreover, the evaluation of alternative lattice topologies, especially body-centered cubic (BCC) configurations, could provide additional insights into the relationship between channel geometry, pressure loss, and heat transfer enhancement in conformal cooling applications.

Acknowledgements This work was carried out as part of the activities of the DEMIBAV Project—Demonstrator of molds for low-volume plastic part injection (27194*33), Line IV—Ferramentarias Brasileiras Mais Competitivas, under the MOVER Program (Rota 2030). The program is managed by FUNDEP and funded by the Federal Government of Brazil, to whom the authors express their gratitude. The authors also acknowledge the Federal Universities of Bahia and Santa Catarina, SENAI CIMATEC, as well as FEESC, FEP, and all other partners and research and technology institutions (ICTs) involved in the DEMIBAV Project.

Responsibility Notice The authors are the only responsible for the material included in this chapter.

References

Cheloni JPM, Zluhan B, Silveira ME, Fonseca EB, Valim DB, Lopes ESN (2025) Mechanical and failure mode of body-centered cubic, gyroid, diamond and Voronoi functionally graded additively manufactured biomedical lattice structures. J Mech Behav Biomed Mater 163, 106796. URL: https://www.sciencedirect.com/science/article/pii/S1751616124004284

Dutkowski K, Kruzel M, Rokosz K (2022) Review of the state-of-the-art uses of minimal surfaces in heat transfer. Energies. https://doi.org/10.3390/en15217994

Kaur I, Singh P (2021) State-of-the-art in heat exchanger additive manufacturing. Int J Heat Mass Transf. https://doi.org/10.1016/j.ijheatmasstransfer.2021.121600

Mahmoud D et al (2023) Enhancement of heat exchanger performance using additive manufacturing of gyroid lattice structures. Int J Adv Manuf Technol 126(9):4021–4036

Marques S, Souza AF, Miranda J, Yadroitsau I (2015) Design of conformal cooling for plastic injection moulding by heat transfer simulation. Polímeros. URL: https://www.researchgate.net/publication/289488085

Meltio (2023) Mold-making with LMD and dual wire. Whitepaper

Reynolds BW, Fee CJ, Morison KR, Holland DJ (2023) Characterisation of heat transfer within 3D printed TPMS heat exchangers. Int J Heat Mass Transf 212:124264

Soares A, Marconi P, Carvalho G, Ribeiro Jr A, Beal V, Ferreira C (2023) Performance evaluation of cooling systems in injection moulds made of steel and aluminium through numerical simulations. In: 12th Brazilian congress of manufacturing engineering

Wang Y, Lee C (2023) Design and optimization of conformal cooling channels for increasing cooling efficiency in injection molding. Appl Sci 13:7437

Effects of Electrolyte Temperature on Jet ECD of Copper Monolayers in Electrochemical Additive Manufacturing

Thiago Ferreira Querino, Letícia Bueno do Amaral, Rafaela Evangelista da Silva, and Arthur Alves Fiocchi

Abstract Electrochemical Additive Manufacturing (ECAM) represents a promising approach for fabricating metallic components at room temperature. However, challenges such as low material deposition rates (MDR), geometric complexity, and slow processing speeds hinder its broader adoption. This study examines the impact of electrolyte temperature on copper monolayers deposited by the Jet Electrodeposition (Jet ECD) technique. A copper sulfate and potassium chloride electrolyte was applied at five distinct temperatures (17 °C, 21 °C, 25 °C, 29 °C, and 33 °C). The experiments were conducted using a CNC hybrid multifunction machine tool, focusing on the ECAM module with Jet ECD. Morphological analysis, 3D profilometry, MDR, and surface roughness (R_a) were evaluated using Scanning Electron Microscopy (SEM) and laser interferometry. The results revealed columnar grains with an average length of 2.477 ± 0.493 μm in the monolayers deposited at 33 °C. The height of the deposited monolayers, measured by 3D profilometry, was consistent across all tested temperatures, remaining within the 30–40 μm range. While no statistically significant change in the MDR was observed due to high experimental variability, the mean values indicated a clear trend, with the MDR at 33 °C ($3.5 \times 10^{-5} \pm 1.4 \times 10^{-5}$ mg min^{-1}), approximately double that at 17 °C. The temperature increases also led to a 52% rise in R_a values compared to the lowest recorded roughness, which was 1.12 ± 0.09 μm. In conclusion, electrolyte temperature is a critical parameter in Jet ECD that presents a fundamental trade-off. While higher temperatures promote a trend of increased MDR and columnar grain growth, they do so at the significant cost of higher surface roughness.

Keywords Electroplating · Electrochemical 3D printing · Copper · Electrochemical fabrication · Metallic additive manufacturing

---

T. F. Querino (✉) · L. B. do Amaral · R. E. da Silva · A. A. Fiocchi
Federal University of Uberlândia, Uberlândia, Minas Gerais, Brazil
e-mail: tquerino@ufu.br

V. E. Beal and A. S. Ribeiro Junior (eds.), *Advances in Manufacturing Engineering Research*, Lecture Notes in Mechanical Engineering,
https://doi.org/10.1007/978-3-032-09858-0_7

1 Introduction

Although seemingly recent in industry, additive manufacturing of metal components dates to the 1920s, with US patent 1533300 describing electric arc deposition for decorative items. Later, submerged arc welding (SAW) was adopted in West Germany for large-scale metal components (Kazanas et al. 2012).

Dutta et al. (2019) report that Additive Manufacturing (AM) emerged during the Third Industrial Revolution with the promise of greater production flexibility and product customization.

Since 2019, AM has enabled the manufacture of jet engine components, prosthetics, and orthotics, replacing their traditionally manufactured counterparts. Although its productivity remains low compared to conventional processes (Kumar et al. 2021), AM adoption in metal manufacturing is growing rapidly as production costs decline.

Typical limitations of AM processes, such as Powder Bed Fusion (PBF) and Direct Energy Deposition (DED), include geometric distortions, post-process heat treatment requirements, costly equipment and raw materials, oxidation, and restriction to high-value-added applications (Li et al. 2021; Dutta et al. 2019; Bambach et al. 2017). These issues arise from the high thermal energy required, which creates steep thermal gradients during cooling. The resulting residual stresses can reach several hundred MPa (Sundaram et al. 2015).

In this context, Electrochemical Additive Manufacturing (ECAM) or electrochemical 3D printing emerged, based on depositing thin and highly adherent layers of metal onto the surface of a conductive substrate through the reduction of metal ions in a electrolyte solution. Owing to atomic-scale material transfer, it enables dimensional control from nano- to mesoscales, as well as control over morphology, density, and crystalline structure, allowing precise definition of the components' physicochemical and mechanical properties. Furthermore, ECAM occurs at temperatures close to room temperature (Chen et al. 2017).

Chen (2021) states that ECAM can be used to manufacture components such as helical microstructures, micro gears, integrated circuit boards, surgical tools, and micrometer-scale chemical analysis devices. Similarly, Li et al. (2022) highlight the relevance of the technique for the manufacture of miniaturized aerospace devices, precision machinery components, and microelectromechanical systems (MEMS). Additionally, Bhuiyan et al. (2019) used ECAM to examine the mechanical characteristics of 3D-printed parts in the context of strain gauges, heat sensors, and temperature sensors, producing results that were focused on practical applications.

From 2015 to 2024, research on ECAM has intensified in terms of the number of published works and filed patents, as well as the market opportunities provided by the adoption of the technique (Amaral et al. 2024).

Among the various techniques employed for ECAM, three stand out: (a) Localized Electrochemical Deposition (LED), (b) Meniscus Confined Electrode (MCE), and (c) Jet Electrodeposition (Jet ECD) (Chen et al. 2017; Li et al. 2022; Xu et al. 2020).

Recently, the Jet ECD technique has gained prominence in research, given its higher MDR and lower porosity (Xu et al. 2023).

However, Chen (2021) states that the low MDR of the techniques (typically from 0.008 to 20.400 $\mu m^3 s^{-1}$) constitutes the main barrier to scaling up products from nano or micro to macro. Furthermore, all 3D printed products are limited in the complexity of their geometries. Most products feature wire-based architectures with coarse morphology and detailed porous structures, primarily composed of a single metal, such as copper (Chen et al. 2023) and nickel (Pandey et al. 2024).

The main process parameters for ECAM include: (a) energy field, (b) interelectrode gap (IEG), (c) electrolyte solution concentration, (d) electrolyte temperature, (e) electrode material, (f) electrode motion speed (scan speed), (g) electric current density, and (h) electric potential between electrodes (Ayalew et al. 2023).

In the electrodeposition field, electrolyte temperature has a critical impact on ion deposition, crystal structure, and layer morphology, with higher temperatures promoting larger crystal sizes (Saitou et al. 2008). Higher electrolyte temperatures enhance metal ion adsorption on the cathode, increasing the MDR. A higher temperature decreases the electrolyte's viscosity, lowering its electrical resistance and facilitating the migration of ions to the cathode (Ayalew et al. 2023).

Although the influence of temperature is well established in electroplating, its application in ECAM has not been addressed in the literature. Consequently, electrolyte temperature may play a critical role in enhancing the MDR and affecting surface roughness and morphology of the deposited material, particularly for the Jet ECD technique.

Considering the importance of the solution temperature on the speed of a chemical reaction, and the lack of studies in ECAM, this work aims to investigate the effects of the temperature of the electrolytic solution on the copper deposition rate, surface finish, and morphology of the deposited layer, using the Jet ECD technique on copper substrates.

2 Materials and Methods

2.1 Experimental Setup

A CNC hybrid multifunction machine tool, hereinafter referred to as an additive manufacturing machine (AMM) was specially designed, built, and validated with a focus on the ECAM module using Jet ECD. The milling, laser cutting, material extrusion (MEX), and metrology modules were omitted as they are not part of this research. These systems can operate independently or simultaneously, in full integration with Industry 4.0 frameworks. In Fig. 1, the Electrochemical Additive Manufacturing process using Jet ECD is schematically presented.

The structure consists of welded carbon steel tubing filled with synthetic granite to dampen internal and external vibrations, thereby improving dimensional and

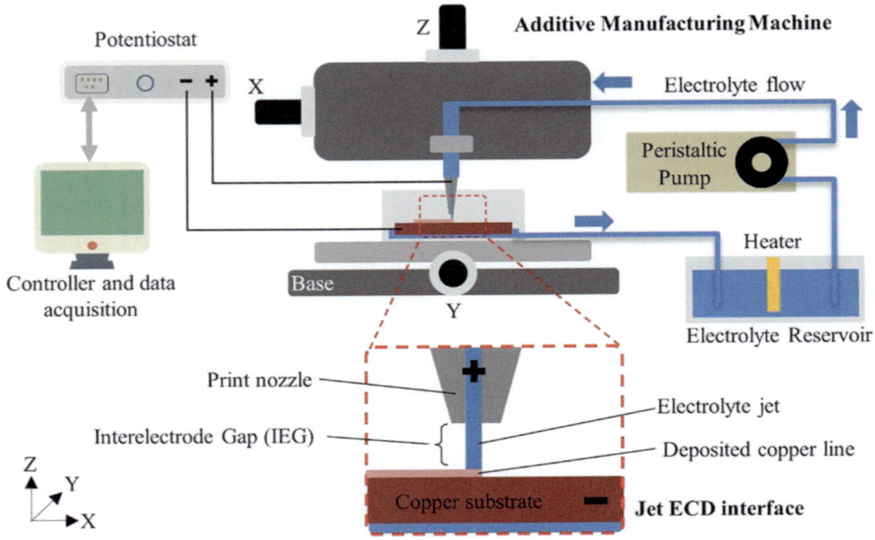

Fig. 1 Schematic representation of the jet electrochemical deposition (jet ECD) and its additive manufacturing machine (AMM)

geometric stability. The machine features a three-axis orthogonal Cartesian system employing supported cylindrical linear guideways, single-nut recirculating ball screws with preloaded elements, supported on angular contact bearings near the motors and rigid radial bearings at the opposite ends, driven by NEMA 23 stepper motors to provide three degrees of freedom. The working envelope measures 190 mm (X), 315 mm (Y), and 100 mm (Z). The build table can accommodate a heated platform for AM processes, with an operating range up to 150 °C. The hybrid machine also allows manual operation through graduated handwheels on all three axes, facilitating rapid referencing and efficient execution of manual tasks or preliminary testing procedures.

The computer controls the electric current supplied by the potentiostat and the movement of the AMM according to the desired length of the monolayer, set at 6 mm. The distance between the nozzle and the substrate is defined as Interelectrode Gap (IEG). It is necessary to apply the electrical potential difference for the reduction reaction directed at the deposition point. The IEG is established by electrically probing the nozzle on the substrate. Once the nozzle and substrate are in touch, the nozzle is moved away by the desired IEG. The electrolyte started flowing and Jet ECD began, moving the nozzle positively along the X axis. The print nozzle (anode) was electrically connected to the positive pole of the potentiostat by an alligator-type test clip. A glass vat contains the substrate (cathode), where the ion reduction occurs. The substrate was connected to the negative pole of the potentiostat by a copper wire and an alligator-type test clip. For electrodeposition to occur, the electrolyte solution is transported into the print nozzle by a peristaltic pump. With the anode and cathode connected to the potentiostat, the electrolyte jet providing contact between them, the AM process begins.

The print nozzle consisted of a 3 mL syringe and a 0.492 mm inner diameter needle made of AISI 304 stainless steel. This assembly was mounted onto the AMM using an aluminum holder. A coupling sleeve, fabricated from resin (Model Brutal, 3DFila) via Masked Stereolithography (MSLA), was produced to connect the pump to the print nozzle via silicone piping.

A peristaltic pump was designed to have a fan and air filter for adequate cooling of the internal electric components (power supply, stepper motor driver, stepper motor, speed controller, and circuit breaker) during prolonged Jet ECD tests. The stepper motor drives three rollers that compress silicone piping, creating pressure differentials for electrolyte transport, while a check valve reduces air bubbles and stabilizes pressure. All components are enclosed in a surface-mounted housing for protection and isolation.

A 25 W submersible heater (Model H-386, Hopar Aquarium Heater) was used to control the electrolyte temperature with a range of 17–35 °C, with a resolution of 2 °C. This submersible heater has temperature control via a bimetallic contact, which is an efficient way to adjust the temperature mechanically.

A Vertex 5A potentiostat from Ivium was used as a power source in CronoPotentiometry mode (constant electric current). The electric current was controlled by IviumSoft® software via USB and a notebook.

2.2 *Materials and Preparation*

The cooper electrolytic solution was prepared with equimolar concentrations (0.6 mol L^{-1}) of copper sulfate pentahydrate ($CuSO_4 \cdot 5H_2O$) and potassium chloride (KCl). The addition of KCl aimed to improve ionic conductivity, reduce grain size, and decrease surface roughness.

The layers were printed on parallelepiped electrolytic copper substrates measuring 15 mm × 15 mm × 2 mm. The surfaces were polished before testing with #1500 and #2000 grit sandpapers, cleaned with a paper towel moistened with acetone and then wiped with isopropyl alcohol.

2.3 *Deposition Procedure*

A clamp system was made with PLA to hold the substrate and provide the electric contact with copper wire.

The contact between electrodes was determined using a multimeter (Model 115, Fluke) to determine the zero reference point of the substrate. Manually, using the graduated handwheels of the AMM, the anode was moved 400 μm apart as IEG, as suggested by (Chen et al. 2023).

Table 1 List of input and output parameters for deposition of copper monolayers by jet electrochemical deposition (Jet ECD)

Input		Output
Variable	Constant	
Electrolyte temperature: 17 °C 21 °C 25 °C 29 °C 33 °C	0.4 mm interelectrode gap (IEG) 0.32 mL/s electrolyte flow rate 4 mA electric current 0.492 mm nozzle diameter 238.6 A/dm^2 electric current density 0.8 mm/min scan speed	Morphology 3D profilometry Material deposition rate (mg/min) Surface roughness R_a (μm)

The electrolyte temperature was measured by a type-K thermocouple and a digital instrument (Model POL-41-A, Politerm). The ambient temperature during the experiments was 16 °C and 50% relative humidity, monitored with a thermo-hygrometer (Model FY-10, Gaqqee). The environment was air-conditioned.

For copper deposition, the selected current intensity value was 4 mA, according to the studies by (Arshad et al. 2022; Chen et al. 2022).

Considering a nozzle inner diameter of 0.492 mm, the current density was calculated to be 238.6 A dm^{-2}.

The nozzle scan speed was set to 0.8 mm min^{-1}, following the methodology of (Amaral et al. 2024). This value was selected to promote better surface roughness.

The established electrolyte flow rate was 0.32 mL s^{-1} to ensure the highest possible solution flow to the deposition region, following the capacity of the available experimental apparatus.

2.4 Experimental Design

To carry out this work, the layers were manufactured according to the data described in Table 1. For each temperature value, three layers for each condition were deposited parallel to each other on each of the five copper substrates, resulting in a total of 15 samples. The samples were analyzed according to four parameters: morphology, 3D profilometry, material deposition rate (MDR), and surface roughness.

2.5 Characterization

To evaluate morphology, the samples were analyzed by Scanning Electron Microscopy (SEM), model EVO MA 10, Carl Zeiss. A working distance of 9.0 mm (on average), an electron acceleration applied voltage of 20 kV, and a shadow imaging mode were used. The image measurements were performed using ImageJ software (Version 1.54 g).

The layer profile was evaluated using a UBM Microfocus Expert IV optical laser interferometer with a lateral resolution of 1000 points/mm and a measurement range of 0.01–1000 μm. After collecting the layer profiles, they were evaluated using the Mountains Map® computational tool.

The material deposition rate (MDR), or the amount of material deposited at each time, is expressed by Eq. (1).

$$\text{Material Deposition Rate} = \frac{m}{t} \tag{1}$$

The deposited mass m was calculated by the product of the copper density 8.94 g cm^{-3} and the deposited volume, obtained using the Mountains Map® software. t is the time of each test, with the mass expressed in milligrams (mg) and the time expressed in minutes (min).

Surface roughness parameter (R_a) was determined using Mountains Map® software, using a 0.8 mm cut-off and a Gaussian filter, following (ISO 4287 2002).

3 Results and Discussion

3.1 Morphology

SEM of copper monolayers in the shape of a line segment is shown in Fig. 2 as a function of electrolyte temperature ranging from 17 to 33 °C. The line segments were intended to be 0.5 mm wide and 6.0 mm long.

The morphology of the SEM copper deposited monolayers qualitatively reveals some differences regarding dimensional and geometric tolerances among the tested temperatures. The highest temperature (33 °C) resulted in a more heterogeneous surface finishing, as proven by the results of surface roughness measurement R_a (1.70 ± 0.06 μm) in Fig. 5.

Magnification of the deposition reveals the formation of a crystalline structure of columnar grains with random orientation, with an average length of approximately 2.477 ± 0.493 μm. The increased surface area of that rougher finishing is not ideal for AM parts requiring dimensional and geometric accuracy. On the other hand, it can be interesting for electrochemical sensors, which might increase sensitivity.

The elongated columnar grains found at the highest temperature (33 °C) are also a result of the highest MDR ($3.5 \times 10^{-5} \pm 1.4 \times 10^{-5}$ mg min^{-1}). These geometries are related to areas of higher surface energy where depositions prefer to occur. This behavior is strongly influenced by mass transport limitations and the non-uniform distribution of copper ions near the surface. As a result, preferential growth occurs along specific crystallographic directions. Nevertheless, such a structure can improve electrical conductivity due to the reduced presence of grain boundaries, which act

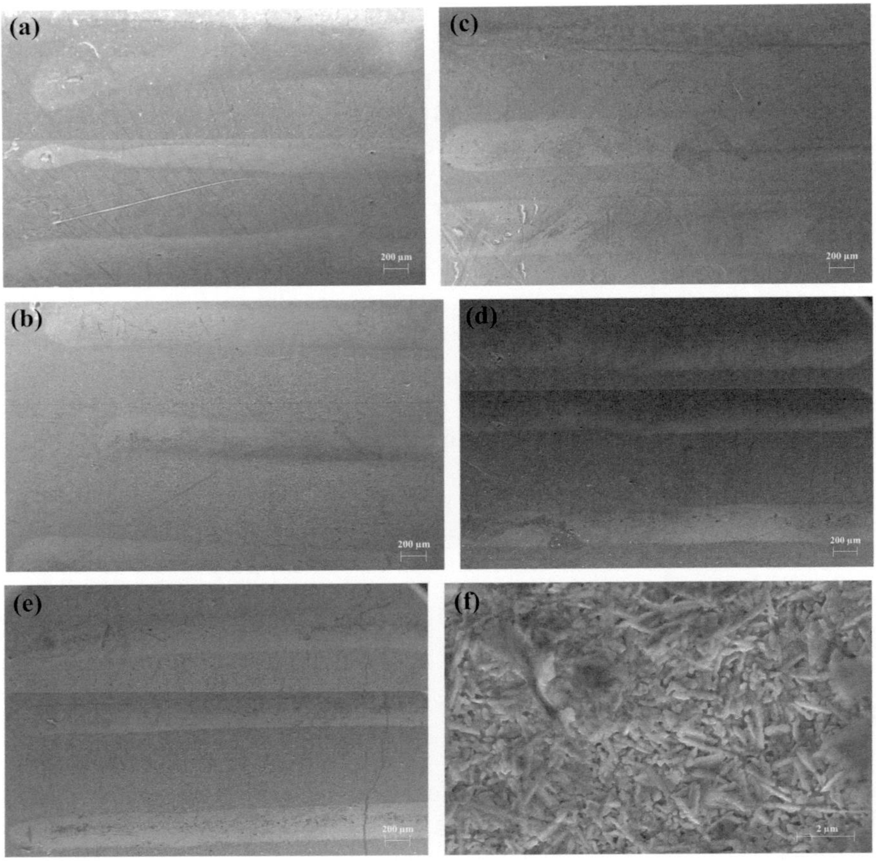

Fig. 2 SEM depositions of copper monolayers by jet electrochemical deposition (jet ECD) at different temperatures **a** 17 °C, **b** 21 °C, **c** 25 °C, **d** 29 °C, **e** 33 °C, and **f** magnification of elongated copper grains at 33 °C

as barriers to electron movement due to their higher electric resistance. Paunovic (2006) and Chen et al. (2020) reported similar structural features and emphasized that these morphologies are associated with elevated surface roughness, but do not quantify it.

3.2 3D Profilometry

The 3D profilometry deposits of the line segments of copper monolayers are presented in Fig. 3 as a function of electrolyte temperature. The 3D profiles have different heights ranging from 30 to 40 μm, regardless of the adopted temperature value. However, the behaviors of the layers' growth rate were different, as shown in Fig. 4.

Three-dimensional profilometry shown in Fig. 3 revealed the topographical differences in the tested deposition conditions with greater detail than SEM (Fig. 2). This demonstrates the superiority of 3D profilometry over SEM as a metrological tool for dimensional and geometric analysis. In all evaluated conditions, island formations were observed, indicating heterogeneous deposition. The number of islands increased with temperature, likely due to the accelerated reaction rate. Nevertheless, these localized deposits were attributed to issues related to the Jet ECD nozzle scan speed.

The initial driver settings for the stepper motors caused acceleration and deceleration during nozzle translation. The goal was to achieve a constant scan speed of

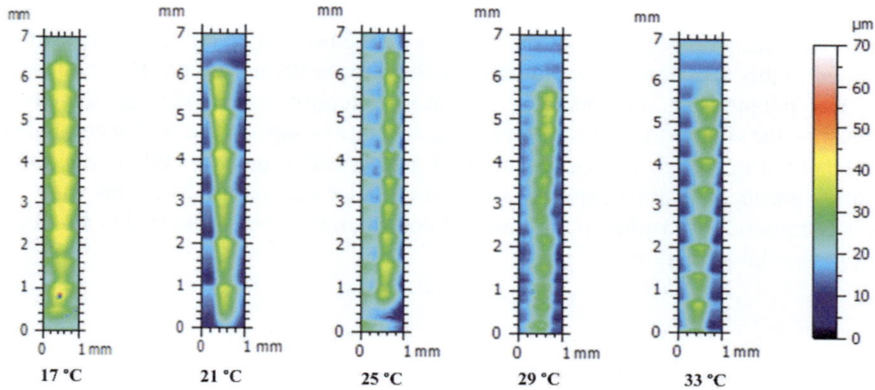

Fig. 3 Top views of 3D profilometry of the depositions of copper monolayers by jet electrochemical deposition (jet ECD) at different temperatures

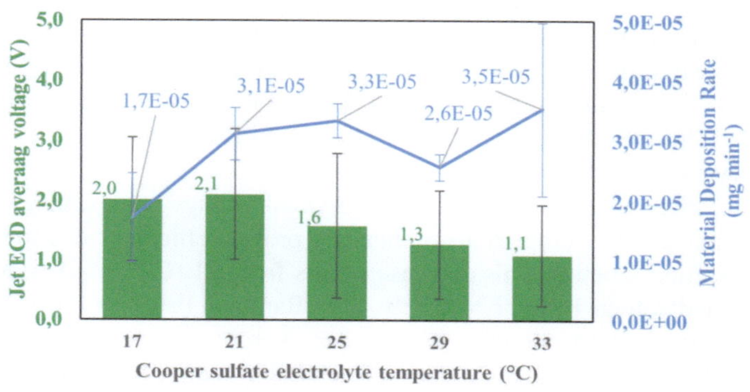

Fig. 4 Graph showing material deposition rate (MDR) and average applied voltage results for depositions of copper monolayers by jet electrochemical deposition (jet ECD) at different temperatures

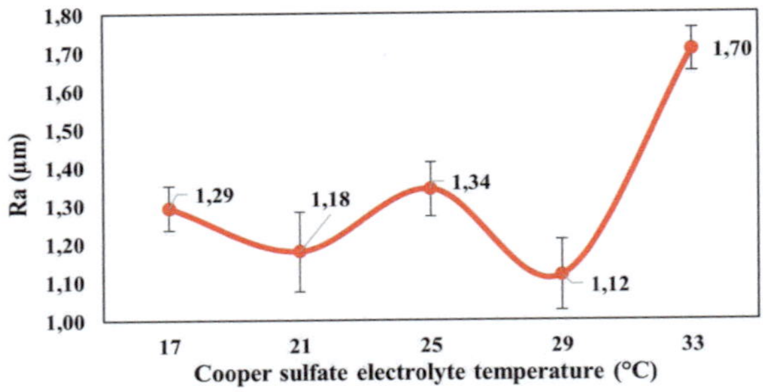

Fig. 5 Graph showing R_a surface roughness results for depositions of copper monolayers by jet electrochemical deposition (jet ECD) at different temperatures

0.8 mm min^{-1}. However, the undesired pulsed motion intensified localized deposition when the scan speed decreased. Increased deposit thickness reduced the IEG. When the IEG was minimal, deposition concentrated at a single point (tip effect) until the nozzle moved to a new, more favorable region. At that point, the deposition shifted to this subsequent area, located downstream in the nozzle travel direction.

The potentiostat was operated in chronopotentiometry mode, continuously adjusting the voltage to stabilize the electric current, as shown in Fig. 4. The combined effects of temperature variation, undesired pulsed nozzle motion, and voltage fluctuation produced distinct deposition patterns. Under the conditions of this study, it was not possible to isolate the individual contributions, which highlights the need for further investigation.

3.3 Material Deposition Rate

Jet ECD voltage and material deposition rate (MDR) are presented in Fig. 4 as a function of the copper sulfate electrolyte temperature. No significant increase was observed in the MDR when considering the experimental deviations. Meanwhile, a non-linear behavior of the average applied voltage was observed.

This high variability suggests that the process, under these conditions, is susceptible to minor fluctuations in parameters such as jet flow and positioning relative to the cathode. This observation corroborates the island formations detected in the 3D profilometry analysis (Fig. 3).

It can be observed that, for the average applied voltage values, they decreased with the increase of the temperature, enhancing process efficiency due to improved reduction reaction kinetics. This decrease occurs from 21 °C onward, reducing the average applied voltage by approximately 48%. Bard and Faulkner (2001) state that the Chronopotentiometry mode requires lower overpotential as temperature rises, due to the reduction in ohmic resistance and activation energy for the Cu^{2+} reduction reaction.

Therefore, controlling the electrolyte temperature is beneficial for the energy consumption of the Jet ECD. For components with long processing times, this can represent a significant economic advantage. It is noteworthy that greater dispersion in voltage values was observed at higher temperatures, suggesting some instability in the electrochemical control.

Nonetheless, the mean values suggest a trend of increased MDR at higher temperatures, with the highest average MDR recorded at 33 °C. This trend is consistent with the theory described by Gamburg and Zangari (2011), where raising the solution temperature enhances ionic diffusion, leading to a greater availability of Cu^{2+} ions at the cathodic interface. It can be inferred that while the underlying chemical kinetics are accelerated by temperature, process instabilities in the current setup masked a statistically significant effect. One of the sources of instability identified was the electrical connection made by an alligator-type test clip. As the experiment progressed, a layer of copper and oxides formed on the clip. This undesirable layer increased the electrical resistance and impaired the deposition on the substrate. For future experiments, it is necessary to replace this connection with a more rigid model, free from the influence of the electrolyte.

3.4 Surface Roughness

The results of the roughness parameter R_a as a function of the copper sulfate electrolyte temperature are presented in Fig. 5.

The average R_a roughness values did not exhibit a linear behavior with increasing temperature. Deviations remained on the order of approximately 10%. The variations in R_a between 17 and 29 °C were minimal and within the bounds of experimental uncertainty. This fact indicates that, within this range, the deposition conditions favored homogeneous nucleation and a more uniform layer formation.

However, at 33 °C, the R_a value increased 52% ($R_a = 1.70 \pm 0.06$ μm). This increase suggests a mismatch between ion transport and the reaction rate, leading to irregular growth and higher roughness at elevated temperatures. This phenomenon is also evident in Fig. 2e for the same temperature. In the context of ECAM, such behavior can be problematic, as the highest points of the growing layer tend to receive a greater ion flux due to the lower local resistance. This can result in a cumulative effect, potentially compromising the final geometry of the manufactured component.

The R_a values obtained in this research are approximately three to six times higher than those reported in similar studies, such as that of Arshad et al. (2022). Notably, these authors did not report any method of controlling the electrolyte temperature during the deposition process.

This result, combined with the results presented in Fig. 4, underscores the need to adjust the electrolyte temperature according to the specific stage of the process.

4 Conclusions

This study aimed to evaluate the influence of electrolyte solution temperature on the Electrochemical Additive Manufacturing (ECAM) process of copper monolayers via Jet Electrochemical Deposition (Jet ECD). The findings indicate that:

- At 33 °C, randomly oriented columnar grains ($2.477 \pm 0.493\ \mu m$) were produced, a microstructure that can enhance electrical conductivity by reducing grain boundaries.
- 3D profilometry revealed the formation of island-like deposition patterns, which became more pronounced with increasing temperature. The 3D profiles have different heights ranging from 30 μm to 40 μm.
- The material deposition rate (MDR) measured at 17 °C was $1.7 \times 10^{-5} \pm 7.5 \times 10^{-6}$mg $\times$ min^{-1}, representing a reduction of approximately 51% compared to the value $3.5 \times 10^{-5} \pm 1.4 \times 10^{-5}$mg $\times$ min^{-1} obtained at 33 °C. Although the mean values suggest a positive correlation between MDR and electrolyte temperature, the relatively large standard deviations indicate substantial variability in the process. This variability is likely associated with inconsistencies in jet flow dynamics and nozzle alignment during the experiments.
- From 17 to 29 °C, R_a variations remained around 10%. At 33 °C, the R_a value increased 52% ($R_a = 1.70 \pm 0.06\ \mu m$). This suggests that temperature influenced ion transport and reaction rates, leading to irregular growth and higher surface roughness.
- Conclusively, lower processing temperatures are advantageous for applications demanding high-dimensional fidelity. Conversely, elevated temperatures can be leveraged for applications where manufacturing throughput and tailored microstructural features are paramount.

Acknowledgements The authors would like to thank National Council for Scientific and Technological Development (CNPq); Coordination for the Improvement of Higher Education Personnel (CAPES); FINEP (grants 2209/22 and 2127/22), Minas Gerais State Research Support Foundation (FAPEMIG) (grants APQ-01759-17 and APQ-00403-24); Multisectoral Association of River Basin Users (ABHA) (grant 23117.006727/2025-55); FEMEC Maker-UFU funding allocated by the Brazilian National Congress (grants 23117.039298/2021-79 and 23117.024929/2023-17); Post Graduate Program in Mechanical Engineering (PPGEM-UFU); FEMEC Maker-UFU Advanced Manufacturing Laboratory; Institute of Chemistry (IQ-UFU), Electro analytics Research Center (NuPE-UFU); Laboratory of Tribology and Materials (LTM-UFU); Laboratory of Scanning Electron Microscopy (LAMEV-UFU); Multi-User Laboratory Network (RELAM-UFU), Laboratory for Teaching and Research in Machining (LEPU-UFU); Center for Research and Development of Welding Processes and Additive Manufacturing (LAPROSOLDA); Laboratory of Friction and Wear Technology (LTAD-UFU); Brazilian Reference Center for Technological Innovations for Paralympic Sports (CINTESP.Br); and EMBRAPII-FEMEC-UFU.

Acknowledgment of AI-Assisted Translation Review The authors acknowledge the use of generative artificial intelligence tools—ChatGPT 5.0 (OpenAI), Gemini 2.5 Pro (Google DeepMind), and Copilot (Microsoft)—to assist in the linguistic and stylistic revision of the English translation of this manuscript. These tools were employed to enhance clarity, technical accuracy, and consistency in terminology. All content generated or suggested by these tools was critically reviewed, validated, and edited by the authors to ensure scientific integrity and compliance with academic standards. The authors remain fully responsible for the accuracy and originality of the information presented.

References

Amaral LB, Querino TF, Fiocchi AA (2024) Influência da Velocidade de Translação do Bocal de Impressão e da Densidade de Corrente Elétrica na Deposição de Manufatura Aditiva Eletroquímica por Jet-ECD. Trabalho apresentado no 2° Congresso Brasileiro de Manufatura Aditiva (CBMADi), Uberlândia, MG

Arshad MH, Saxena KK, Smith R, Qian J, Reynaerts D (2022) Localized and mask-less copper deposition with free-flow jet micro-electrochemical additive manufacturing. Int J Electr Mach 27:16

Ayalew AA, Han X, Sakairi M (2023) A critical review of additive material manufacturing through electrochemical deposition techniques. Add Manuf 103796

Bambach MD, Bambach M, Sviridov A, Weiss S (2017) New process chains involving additive manufacturing and metal forming—a chance for saving energy? Proc Eng 207:1176–1181

Bard AJ, Faulkner LR (2001) Fundamentals and applications: electrochemical methods. Electrochem Methods 2(482):580–632

Bhuiyan MEH, Behroozfar A, Daryadel S, Moreno S, Morsali S, Minary-Jolandan M (2019) A hybrid process for printing pure and high conductivity nanocrystalline copper and nickel on flexible polymeric substrates. Sci Rep 9(1):19032

Chen KJ, Wu JA, Chen C (2020) Effect of reverse currents during electroplating on the⟨111⟩-oriented and nano twinned columnar grain growth of copper films. Cryst Growth Des 20(6):3834–3841

Chen X, Liu X, Childs P, Brandon N, Wu B (2017) A low-cost desktop electrochemical metal 3D printer. Adv Mater Technol 2(10):1700148

Chen X (2021) Electrochemical metal 3D printing. Doctoral dissertation, Imperial College London

Chen X, Chen J, Saxena KK, Zhu J, Gu X, Guo Z (2022) Localization of jet electrochemical additive manufacturing with a liquid confinement technique. J Manuf Proces 81:48–64

Chen X, Gu X, Saxena KK, Arshad MH, Huang J, Reynaerts D (2023) Enhancing shape precision in jet electrochemical additive micro-manufacturing process through confined electrolyte. Precis Eng 84:69–80

Dutta B, Babu S, Jared BH (2019) Science, technology, and applications of metals in additive manufacturing. Elsevier, Netherlands

Gamburg YD, Zangari G (2011) Theory and practice of metal electrodeposition. Springer Science & Business Media

Kazanas P, Deherkar P, Almeida P, Lockett H, Williams S (2012) Fabrication of geometrical features using wire and arc additive manufacture. Proc Inst Mech Eng Part B: J Eng Manuf 226(6):1042–1051

Kumar SP, Elangovan S, Mohanraj R, Ramakrishna JR (2021) Review on the evolution and technology of State-of-the-Art metal additive manufacturing processes. Mater Today: Proc 46:7907–7920

Li X, Ming P, Ao S, Wang W (2022) Review of additive electrochemical micro-manufacturing technology. Int J Mach Tools Manuf 173:103848

Li T, Yan X, Fang X, Jin P, Li J, Rabbi KF, Miljkovic N (2021) In situ jet electrolyte micromachining and additive manufacturing. Appl Phys Lett 119(17)

Pandey AM, Kapil S, Das M (2024) A novel method of effectively fabricating the micro-objects by selective jet electrodeposition (SJED). Rapid Prototyping J 30(6):1105–1120

Paunovic M (2006) Fundamentals of electrochemical deposition. Wiley, USA

Saitou M, Oshiro S, Asadul Hossain SM (2008) Effect of temperature on nickel electrodeposition from a nickel sulfamate electrolyte. J Appl Electrochem 38:309–313

Sundaram MM, Kamaraj AB, Kumar VS (2015) Mask-less electrochemical additive manufacturing: a feasibility study. J Manuf Sci Eng 137(2):021006

Xu J, Ren W, Lian Z, Yu P, Yu H (2020) A review: development of the maskless localized electrochemical deposition technology. Int J Adv Manuf Technol 110:1731–1757

Xu J, Zhang G, Lv J, Guo Z, Wang F, Zhang Y, Xu J, Cheng G (2023) High-efficiency localized electrochemical deposition based on ultrafast laser surface modification. Surf Coat Technol 471:129923

Permanent Mold Casting of A356 Alloy to Simulate Critical Parts of Lightweight Wheels in Laboratory Scale

Bruna Callegari, Luis Fernando Folle, Leonardo Oliveira Passos da Silva, Tiago Nunes Lima, and Rodrigo Santiago Coelho

Abstract Aluminum alloys, especially the A356 alloy, containing aluminum, silicon and magnesium as its main elements, are the material of choice to manufacture lightweight automotive wheels, due to their good castability and mechanical behavior associated with its low density. However, manufacturers are constantly seeking new, improved alloys capable of minimizing weight while maintaining or enhancing the performance of wheels. For resource saving, the development of standardized procedures for laboratory testing of developed materials that can reliably replicate the material's behavior in operation is highly interesting for the industry. Therefore, an ASTM-standardized mold was used to cast ingots of the A356 alloy at low pressure. Such mold is used to produce specimens for tensile testing. In this case, focus was given to the auxiliary distribution channels, which are considerably bulky and can be used to provide material for further characterization, such as impact testing, which is highly relevant, provided that toughness is a critical property for automotive wheels. Numerical simulations were carried out to define different pre-heating temperatures of the casting mold, so that typical thermal regimes of different wheel regions could be reproduced. Then, the solidified distribution channel was sectioned to have its microstructure analyzed in terms of secondary dendritic arm spacing (SDAS) and porosity. Based on these results, the most adequate position for placement of the V-notch of Charpy test specimens was defined, and specimens were subsequently machined. Impact test results showed that, although the chosen region is highly representative of different regions microstructurally speaking, as long as the thermal regime is correctly designed, toughness values obtained were well below those informed by manufacturers as the threshold for guaranteed safety upon wheel

B. Callegari (✉) · L. O. P. da Silva · T. N. Lima · R. S. Coelho
Instituto SENAI de Inovação em Conformação e União de Materiais, SENAI CIMATEC, Salvador, Brazil
e-mail: bruna.callegari@fieb.org.br

L. F. Folle
Universidade Federal de Santa Maria, Santa Maria, Brazil

R. S. Coelho
Universidade SENAI CIMATEC, Salvador, Brazil

V. E. Beal and A. S. Ribeiro Junior (eds.), *Advances in Manufacturing Engineering Research*, Lecture Notes in Mechanical Engineering,
https://doi.org/10.1007/978-3-032-09858-0_8

application. This means that laboratory casting practices might need refinement to avoid additional issues and make sure that results can be deemed representative and reliable.

Keywords A356 Alloy · Lightweight wheels · Permanent mold casting · Simulation · Microstructure

1 Introduction

Aluminum (Al) alloys have a wide range of industrial applications, typically taking advantage of their lightweight nature combined with good mechanical, corrosion, electrical, and thermal properties. Their thermal conductivity, for instance, is only matched by that of copper, a much more expensive and heavier material. Although not very strong, the strength-to-weight ratio of aluminum alloys surpasses that of most metals. Aluminum is one of the most versatile metals for casting, being extensively used in the automotive industry (Krall et al. 2024; Rams and Whittaker 2022).

The advantages of aluminum for casting lie in its low melting point, negligible gas solubility (except for hydrogen), and the high-quality surface finish achievable in cast products. However, its main drawbacks include high shrinkage levels during the solidification of certain alloy classes. Another issue is the variability in mechanical properties and the fact that these properties are generally inferior to those of mechanically worked alloys. The most commonly used aluminum alloys are based on the Al–Si, Al–Si–Mg, and Al–Si–Cu systems, such as the A356 (Polmear et al. 2017; Saini and Jha 2023).

The A356 alloy, in particular, is widely used in the manufacturing of automotive wheels. Cast wheels, despite having inferior mechanical properties compared to forged ones, still present a sufficient mechanical performance, and are preferred by industries due to their greater stylistic versatility and dimensional accuracy. Alloys used for wheel casting must exhibit good casting characteristics, including reduced susceptibility to hot cracking and shrinkage, as well as high resistance to impact, fatigue, and corrosion. As it meets all these requirements, the A356 alloy has become the standard material for automotive wheel manufacturing (Hamasaiid et al. 2019; Kaba et al. 2018; Wu et al. 2022).

In aluminum casting alloys, defects like porosity formation and microstructural refinement significantly influence the material's mechanical properties, especially toughness, which is a critical property for automotive wheels in situations such as collisions, and high-speed impact against curbs, potholes, speed bumps, and cross-street drains. Wheel manufacturers carry out full-scale impact tests in fabricated wheels. However, when it comes to alloy development or improvement, it is useful to make use of laboratory-scale tests using standardized specimens which are still able to predict the alloys real behavior in an adequate and reliable manner, thus saving valuable resources such as feedstock, energy and time.

Table 1 Target composition of the studied alloy

Element	Al	Si	Mg	Fe	Sr	Ti
Amount (wt%)	Bal	6.8–7.2	0.25–0.30	0.15 max	0.015–0.020	0.15–0.20

The aim of this paper is to conduct a study to evaluate the suitability of the laboratory casting of a standardized ingot to predict the behavior of the A356 alloy and define how representative it would be for assessment of the impact toughness properties of the alloy. The study was supported by numerical simulations, to define different mold pre-heating temperatures that would lead to different thermal regimes representing different wheel regions.

2 Methodology

2.1 *Material*

The target composition of the A356 alloy studied in this work is provided in Table 1.

2.2 *Casting Procedure*

Alloy casting was carried out in a permanent mold designed according to the ASTM B108 standard (American Society for Testing and Materials 2019). This mold design provides two specimens for tensile testing, while specimens for impact testing can be machined from the adjacent distribution channels. The mold was fabricated in H13 steel with a cavity machined to a surface roughness (R_a) between 0.8 and 1.6 μm, as shown in Fig. 1. The A356 alloy was molten in a SiC crucible with additions of alloys A413, P1020 (99.7% Al + 0.1% Si + max. 0.2% Fe), Al–90% Fe, Al–15% Sr, Al–5% Ti–1% B, metallic Si and pure Mg for composition adjustment, which was done by taking samples of the molten material and analyzing the composition by optical emission spectroscopy in a Bruker Q2 ION analyzer.

The mold was pre-heated to two different temperatures—510 and 560 °C (hereafter referred to as "LT" and "HT," respectively)—which were defined by numerical simulations of the casting process to replicate typical cooling times of rim and hub regions of lightweight wheels, respectively, using results of real wheel castings as input. The pouring temperature ranged between 700 and 720 °C. Commercial Degasser 200 was used for degassing, and Coveral flux was used to protect the melt from oxidation. Powdered graphite was applied to the cavity's surface to facilitate the release of the ingot. The ingots were characterized in the as-cast state, without any heat treatment.

Fig. 1 Permanent mold fabricated in H13 steel according to ASTM B108 standard

2.3 Numerical Simulation

Numerical simulation was carried out in ProCAST software. The casting material was the aluminum alloy A356 (found under the name AlSi7Mg0.3 in the software's database) with a chemical composition of 7.0 wt% Si, 0.3 wt% Mg and the remainder as aluminum. For the simulation, the pouring temperature was set to 720 °C, and the mold material was set to H13 steel. The mold filling time was defined as 10 s, according to preliminary casting try-outs. The part/mold and air/mold heat transfer coefficients were 2000 and 10 W/m^2K, respectively, as extracted from the ProCAST library as well. Finally, the chosen element size was 2.0 mm, obtained through a convergence and mesh analysis.

2.4 Microstructural Characterization

The distribution channel of the ingot was sectioned in different regions, as shown in Fig. 2a, for microstructural characterization, to define the best position of the V-notch when machining specimens for Charpy tests. Only one side was analyzed due to the symmetry of the ingot. Each section was cold-embedded in resin, ground with sandpapers from mesh 220 to mesh 1200 and polished with a 1-μm diamond paste. When necessary, they were etched with a 5% NaOH solution to reveal the microstructure. A real cast wheel was used as reference, and cross sections of its

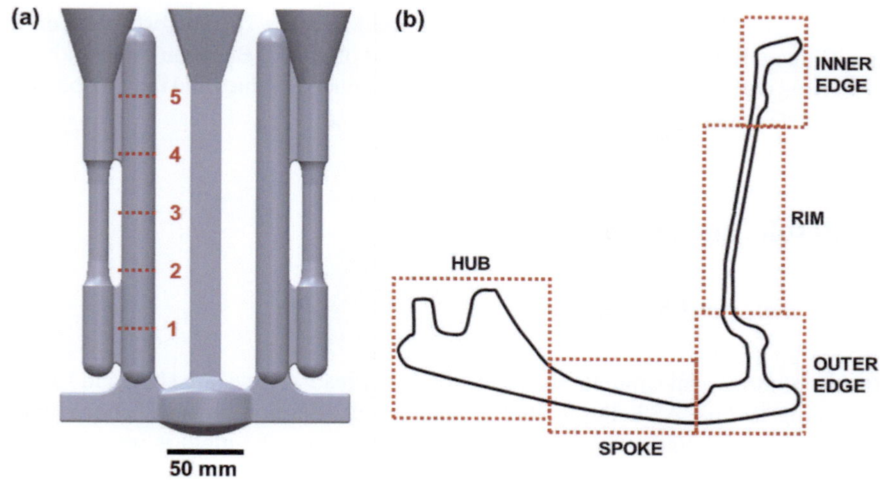

Fig. 2 3D drawing of the ingot with indication of regions for SDAS and porosity measurement (**a**); schematic representation of the cross section of the reference wheel with indication of its different regions

different regions—hub, spoke, rim, and inner and outer edges—were also cut and prepared in the same manner (Fig. 2b).

Scanning electron microscopy (SEM) was done to analyze the microstructure. A JEOL® JSM-6510LV tungsten microscope was used for image acquisition using backscattered electrons' signal. Images for SDAS and porosity measurements were acquired in a Zeiss® Axio Scope.A1 optical microscope equipped with a Zeiss® Axiocam ERc 5 s camera. For SDAS calculation, four images were taken from etched specimens with a magnification of 50*x*, allowing from 4 to 10 measurements in each micrograph. Measurements were done using ImageJ software according to Method D of the work by Vandersluis and Ravindran (Vandersluis and Ravindran 2017), described by Eq. (1), where "L" is the length of the primary dendritic arm and "N" is the number of secondary branches. Percent area porosity was calculated using ImageJ software based on single images taken at the lowest magnification (6.4*x*) of non-etched samples.

$$\text{SDAS} = \frac{L}{N-1} \tag{1}$$

2.5 *Impact Testing*

Specimens for Charpy impact testing were cut from the ingots by electrical discharge machining (EDM). Specimen dimensions and test procedures followed the ASTM

E23 standard (American Society for Testing and Materials 2023). Tests were done in a Tinius Olsen machine at a room temperature of 20 °C. Between three and five specimens were tested for each condition. Average energy values and their respective standard deviations were calculated.

3 Results and Discussion

3.1 Numerical Simulation

Results of numerical simulations, regarding solidification time, are shown in Fig. 3. According to ProCAST, the feature of interest (distribution channel) should take between 700 and 400 s, approximately, to solidify when the mold is pre-heated to 560 °C. Upon pre-heating to 510 °C, on the other hand, the same region takes from around 90–160 s to cool down. The farther the region is from the lower feeder, the longer it takes to solidify, due to the direction of melt flow inside the mold starting from the center.

Another feature that should be noted is that the thermal regime faced by the distribution channel is different than the one experienced by the gauge length of the tensile test specimen, which cools down faster. Therefore, attention should be given to the correlation between test results and the respective microstructure of the specimen. However, tensile behavior is not the focus of the present work.

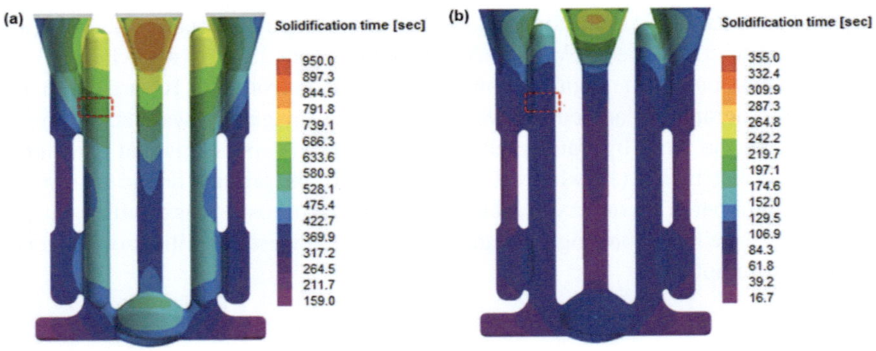

Fig. 3 Numerical simulation results showing thermal profiles during solidification in ingots pre-heated at 560 °C—HT (**a**) and at 510 °C—LT (**b**). The dashed red lines indicate the final region chosen for notch placement

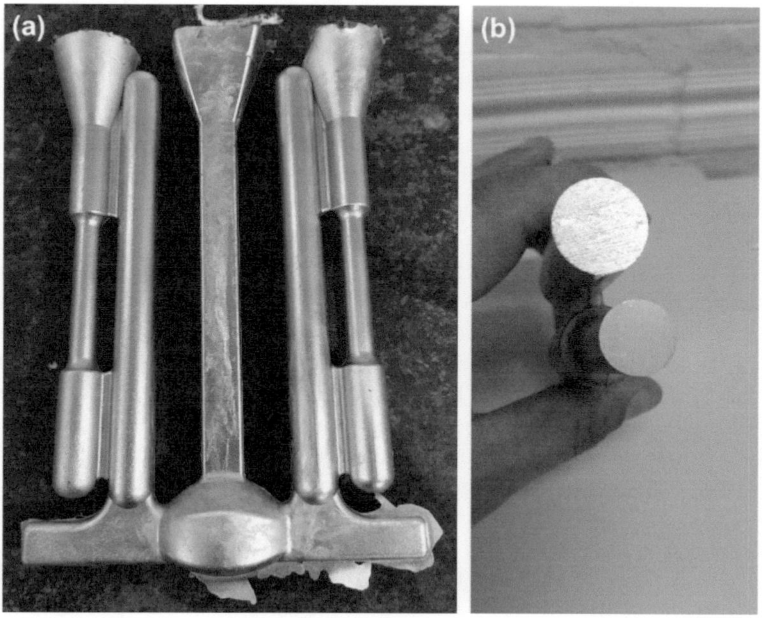

Fig. 4 Example of cast ingot (**a**); cross section of the central region of tensile test specimen + distribution channel

Table 2 Average composition of the cast alloy with the respective standard deviations

Element	Al	Si	Mg	Fe	Sr	Ti
Amount (wt%)	Bal	7.01 ± 0.14	0.30 ± 0.04	0.14 ± 0.01	0.015 ± 0.003	0.18 ± 0.03

3.2 *Casting Quality*

The quality of produced ingots can be seen in Fig. 4. No critical defects, such as coarse contraction pores, were observed in the regions of interest, being limited to feeders and risers. The average composition of all ingots cast in this work, measured in samples taken from the respective melts, is presented in Table 2.

3.3 *Microstructure*

Results of SDAS measurements in different regions of a cast wheel are shown in Fig. 5. The hub and the spoke present similar degrees of dendritic refinement, with values ranging between 40 and 55 μm, and average values of 51 and 45 μm, respectively. Edges and rim, in their turn, present SDAS values ranging from 25 to 35 μm, with very close averages (around 30 μm). Therefore, it can be said that the set hub + spoke faces similar thermal regimes, with slower cooling rates, while the set rim + edges faces another regime that leads to a faster solidification, most likely due to the smaller thickness and primary melt feeding, while the hub and spoke, in addition to being larger, are the last regions to be filled and to solidify (Duan et al. 2020; Merlin et al. 2009).

Figure 6a shows the measured SDAS values in the different regions analyzed in the distribution channel. Based on Fig. 5, it can be observed that the regions whose thermal regimens lead to dendritic sizes closer to those from the hub and the rim when facing the two different cooling rates applied are regions 4 and 5. In both of them, the SDAS in condition "HT" is very close to 50 μm. In "LT", the spacing is slightly overestimated, lying between 35 and 40 μm, while the typical values observed in the rim were a bit lower, closer to 30 μm. Nonetheless, considering deviations in both cases (rim in Fig. 5 and regions 4 and 5 in Fig. 6a), the temperature of 510 °C was considered appropriate to simulate this part of the wheel. Moreover, no appreciable variation of the SDAS was observed among different regions of the part in condition "LT," evidencing that the effect of the cooling rate is reduced as this rate increases, while condition "HT" shows a direct correlation between solidification order (distance from the feeder) and dendritic spacing.

Fig. 5 Typical SDAS values obtained in different regions of a cast wheel

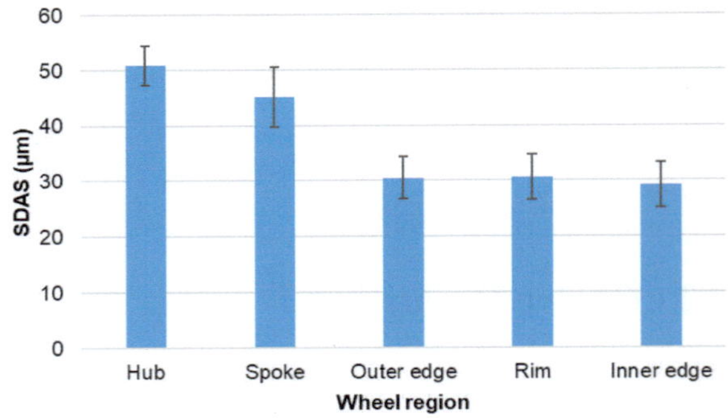

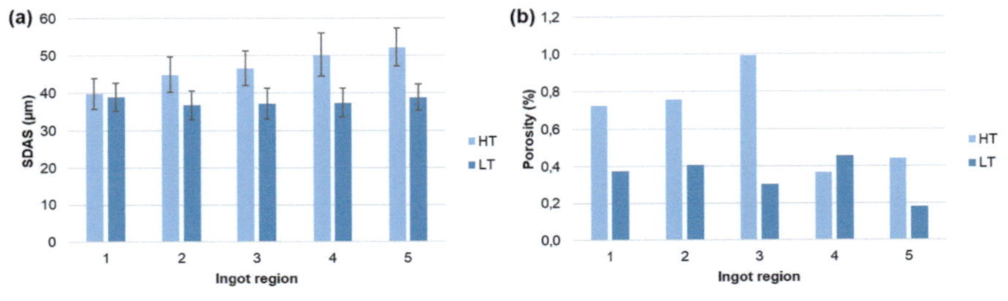

Fig. 6 SDAS **a** and porosity **b** values measured in each region of the cast ingot

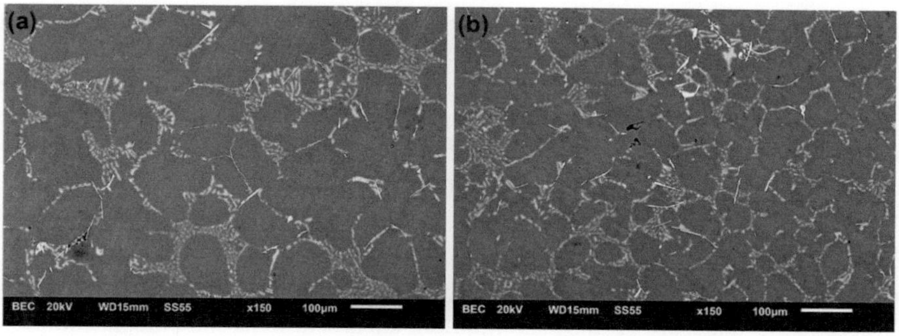

Fig. 7 SEM images showing the microstructure of region 4 in conditions "HT" **a** and "LT" **b**

Porosity (Fig. 6b), on the other hand, does not follow straightforward tendencies. However, since calculations were done with basis on one single image for each region, it was not possible to calculate associated errors and deviations. Regardless, the only case in which a high porosity level, as per specifications of the wheel manufacturer, was observed was in region 3 of condition "HT." Regions 4 and 5 presented low porosity in both conditions. Thus, considering the size of the impact test specimen, region 4 was chosen for the placement of the central V-notch (highlighted by the dashed red line in Fig. 3). The microstructures of this region for both conditions are shown in Fig. 7, comprising mainly the eutectic silicon phase dispersed in the Al matrix. Besides the eutectic phase, needles of the β-AlFeSi phase are also observed. However, these are typical of the alloy's microstructure due to the presence of iron (Mbuya et al. 2003; Taylor 2012).

3.4 *Impact Toughness*

After machining with the V-notch positioned in region 4, impact tests were done, leading to average absorbed energy values of 3.92 ± 0.00 J/cm^2 in condition "HT" and 6.28 ± 0.54 J/cm^2 in condition "LT." As expected, the refined microstructure presented an enhanced toughness. These values are close to the ones obtained by Kaba et al. permanent mold specimens (Kaba et al. 2018) and even higher than those obtained in different regions of cast wheels by Merlin et al. (2009), with comparable SDAS values. However, they are much lower than the required for different wheel regions, of 15 J/cm^2, according to manufacturer.

To evaluate if the heat treatment could help improve toughness, raising it closer to the standard value, impact tests were also done in specimens submitted to a typical T6 treatment, involving solution treatment at 540 °C for 3 h and aging at 125 °C for 2 h, followed by a simplified thermal cycle to represent the painting procedure applied during actual wheel fabrication (165 °C/1 h). However, impact toughness presented a negligible improvement, reaching 4.71 ± 0.44 J in condition LT and 6.87 ± 0.00 J/cm^2 in condition HT.

Therefore, although the thermal regime of the channel is representative of the thermal regime of the wheel, in terms of microstructural refinement (dendritic spacing), and although porosity levels are relatively low, and meet manufacturer specifications, other factors should be further evaluated to understand why toughness results are not representative of the specified wheel regions. However, some hypotheses can be raised:

- The casting procedure was done ensuring that all elements would like within the acceptable range. For Fe, the maximum amount is 0.15 wt%, and the concentration of this element was close to this value in all cast ingots. In the industry, although this tolerance exists, casting practices are defined in a way that Fe concentration is always kept as low as possible, because the presence of the needle-shaped β-AlFeSi phase is one of the most detrimental features to mechanical performance, especially in terms of ductility and toughness (Mbuya et al. 2003; Taylor 2012). Therefore, the amount of this phase in the ingot could be somewhat higher than what is observed in commercial wheels. However, a deeper microstructural analysis of the ingot and of actual wheels would be necessary, including phase quantification, to evaluate this hypothesis.
- Although porosity levels were sufficiently low, and the observed porosity was mainly interdendritic, meaning that it is more associated with material flow through interdendritic regions than with gas entrapment, the degassing procedure used in the laboratory was different than the one used industrially. Therefore, there could be gases such as oxygen and hydrogen in levels higher than expected. However, a thorough chemical analysis of the ingots using elemental analyzers would be necessary to quantify these elements.

4 Conclusions

This study was carried out aiming to evaluate the suitability of a procedure to define the most adequate region to analyze the impact toughness behavior of a low-pressure cast A356 alloy used in automotive wheels. A procedure was established based on numerical simulation to define mold pre-heating temperatures that would lead to thermal regimes similar to those observed in different parts of the wheel; in this case, hub and rim regions were chosen. The target thermal behavior was defined in the auxiliary distribution channel, from which specimens can be taken for impact tests. Results show that numerical simulations are useful to define adequate pre-heating temperatures, and one region of the distribution channel achieved microstructural features—SDAS and porosity—considered adequate to represent different wheel regions upon such pre-heating. However, although the microstructure is very similar to the typical one of cast A356 alloy, toughness results were much lower than those expected based on wheel application requirements. Some hypotheses were pointed out to explain such differences, but further studies are necessary to assertively define the actual reason.

Acknowledgements The authors would like to acknowledge the Foundry Laboratory (LAFUN) of the Federal University of Rio Grande do Sul for carrying out the casting tests.

Competing Interests The authors have no conflicts of interest to declare that are relevant to the content of this chapter.

Responsibility for Information The authors are solely responsible for the information included in this work. No AI systems were used in the translation or writing of the text.

References

American Society for Testing and Materials (2019) ASTM B108/B108M-19: Standard specification for aluminum-alloy permanent mold castings. In: Annual book of ASTM standards. ASTM International, p 23. https://doi.org/10.1520/B0108_B0108M-19

American Society for Testing and Materials (2023) ASTM E23-18: Standard test methods for notched bar impact testing of metallic materials. In: Annual book of ASTM standards. ASTM International, p 26. https://doi.org/10.1520/E0023-18

Duan Y, Zhang F, Yao D, Tian L, Yang L, Guan Y, Hu J (2020) Numerical prediction of fatigue life of an A356-T6 alloy wheel considering the influence of casting defect and mean stress. Eng Fail Anal 118:104903. https://doi.org/10.1016/j.engfailanal.2020.104903

Hamasaiid A, Dargusch MS, Dour G (2019) The impact of the casting thickness on the interfacial heat transfer and solidification of the casting during permanent mold casting of an A356 alloy. J Manuf Process 47:229–237. https://doi.org/10.1016/J.JMAPRO.2019.09.039

Kaba M, Donmez A, Cukur A, Kurban AF, Cubuklusu HE, Birol Y (2018) AlSi5Mg0.3 alloy for the manufacture of automotive wheels. Int J Metalcast 12(3):614–624. https://doi.org/10.1007/s40962-017-0191-2

Krall P, Weißensteiner I, Pogatscher S (2024) Recycling aluminum alloys for the automotive industry: breaking the source-sink paradigm. Resour Conserv Recycl 202:107370. https://doi.org/10.1016/J.RESCONREC.2023.107370

Mbuya TO, Odera BO, Ng'ang'a SP (2003) Influence of iron on castability and properties of aluminium silicon alloys: literature review. Int J Cast Metals Res 16(5):451–465. https://doi.org/10.1179/136404603225006468

Merlin M, Timelli G, Bonollo F, Garagnani GL (2009) Impact behaviour of A356 alloy for low-pressure die casting automotive wheels. J Mater Process Technol 209:1060–1073. https://doi.org/10.1016/j.jmatprotec.2008.03.027

Polmear I, StJohn D, Nie J-F, Qian M (2017) Light alloys: metallurgy of the light metals, 5th edn. Elsevier.

Rams, J., & Whittaker, M. (Eds.). (2022). Light alloys and composites - High temperature extreme alloys and composites (Vol. 1). Elsevier, Netherlands

Saini DK, Jha PK (2023) Fabrication of aluminum metal matrix composite through continuous casting route: a review and future directions. J Manuf Process 96:138–160. https://doi.org/10.1016/J.JMAPRO.2023.04.041

Taylor JA (2012) Iron-containing intermetallic phases in Al-Si based casting alloys. Proc Mater Sci 1:19–33. https://doi.org/10.1016/j.mspro.2012.06.004

Vandersluis E, Ravindran C (2017) Comparison of measurement methods for secondary dendrite arm spacing. Metall Microstruct Anal 6:89–94. https://doi.org/10.1007/s13632-016-0331-8

Wu J, Sui D, Han Q (2022) High quality plate-shaped A356 alloy casting by a combined ablation cooling and mold heating method. J Mater Process Technol 303:117536. https://doi.org/10.1016/J.JMATPROTEC.2022.117536

Determination of the Mechanical Properties of Hybrid Composite Material for Aeronautical Structures

Evelyn Coelho Dias and Jhonatan Acácio Silva

Abstract With the development of technology, industries have observed that hybrid materials (composites) are a promising solution due to the combination of their distinct mechanical properties. However, one of the major challenges to be mastered is the manufacturing process and the bonding interface between the materials. Numerical simulation analyses have been widely applied and studied as basic tools for predicting the behavior of hybrid composite materials. However, for simulation to be effective, a broad understanding of the mechanical properties of the material under study is first required. To assess the potential for improvement in the materials' characteristics, a total of 15 test specimens were developed in the form of sandwich panels, composed of two layers of 334HM carbon fiber impregnated with epoxy resin and a core made of 6063-T6 aluminum. The core material was prepared with three distinct surfaces: five natural tests, five sanded tests with No. 60 sandpaper, and five blasted tests with microspheres. This test was performed to predict potential improvements in adhesion between the core and the composite. After lamination, the specimens were subjected to three-point flexural testing using a universal testing machine. By applying different forces and loads, the material responses and their ability to withstand stresses and deformations were evaluated, as well as the adhesion characteristics between the composite and the hybrid core. This process provided essential data for calculating the material's ultimate loads, including bending stress, elastic modulus, tensile stress, and yield strength. After comparing the results of the different surfaces, it was concluded that the preparation of the sandblasted aluminum surface had a superior performance to the others, given that its roughness allows for better adhesion of the resin/fiber to the material. In addition, it also presented the best mechanical behavior, providing an improvement of approximately 44.40% in its characteristics compared to pure 6063-T6 aluminum.

E. C. Dias (✉) · J. A. Silva
Universidade do Vale do Itajaí—UNIVALI, Itajaí, Santa Catarina, Brazil
e-mail: evelyn.dias1301@gmail.com

J. A. Silva
e-mail: jhonatanacacio@univali.br

V. E. Beal and A. S. Ribeiro Junior (eds.), *Advances in Manufacturing Engineering Research*, Lecture Notes in Mechanical Engineering,
https://doi.org/10.1007/978-3-032-09858-0_9

Keywords Composite materials · Material properties · Mechanical strength · Aircraft fuselage

1 Introduction

After the space race resulting from the Cold War, composite materials have gained increasing importance in the aerospace industry (Zanatta 2012). Their high specific properties, as well as their high flexibility in obtaining modified hybrid materials according to the unique requirements of a project, have led to an increasingly high demand for these materials (Freitas 2018).

Metals are widely used materials in all types of applications because their mechanical properties and behaviors have been studied and understood since the beginning of industrialization. Due to the strength, weight, and molding flexibility of composite materials, they have become invaluable for industrial studies and applications. However, the high cost, specialized labor, and time-consuming processes associated with the manufacture of composite materials limit their use (Monge 2018).

The *Carbon Fiber Reinforced Polymer (CFRP)* for the manufacture of aircraft such as the Airbus A350 and Boeing 787, where its use can reach 50% of its entire structure (Morgado et al. 2020; Pérez et al. 2019). In the current industry, CFRP is being combined with metal alloys to form a hybrid composite material, whose objective is to take advantage of the best characteristics of each: the metal will provide ductility to the assembly, providing a more flexible manufacturing that allows greater ease in joining with other components, in addition to the composite helping in fatigue resistance and stiffness with a reduced increase in mass (Monge 2018; Arputhabalana et al. 2019).

To further enhance Monge's existing research in this area, the study aims to model test specimens to validate the characteristics obtained from joining these materials using a sandwich panel, thereby providing insight into the performance and behavior of hybrid composites. These hybrid composites still face significant challenges, such as adhesion between the composite and the core, the presence of bubbles during lamination, and failures resulting from the manufacturing process.

2 Materials and Methods

This study will use 334 HM carbon fiber, 2004 epoxy resin, and 6063-T6 aluminum, which will be joined using a sandwich panel. The specimens will be manufactured using a combination of manual lamination for the composite and a vacuum bag process for joining the layers. The vacuum application is intended to reduce bubble formation and ensure greater adhesion between the composite and the aluminum reinforcement, as studied by Souza (2019).

To carry out this study, approximately 15 test specimens measuring 150 × 50 mm will be manufactured in order to obtain more assertive sampling and a better structural analysis of the composites, in accordance with ASTM D790 and ISO 178 (2022) standards, which indicate the minimum number of 5 tests required to validate the results of each experiment.

2.1 Selection of Hybrid Composite Materials

Unidirectional carbon fiber 334 HM was used to manufacture the specimens, which will be responsible for the reinforcement of the composite used in the tests. Table 1 taken from the supplier Texiglass (2024) contains information on the parameters of the selected material.

For the matrix, 2004 epoxy resin was used, a polymeric matrix commonly used in the manufacture of composites according to Costa and Pereira (2020). In addition to the materials that make up the two phases of the composite, this study will add a hybrid 6063 aluminum core, which aims to maximize the characteristics of the material and improve its properties. The parameters of the aluminum alloy were studied by Decesh et al. (2020) and can be found in Table 2.

Three core preparation models were made, being 5 aluminum sheets with a natural surface, 5 sheets with a sandblasted surface treatment and 5 sheets sanded with 60 grit sandpaper. In Fig. 1, we can see the aluminum reinforcements sandblasted with microspheres (Fig. 1a), sanded with 60 grit sandpaper (Fig. 1b) and natural (Fig. 1c) respectively, already cut to the measurements of 150 × 50 mm, which were used in the present research.

Table 1 Parameters of carbon fiber 334 HM

Parameters	Nominal
Weight (g/m^2)	320
Number of threads (Threads/cm)	3.8
Number of wefts (Threads/cm)	1.5
Thickness (mm)	0.5
Warp breaking load (kgf/cm)	779
Weft breaking load (kgf/cm)	3.6
Elastic modulus (GPa)	240
Fabric tensile strength (MPa)	4900

Table 2 Parameters of aluminum 6063-T6

Parameters	Nominal
Density (g/cm^3)	2.71
Modulus of elasticity (MPa)	69,000
Melting point (ºC)	600–650
Thickness (mm)	3.17
Length (mm)	150
Width (mm)	50
Tensile strength (MPa)	210–240
Yield strength (MPa)	10
Hardness (HB)	60

Fig. 1 Aluminum reinforcement core

2.2 *Test Specimens Manufacturing*

The specimens were manufactured in a single mold using a combination of manufacturing processes. The fiber/resin composite was manually laminated, and after the panels were assembled, the hybrid composite material was subjected to a vacuum curing process. This method reduces the formation of air bubbles and aids in adhesion between the core and the composite.

First, a glass plate was prepared with three layers of release wax, serving as the lower mold for the experiment. The carbon fiber was cut to the dimensions required to create the test specimens (200 × 60 mm) with a rim. The aluminum plates were also previously cut and prepared to 150 × 50 mm, requiring surface cleaning of the metal before rolling. Only water and mild soap were used to clean the aluminum.

The resin-impregnated fiber layers were then sandwiched in two sequential layers, forming the composite material used. After lamination was complete, an aluminum core was added, followed by the same proportion of composite layers, as shown in Fig. 2, according to the sandwich method adopted in this research.

Finally, the peel ply layers were installed and perforated film that aid in demolding the material, followed by the vacuum bag layer, which was secured with sealing tape to seal the entire mold. In this study, a longitudinal vacuum channel was used to better absorb resin buildup, as shown in Fig. 3.

After the mold was completely sealed, the vacuum pump was activated for approximately 1 h and 30 min until the pressure in the system was stable in order to remove possible bubbles resulting from manual lamination and consequently improve the adhesion of the composite with the metallic hybrid.

Fig. 2 Panel assembly

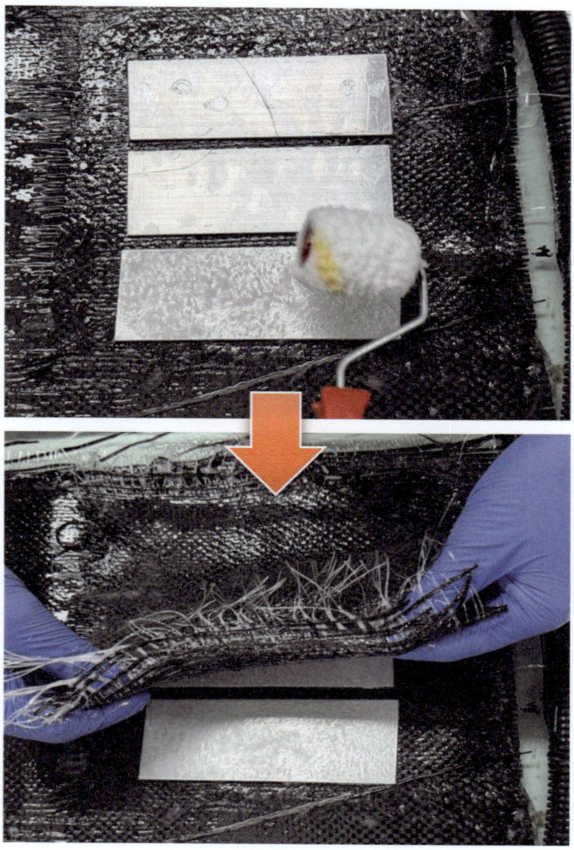

Fig. 3 Vacuum applied to the mold

The curing process was estimated to take 24 h, but the plate was demolded six days after the experiment. The protective layers were carefully removed from the mold, cut to the correct size, and the burs were removed using a rotary sander.

2.3 Mechanical Tests

For this research, the 3-point bending test was carried out with the aid of a universal testing machine, following the parameters listed in Table 3, with a total of 15 tests.

Table 3 Experimental parameters

Parameters	Nominal
Test speed (mm/min)	1
Distance between supports (mm)	120
Support diameter (mm)	16
Air corps (mm)	150 × 50
Average body Thickness (mm)	5.6

Fig. 4 Support of test specimens

The tests were conducted with a two-support support where the test specimen was centered on the base and the applied force was centered in the middle of the piece, with the predetermined spacing, as shown in Fig. 4.

The test report provided us with the deformation graph based on time and applied force, obtaining the essential data that was used to calculate the limit loads of the materials, including the yield limits and tensile strength of each set later.

3 Results and Analysis

After performing the flexural tests, it was possible to perform several calculations to characterize the properties of the hybrid composite material in addition to observing failures between the adhesion and cohesion of the composite material in relation to the core.

Nastran software (Fig. 5) was used, following the instructions and guidelines provided by Autodesk itself (2025), to analyze the main forces and stresses applied during the bending tests, considering that the aluminum lamination is in the same direction as the carbon fiber strands. From these results, it can be observed that the body undergoes compression stress at the top of the test and tension stress at the bottom, while bending is applied to the cross-section.

To analyze the behavior of the composite material in relation to the core, the 3DEXPERIENCE® Simulia software was used (Fig. 6), which provided a better understanding of the adhesion problems of the hybrid composite, as shown in

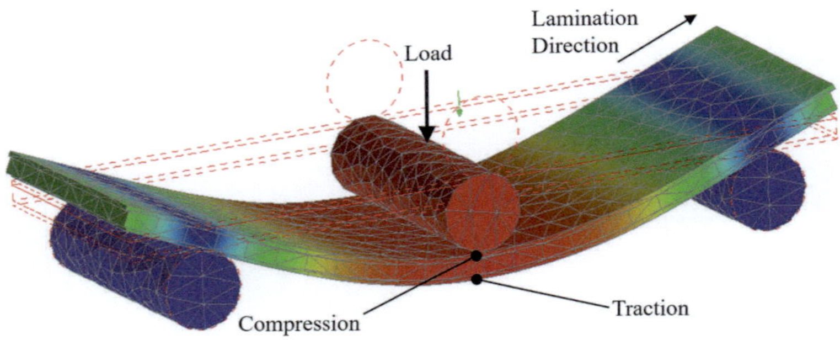

Fig. 5 Specimens stress distribution simulated for the bend test

Fig. 6a, which shows the detachment of the composite material in relation to the core. Figure 6b shows the von Misses stress distribution during the bending tests.

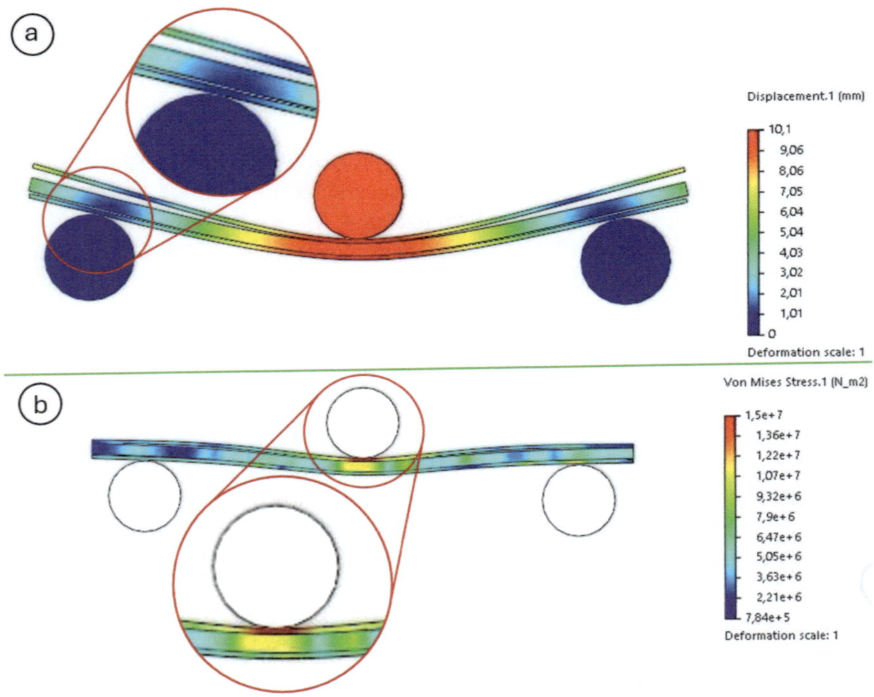

Fig. 6 Displacements (**a**) and von Mises stress (**b**) distributions

3.1 Material Adhesion Analysis

When performing the analysis using a stereoscope in the tests performed, it was possible to observe adhesion points between the core and the composite material in the test specimens that had a sanded (CP6-CP10 tests) and sandblasted (CP10-CP15 tests) surface treatment, as shown in Fig. 7.

In the test specimens that were in natura (CP1-CP5 tests), several failures were found due to fiber rupture, as shown in Figs. 8 and 9.

Additionally, some adhesion failures resulting from the lamination process were found. Figure 10 shows a comparison between a lamination with good adhesion and a failure due to lack of adhesion.

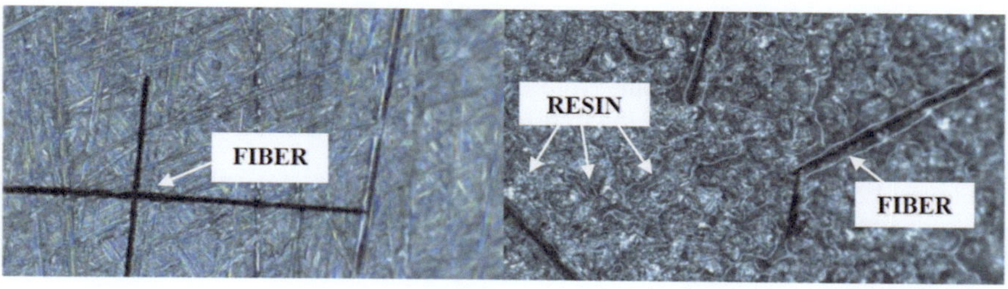

Fig. 7 Hybrid material adhesion points with the core

Fig. 8 Failure by rupture test 3

Fig. 9 Failure by rupture test 4

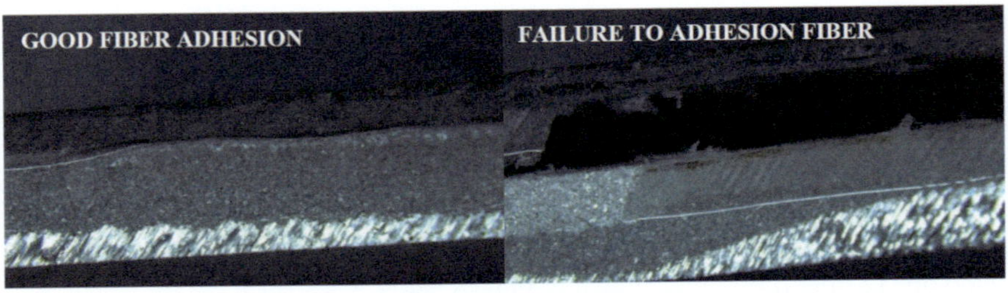

Fig. 10 Comparison between good adhesion and failure

3.2 *Analysis of Mechanical Test Results*

Figure 11 shows a graph with the force versus deformation results obtained with the tests without core treatment. Note that there were 3 studies (CP3, CP4, CP5) with failure in the adhesion of the hybrid composite, having greater deformation at maximum force.

Figure 12 shows a graph with the results of force in relation to deformation obtained with the tests treated using No. 60 sandpaper. Note that there were 2 studies (CP6, CP7) with failure in the adhesion of the hybrid composite, having greater deformation at maximum force.

Figure 13 shows a graph showing the force versus deformation results obtained from the blasting tests. It should be noted that there were no flaws in the studies,

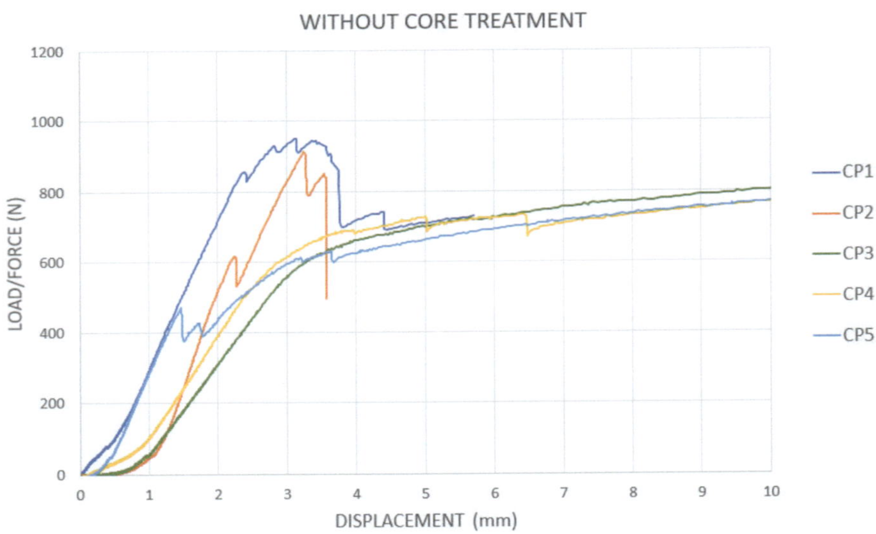

Fig. 11 Load versus displacement graph for natural specimens

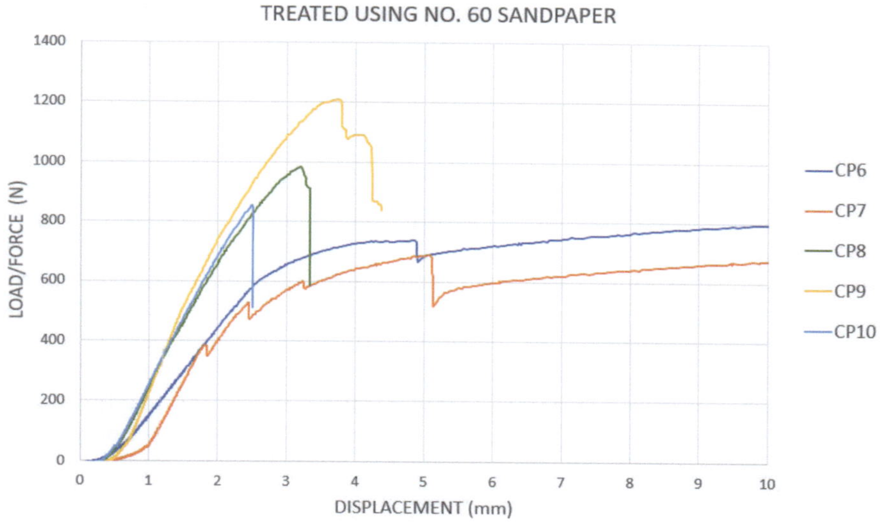

Fig. 12 Load versus displacement graph for sanded specimens

ensuring a much lower standard deviation than the other tests, as these had little variation in their results.

To characterize the properties obtained through the tests, the bending stresses, tensile stresses and elastic modulus of each study were calculated, using, respectively, Eqs. (1–3) which were taken from Hibbeler (2010).

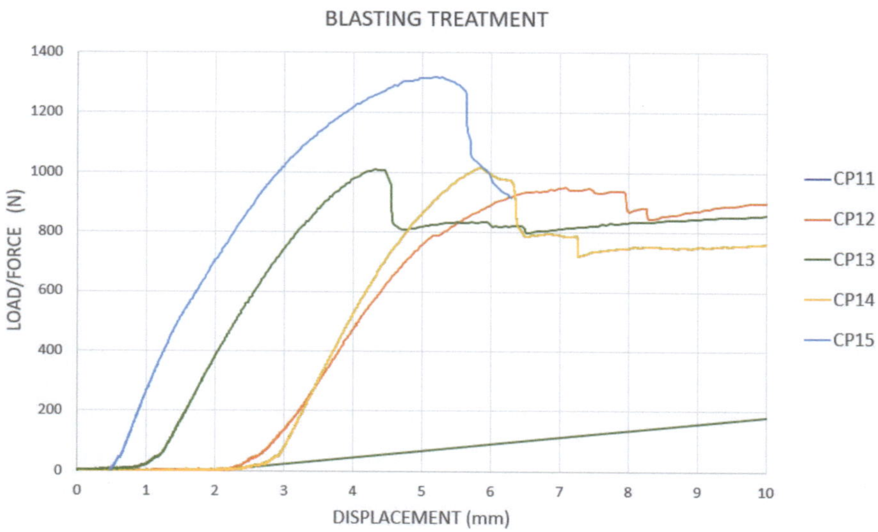

Fig. 13 Load versus displacement graph for sandblasted specimens

$$\sigma_{\text{bending}} = \frac{3 \times F \times L}{2 \times w \times h^2} \tag{1}$$

$$\sigma_{\text{Traction}} = \frac{F \times L}{w \times h^2} \tag{2}$$

$$\varepsilon = \frac{F \times L^3}{48 \times f \times I} \tag{3}$$

where σ_{bending} refers to the maximum bending stress (MPa); σ_{traction} refers to the maximum tensile stress (MPa); ε refers to the modulus of elasticity (GPa); F refers to the maximum force applied (N); L refers to the distance between the test supports (mm); w refers to the width of the test specimen (mm); h refers to the thickness/height of the test specimen (mm); and I is the moment of inertia of the body (kg m^2).

After obtaining the main characteristics of the tests, the averages between the results obtained for each study were analyzed. Table 4 shows the results obtained through the tests without core treatment, reaching an average tensile stress of 263.73 MPa and an average bending stress of 395.6 MPa. Its deformation varied between 3.13 and 18 mm, generating an average deformation of 11.21 mm.

Table 5 presents the results obtained through tests with the core treated with sandpaper, reaching an average tensile stress of 281.46 MPa and an average bending stress of 422.14 MPa. Its deformation varied between 2.48 and 20.38 mm, generating an average deformation of 10.05 mm.

Finally, Table 6 presents the results obtained through tests with the blast-treated core, reaching an average tensile stress of 328.19 MPa and an average flexural stress

Table 4 Results of untreated tests

Parameters	CP01	CP02	CP03	CP04	CP05	Average
Maximum force (N)	952.47	913.10	823.15	814.02	804.89	861.53
Bending stress (MPa)	437.36	419.28	377.98	373.78	369.59	395.6
Tensile stress (MPa)	291.57	279.52	251.98	249.19	246.39	263.73
Deformation (mm)	3.13	4.31	15.56	15.05	18.02	11.21
Elastic Modulus (GPa)	525.84	366.09	91.41	93.46	77.18	230.8

Table 5 Results of sanding tests

Parameters	CP06	CP07	CP08	CP09	CP10	Average
Maximum force (N)	823.14	725.85	983.98	1209.3	854.38	919.33
Bending stress (MPa)	377.97	333.3	451.833	555.29	392.32	422.14
Tensile stress (MPa)	251.98	222.2	301.22	370.19	261.54	281.46
Deformation (mm)	20.38	20.44	3.20	3.74	2.48	10.05
Elastic Modulus (GPa)	69.79	61.36	531.35	558.74	595.31	363.31

of 492.28 MPa. Its deformation ranged from 3.42 to 7.08 mm, this being the study with the smallest variation in values.

Based on the analysis of the results, Fig. 14 presents the comparison between the averages of the analyses obtained, with the blue line representing the commonly known characteristics of 6063-T6 aluminum, the orange line representing the tests with the natural core, the yellow line representing the sanded tests and the green line representing the sandblasted tests.

Table 6 Blasting test results

Parameters	CP11	CP12	CP13	CP14	CP15	Average
Maximum force (N)	949.93	1009.9	1016.6	1318.1	1065.9	1072.09
Bending stress (MPa)	436.19	463.73	466.81	605.25	489.44	492.28
Tensile stress (MPa)	290.79	309.15	311.20	403.50	326.30	328.19
Deformation (mm)	7.08	4.31	5.86	5.21	3.42	5.18
Elastic Modulus (GPa)	231.85	404.9	299.78	437.43	538.56	382.50

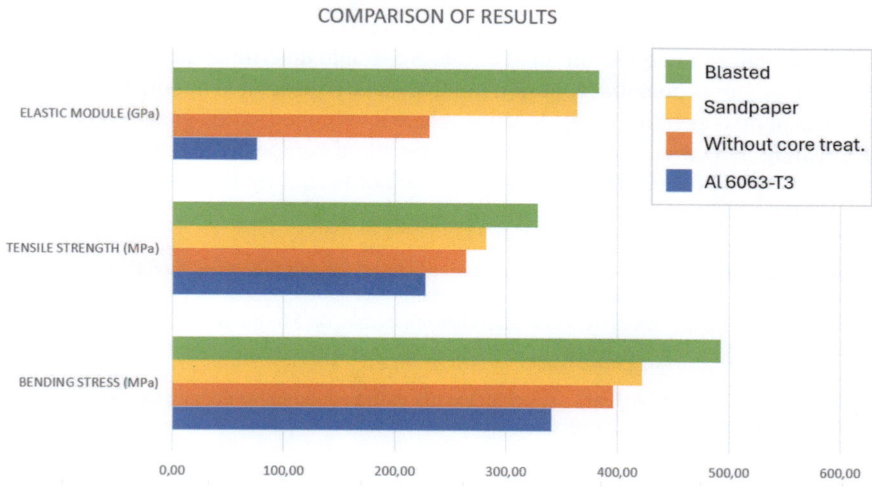

Fig. 14 Comparative results graph

4 Conclusion

Analyzing the test results, it was found that the natural core provided a 16.00% improvement in mechanical properties compared to commercial 6063-T6 aluminum. Tests performed with sanded surfaces achieved a 23.80% improvement, while sand-blasted tests demonstrated a significant 44.40% improvement, also compared to commercial 6063-T6 aluminum.

Roughening the aluminum surface proved to be a key factor in achieving superior results in terms of adhesion and cohesion of the composite material to the core. Among the methods evaluated, sandblasting stood out as the most effective, resulting in a significant improvement over other techniques and providing greater reliability and precision to the experiment.

Overall, when comparing the properties of commercial aluminum with the hybrid composite material, the study revealed promising characteristics for applications in the aeronautical industry. The hybrid composite material proved to be moldable and capable of maintaining its light weight, while demonstrating significantly superior strength to 6063-T6 aluminum, currently widely used in the field.

For future work, we suggest conducting an applied simulation that takes into account the pressures and forces acting on the material, considering the specific altitude of the aircraft in question. In addition, we should determine the applicability of this material, as its potential cohesion failures make it difficult to maintain, becoming a critical point depending on the location where it is applied, whether structural, filler, or finishing. However, the hybrid aluminum-carbon composite material remains an excellent option for aerial applications, as demonstrated by the results obtained in this study.

Acknowledgements I express my sincere gratitude to my research partner, Jhonatan Acacio, for his constant support throughout my university career.

I also thank Felipe Brum Rossato and Gabriel Renan de Moura Bunde for their valuable assistance in developing this research, as well as for their time and dedication to this study and essays.

Special thanks to CREA-Junior, which made my participation in the congress possible and has been a constant supporter of the personal and professional development of its members.

Responsibility for Information The authors are solely responsible for the information included in this work.

References

Arputhabalana J, Prabhu S, Palanikumar K, Venkatesh S, Vijay K (2019) Assay of machining attributes in drilling of natural hybrid fiber reinforced polymer composite. Mater Today: Proc 16:1097–1105

ASTM D790. Standard test methods for flexural properties of unreinforced and reinforced plastics and electrical insulating materials. Available at: https://cdn.standards.iteh.ai/samples/97762/3bfa4bc9fbf54c68ad0a721a86682dff/ASTM-D790-17.pdf. Accessed on 15 Jan 2025

Autodesk (2025) Autodesk inventor Nastran. Available at: https://help.autodesk.com/view/NINCAD/2025/ENU/?guid=GUID-632CF56E-F5A6-4D47-AC00-9DBEBF70B306. Accessed on 10 Jan 2025

Costa L, Pereira M (2020) Composite materials: theory and practice. LTC, Rio de Janeiro

Decesh R, Ankit SA, Sanjay K (2020) Evaluation of Johnson-Cook material model parameters of AA6063-T6. Int Res J Eng Technol (IRJET)

Freitas MAC (2018) Development of a tool for hybrid drilling carbon-steel. Dissertation, Master in Engineering, Faculty of Engineering, University of Porto (FEUP), Porto

Hibbeler RC (2010) Chapter 6—Bending. In: Strength of materials, 7th edn, pp 203. Pearson Prentice Hall, São Paulo

ISO 178 (2022). ISO 178—Specimen size. Available at: https://infinitalab.com/blogs/testing-methods/iso-178-specimen-size/. Accessed on 15 Jan 2025

Monge JMS (2018) Hybrid metal/composite system. Dissertation (Master in Engineering)—Faculty of Engineering, University of Porto (FEUP), Porto

Morgado MA, Carbas RJC, Dos Santos DG, Da Silva LFM (2020) Strength of CFRP joints reinforced with adhesive layers. Int J Adhesion Adhesives 97:102475

Pérez JF, Cantero JL, Álvarez JD, Miguélez MH (2019) Hybrid composite-metal stack drilling with different minimum quantity lubrication levels. Materials 12:448

Souza LHC (2019) Study of aluminum matrix composites. Monograph (Undergraduate Degree in Mechanical Engineering)—School of Mines, Federal University of Ouro Preto, Ouro Preto

Texiglass. Carbon fiber catalog. Available at: https://texiglass.com.br/wp-content/uploads/2024/06/TEXICARBON-334-HM.pdf. Accessed on 12 Oct 2024

Zanatta (2012) Composite materials in aviation. Available at: http://www.aviacao.org/article/materiais-compositos/2/. Accessed on 15 Oct 2024

Development and Manufacturing of a Motor for Mini Rockets with an Apogee of 200 Meters

Maria Nícolle Ferreira Queiroz, João Batista Silva Costa, João Ribeiro Marques, Karen Maysa Dias Oliveira, and Henrique Takashi Idogava

Abstract The field of Rocketry involves the development, construction, and launch of small-scale rockets, serving both educational purposes and competitions, and provides a field of study and practical application that allows the development of reduced-scale technologies simulating rocket behavior under controlled conditions. This work aims to present the design and manufacturing of an engine for mini-rockets with a 200 m apogee, used in the FRMF—*Festival Regional de Minifoguetes*. The research includes the idealization, modeling, and manufacturing of the rocket engine, utilizing tools such as experimental spreadsheets and computer-aided design (CAD) software, in addition to discussing the results obtained during the development process and the engine's performance in the competition. The study focuses on the comparison between theoretical results, generated by simulation software, and practical results obtained from tests and during the competition. The structural design of an engine constitutes the starting point in mini-rocket development, being the most determining component for the vehicle's overall performance. This project seeks to integrate fundamental principles of physics, thermodynamics, and fluid mechanics into a compact and efficient system, with a special focus on safety, a critical aspect in rocketry. The adopted methodology was based on a pre-existing project from the literature, serving as a starting point for the adaptation and optimization of the engine. Dimensioning tables were used to simulate critical parameters, such as internal pressure and material resistance. The introduced modifications, such as replacing traditional screw fittings with threading, aim to increase practicality. The parts were manufactured from SAE 1020 steel, and stress analysis was performed using Autodesk Inventor software, employing the finite element method. Through the analyses obtained from burn test results and static bench tests, it was possible to improve safety and performance by adjusting critical parameters identified during

M. N. F. Queiroz (✉) · J. B. S. Costa · J. R. Marques · K. M. D. Oliveira · H. T. Idogava
UNIVASF, Juazeiro, Bahia, Brazil
e-mail: maria.nicolle@discente.univasf.edu.br

V. E. Beal and A. S. Ribeiro Junior (eds.), *Advances in Manufacturing Engineering Research*, Lecture Notes in Mechanical Engineering,
https://doi.org/10.1007/978-3-032-09858-0_10

testing. After implementing the changes and detailed data analysis, the rocket was successfully launched in the competition, reaching an apogee of 231 m, exceeding initial expectations. This work contributes not only as a practical guide for the manufacturing of mini-rocket engines but also as an in-depth analysis of the tools and techniques employed, evaluating their efficiency and impact on the development of small-scale aerospace projects.

Keywords Model rocketry · Propulsion · Rocket engine · Pressure vessels · Validation tests

1 Introduction

Rocket development has been undergoing a great process of technological advancement, mainly since the mid-twentieth century, covering applications ranging from ballistic missiles to the launch of satellites into Earth's orbit. This evolution highlights the importance of expanding studies and technological development in the area, with a focus on rocketry, a field that combines science, technology, and innovation for the development of reduced-scale aerospace systems, frequently used as didactic and experimental tools in educational institutions and academic competitions, and which presents itself as an important tool for the introduction and consolidation of aerospace research. This field is relevant because it allows for the practical application of engineering concepts, providing students and researchers with the opportunity to work on projects that demand planning, creativity, and a solid technical foundation (Norton 2010).

Small-scale rockets, frequently used in competitions and academic experiments, demand special care in the conception of the engine, which is the component responsible for propulsion. The engine's ability to propel the rocket safely and efficiently depends on an optimized design, grounded in engineering principles. This process involves performing structural simulations in specialized software, practical tests on static benches, and dimensional analyses based on the theory of pressure vessels and material properties (Palmerio 2017).

Solid propellant rocket engines, widely used in academic projects, are composed of three main components: a combustion chamber which receives the solid fuel (propellant), a nozzle that accelerates the gases generated by the propellant's burning, and a bulkhead that ensures the combustion gases do not leak. The operating basis of these engines is the progressive combustion of the propellant, composed of a combination of oxidizer and fuel, generating gases that are accelerated by the nozzle and expelled at high velocity, creating the thrust necessary for the rocket's propulsion.

This principle, based on Newton's third law, is described by Carande (2011). The validation of such engines requires technical rigor and safety, using a static test stand to measure the critical data of the thrust-time curve and the engine's operating pressure throughout the burn, which reflect the structural performance of the engine and the propellant (Nakka 2025a, b).

This article aims to report on the manufacturing process through the design and development of a solid propulsion engine for the FRMF competition, held in Natal in 2024, within the 200 m apogee category. The approach includes the use of tools such as Inventor, SRM and EXALT tables, a mechanical lathe and CNC, in addition to a detailed look at the chosen raw materials—SAE 1020 steel and SAE 1045 steel—the manufacturing method, finite element simulations, and the adjustments made during the process. The project validation occurred through practical tests on a static bench, which allowed for a comparison of theoretical with experimental results and ensured the engine's safety and performance, while also opening up possibilities for analyses of potential improvements. Finally, the engine nicknamed "Charlinho," developed within the scope of this project, represents an example of the potential for innovation in academic rocketry, and the research reinforces the importance of technical rigor and creativity in overcoming the challenges associated with the development of academic rocket engines.

2 Methodology

The development of the motor for mini-rockets with a 200 m apogee, nicknamed "Charlinho" by the team, was carried out based on stages integrating mechanical engineering concepts, material analysis, and computational simulations (Fig. 1). The work began with a detailed definition of the motor's functional and operational requirements, taking into account the technical constraints imposed by the Regional Mini-Rocket Festival (FRMF), the competition for which it was developed, and the predetermined rigorous safety criteria, essential for the project's success. This preliminary stage was crucial to establishing the design foundations and the technical specifications to be met. Following the established project logic, the prototype was then developed.

2.1 Theoretical Project

For the initial design, the team used the A-100 motor, developed by Richard Nakka, as a reference, adapting it to meet the project's and competition's specific requirements. Modifications were made to the original fittings, replacing them with threaded connections, a practice previously employed in the team's earlier motors. This approach primarily ensures practicality during assembly and disassembly, which

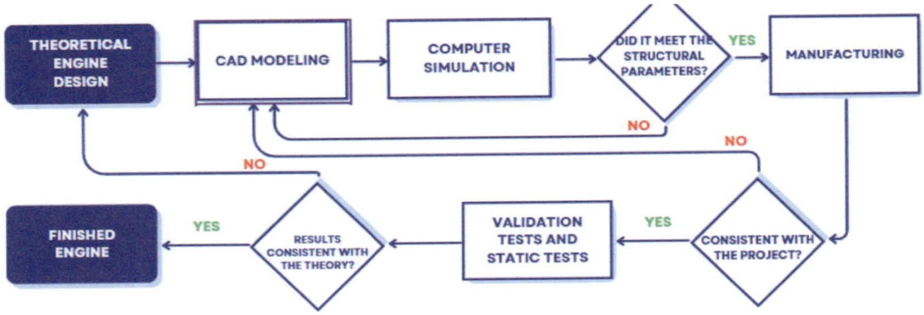

Fig. 1 Flowchart of the methodology steps

are repeated multiple times during testing. It also simplifies the analysis of fittings during manufacturing, as threading allows for faster assembly and eliminates dependency on the mini-rocket's structural alignment—unlike pin holes, which require precise positioning. Following the guidelines of the Brazilian Association of Rocketry (BAR), which mandates six preliminary static tests, the use of the Solid Rocket Motor (SRM) table was critical to simulate parameters such as internal pressure (Fig. 2) and propellant core dimensions. This optimization ensured compliance with the G-class category, where the maximum allowable total impulse is 160 N · s.

After analyzing the results from the SRM simulations, the generated outputs (Table 1) were imported into another spreadsheet tool for further analysis: Ezalt, also developed by Richard Nakka. This spreadsheet is essential for predicting the rocket's apogee and maximum velocity (Table 2). These parameters are critical for validating the chosen motor dimensions and determining whether the engine can deliver sufficient thrust to achieve the target altitude. Additionally, it is important to incorporate a safety margin for drag under varying conditions, as real-world launch performance

Fig. 2 Pressure graph generated by the SRM

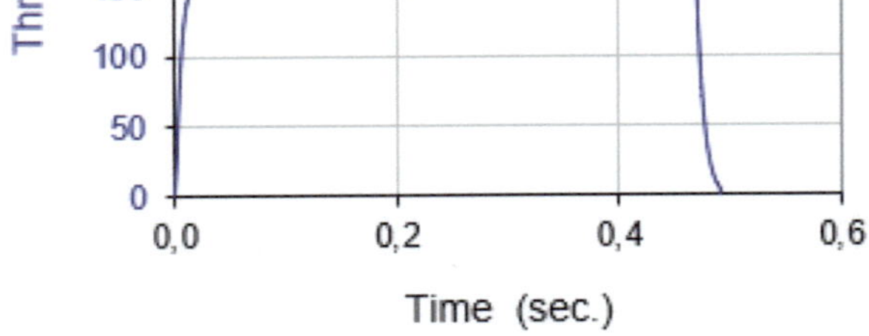

Table 1 Input data for Ezalt

	Input data
Average thrust of the engine (N)	221.1
Total impulse of the engine (N · s)	109.9
Propellant mass (kg)	0.088
Rocket dry mass (kg)	1.400
Maximum rocket diameter (cm)	5
Rocket drag coefficient	0.4

Table 2 Output data from Ezalt

	Output data
Apogee (m)	248
Time to apogee (s)	7.2
Maximum velocity (km/h)	255
Burnout altitude (m)	18

is heavily influenced by aerodynamic drag—which itself depends on wind conditions. This parameter is vital for simulating the rocket's behavior in realistic flight scenarios.

2.2 *CAD and Three-Dimensional Assembly*

Once the theoretical design of the rocket motor was finalized, the team proceeded to computational modeling. This step enabled the fabrication of a prototype with greater precision, reducing costs associated with production, which would otherwise lack prior visualization, and shortening production time by providing advance structural and visual knowledge of the design. The CAD models were developed using Autodesk Inventor, allowing detailed part-by-part design of the motor. The calculated dimensions were iteratively adjusted based on simulation results, striking a balance between structural strength and minimal weight, a critical factor for competition success.

For the nozzle (Fig. 3a), the initial step involved sketching the profile contour. This was followed by a revolved extrusion of the profile. Finally, to create the threading, the spiral command was employed using the "Pitch and Height" mode with a pitch of 1250 mm and a height sufficient to complete 5 full threads.

The bulkhead (Fig. 3b) head was modeled using the same process as the nozzle. First, a profile sketch was created based on the drawing specifications from Richard Nakka's website and the values obtained from the SRM spreadsheet. After completing the sketch, it was revolved. The threading was then created using the spiral command in 'Pitch and Height' mode, with a 1250 mm pitch and sufficient height to complete 5 full threads.

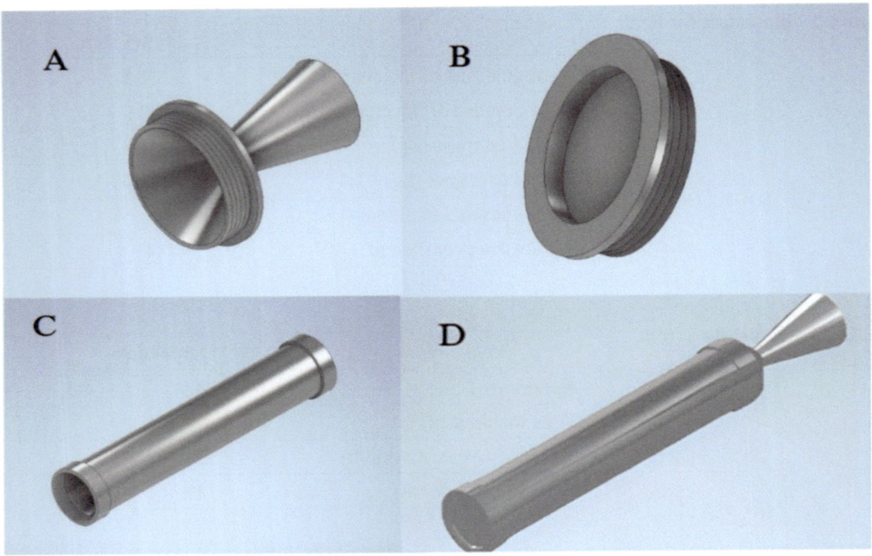

Fig. 3 Nozzle model (**a**), bulkhead model (**b**), chamber model (**c**), and motor assembly (**d**)

For the combustion chamber (Fig. 3c), the dimensions from the SRM projections were used. Initially, for determining the minimum wall thickness t, the circumferential stress equation was applied, given by Eq. (1).

$$t = \frac{P \times R}{S \times E - 0.6 \times P} \tag{1}$$

where P is the estimated operating pressure, R is the chamber radius, S is the maximum allowable stress, E is the weld joint efficiency, substituting these values with those obtained from theoretical predictions in the tables, and using P and S values from Nakka's A-100 model, we obtain the explicit calculus on Eq. (2).

$$t = \frac{7\,\text{MPa} \times 13.5\,\text{mm}}{240\,\text{MPa} \times 1 - 0.6 \times 7\,\text{MPa}} = 0.400763\,\text{mm} \tag{2}$$

However, in actual manufacturing practice, machining parts to such minimal thickness proves extremely challenging and risky. With 0.4 mm as the calculated minimum thickness, we implemented 1.5 mm for the chamber development. This represents a safety factor increase from 1.4 to 5.5.

2.3 Structural and Safety Analysis

Given the reusable nature of this prototype, an increased safety factor was implemented to withstand extensive testing cycles. The threading followed metric standards (1.25 mm pitch, 60° triangular profile), with preliminary strength calculations based on machine element theory, later refined through computational static stress simulations in Autodesk Inventor (Fig. 4). The analysis applied 7 MPa pressure loads with constrained threads (nozzle/head) and chamber ends, yielding the results:

The thread's stress distribution clearly varies along its length, exhibiting cantilever beam behavior (fixed at one end, free at the other). Initial simulations revealed a minimum safety factor of 0.25 in critical regions. To improve simulation accuracy, we implemented mesh refinement at the thread geometry, generating 0.25 mm size elements. This enhanced resolution produced more precise stress concentrations at the thread tips while maintaining safety factors above 6 at the thread base, demonstrating how proper mesh control captures both the critical failure points and structural integrity zones (Fig. 5).

Thread strength simulations were conducted exclusively on the nozzle head, as its simpler geometry experiences the highest pressure loads during motor operation. The results yielded satisfactory and consistent values with the theoretical thread stress analysis, confirming the critical nature of the threading system, the most vulnerable component in the design.

The nozzle and nozzle head employ threaded connections to facilitate static testing operations while achieving significant weight reduction compared to conventional bolt-on designs. This configuration eliminates the need for mandatory O-ring seals, allowing designers to optionally incorporate them based on specific test requirements, a design flexibility that further simplifies assembly and reduces component count.

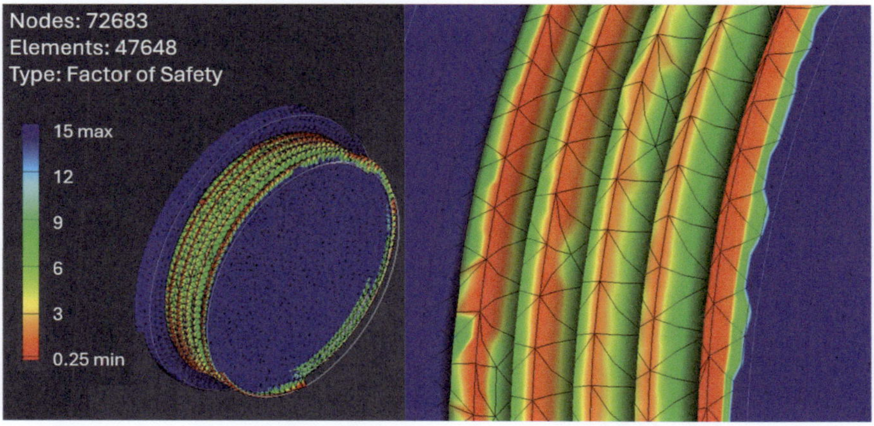

Fig. 4 Safety factor from simulation with detail on the variation in the threads

Fig. 5 Safety factor enhancement after mesh refinement

2.4 *Manufacturing and Static Test*

The selected manufacturing materials were SAE 1020 and SAE 1045 steel, chosen for their mechanical strength, market availability, and cost-effectiveness. Throughout the manufacturing process, including cutting, machining, and threading operations, each component underwent rigorous dimensional inspections to ensure compliance with the computational model specifications.

The manufacturing process begins with creating technical drawings (Fig. 6) for the three components to be produced. As Novaski (2014) emphasizes, well-executed technical drawings are fundamental, and modern 2D/3D digital design tools play a pivotal role in developing components and assemblies with precise fits, significantly enhancing production and assembly processes.

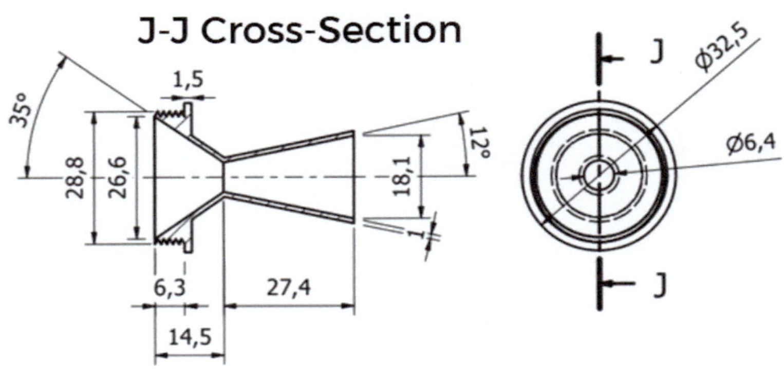

Fig. 6 Technical drawing of the nozzle

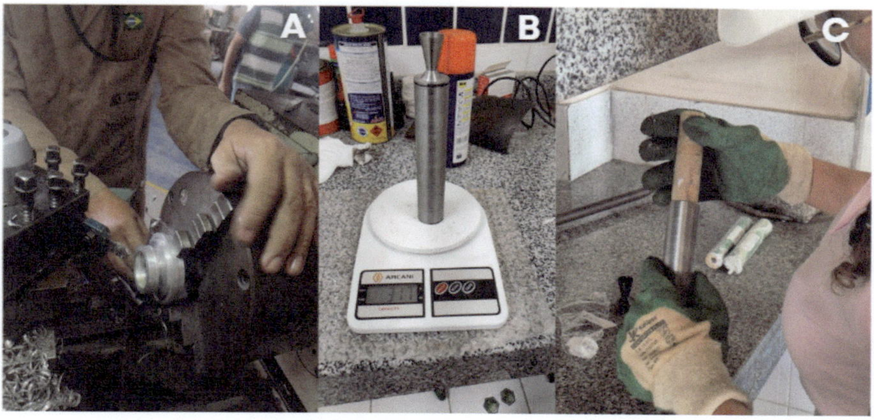

Fig. 7 Manufacturing process of one of the parts (**a**), dead weight of the finished motor (**b**) motor being fueled for static testing (**c**)

The nozzle wall thickness was designed to the minimum possible while maintaining structural integrity, adhering to the competition's 1.5 kg weight limit for the fully-fueled rocket. Manufacturing allowed for subsequent dimensional adjustments to optimize fit, test performance, and accommodate machine limitations. Production commenced using conventional lathes (Fig. 7a) for most components, while CNC lathes provided precision for critical features like nozzle angles. Post-machining, all parts underwent sanding, thread fit verification, and final assembly weighing (Fig. 7b) for performance analysis.

Experimental testing was conducted in two primary phases: burn tests and static test bench evaluations. For combustion trials, the motor was loaded with approximately 98 g of KNSU propellant and ignited using an electronic pyro-technic starter (skib). These tests analyzed critical parameters including chamber pressure, temperature, and combustion stability to validate the motor's preliminary performance, with particular focus on the threaded connection's pressure tolerance. The burn test sequence was repeated three times for verification.

For static testing, we employed a custom-built test stand (Fig. 9) to evaluate the motor under controlled conditions while preventing movement, using the same KNSU propellant proportions as the burn tests. Thrust data was acquired through a P150.45.40.A2-500 load cell mounted to the structure, with measurements processed via an HX711 sensor to record peak thrust and burn duration (Fig. 8). Three static tests were conducted to verify the motor's safe operating parameters, with collected data compared against theoretical simulations. Identified discrepancies, particularly higher-than-expected internal pressures, were resolved by increasing the propellant core diameter, which effectively regulated chamber pressures while maintaining performance.

Following implementation of the improvements identified during testing, the motor was integrated into the mini-rocket for final launch operations. During the

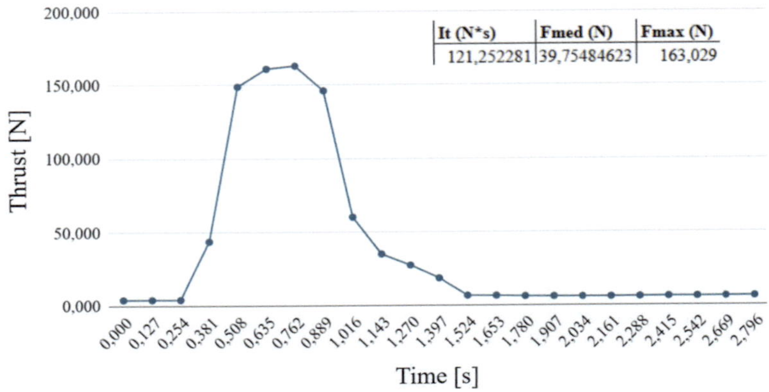

Fig. 8 Graph representing the values obtained from the static test

Fig. 9 Custom-built static test bench used for the experiments

FRMF competition, the motor demonstrated superior performance, propelling the rocket to a 231 m apogee, exceeding initial project expectations. This achievement not only validated the rigorous methodological process employed but also highlighted the effectiveness of the design and manufacturing modifications implemented throughout development.

3 Results and Discussion

The motor development results were analyzed at three key stages: theoretical simulations, static test bench evaluations, and final competition launch performance. This comprehensive approach enabled validation of the design accuracy, identification of discrepancies between predicted and measured values, and revealed improvement opportunities for future motor iterations.

The theoretical-to-experimental comparison revealed a 6.85% deviation between predicted and actual apogee, indicating 93.15% performance relative to Exalt table estimates. This variance can be attributed to thermal losses, manufacturing tolerances, and propellant combustion efficiency variations. Nevertheless, the motor successfully achieved its primary objective by delivering sufficient thrust to reach near-target apogee, demonstrating robust performance within acceptable engineering margins.

Beyond apogee analysis, the motor's total impulse was also evaluated. Static testing recorded 121.25 $N \cdot s$ of impulse, an 11.23% increase over theoretical predictions. This deviation was anticipated, as the motor was designed to guarantee the rocket would reach at least 200 m altitude. The higher impulse likely resulted from: (1) more efficient combustion than theoretical models projected, (2) variations in propellant density/purity, or (3) minor exhaust flow discrepancies, all factors contributing to enhanced performance while maintaining structural safety margins.

A critical design decision was to forego external machining for combustion chamber weight reduction. This choice prioritized extended motor lifespan, enhancing robustness and reliability for multiple test-and-launch cycles. However, the trade-off became evident in the rocket's mass-to-performance ratio, highlighting the need for future studies to optimize chamber geometry while preserving structural durability.

The results demonstrate that the developed motor met all performance expectations, exhibiting high efficiency and reliability within established parameters. The theoretical–experimental discrepancies provide valuable insights for future improvements, including manufacturing optimizations, thermal management refinements, and propellant formulation adjustments, which will inform the development of higher-thrust motors. This study advances the design of efficient small-scale rocket engines, strengthening the technical foundation for future academic projects and competition vehicles.

Acknowledgements The authors express their gratitude to the Interdisciplinary Additive Manufacturing Center (NIMA-UNIVASF) and the Mechanical Engineering Department for providing equipment access, and to the Cactus Rockets Design team for their unwavering support and dedication throughout the manufacturing development process.

Responsibility for Information The authors are solely responsible for the information included in this work.

References

Carande FJF (2011) Solid-fuel rocket motor. Advisor: Prof. Doutor Francisco Miguel Ribeiro Proença Brójo, 189 p. Master's thesis (Master's in Aeronautical Engineering)—Universidade da Beira Interior, Covilhã. Available at: https://ubibliorum.ubi.pt/bitstream/10400.6/3619/1/Projecto_Mestrado_Francisco%20Javier_Fuertes_Carande_m4452.pdf. Accessed on 1 Jan 2025

Nakka R. A-100M rocket motor static test. Available at: https://www.nakka-rocketry.net/A-100M.html. Accessed on 1 Jan 2025

Nakka R. Rocket theory. Available at: https://www.nakka-rocketry.net/rtheory.html. Accessed on 7 Jan 2025

Norton RL (2010) Kinematics and dynamics of machinery, 1st edn., pp 27–34. AMGH, Porto Alegre, Brazil

Novaski O (2014) Introduction to mechanical manufacturing engineering. Editora Blucher, Brazil

Palmerio AF (2017) Introduction to rocket technology. SindCT, São José dos Campos, Brazil, 2nd edn., pp 156–160

Influence of Cryogenic Cooling on Surface Roughness and Cutting Forces During Milling of ASTM A890 GR 6A Super Duplex Stainless Steel

Eduardo Vela Biazi, Ed Cláudio Bordinassi, Nelson Wilson Paschoalinoto, Marcelo Otávio dos Santos, and Gilmar Ferreira Batalha

Abstract Cooling during machining is essential to ensure stable cutting conditions and to extend tool life. The high temperatures generated in the cutting zone affect the properties of both the workpiece material and the cutting tool. Liquid nitrogen acts as an ultra-low-temperature coolant, and when applied to the machining process, it enables the processing of materials with poor machinability, such as super duplex stainless steel, which is widely used across various industrial sectors. Key parameters of interest include surface roughness and cutting forces. This study aimed to investigate the influence of cryogenic cooling on surface roughness and cutting forces during the face milling of ASTM A890 GR 6A super duplex stainless steel. Milling tests were carried out using cryogenic cooling and conventional mineral-based soluble oil, following an experimental design with variations in cutting parameters: cutting speeds of 155 and 125 mm/min, feed rates of 0.10 and 0.05 mm/rev, and depths of cut of 0.60 and 0.30 mm. Cutting forces were measured using a Kistler load cell, and surface roughness was evaluated with a Mitutoyo SJ-301 profilometer. The results showed that cryogenic cooling reduced R_a roughness values by 26.5%, R_z by 25.3%, and R_p by 21.8%. Regarding cutting forces, an increase of 17% in feed force and 40.8% in cutting force was observed when using cryogenic cooling.

Keywords Milling · Super duplex stainless steel · Cryogenic cooling

E. V. Biazi (✉) · G. F. Batalha
University of Sao Paulo, São Paulo, São Paulo, Brazil
e-mail: eduardo.biazi@usp.br

E. V. Biazi · E. C. Bordinassi · N. W. Paschoalinoto · M. O. dos Santos
Maua Institute of Technology, São Caetano do Sul, São Paulo, Brazil

V. E. Beal and A. S. Ribeiro Junior (eds.), *Advances in Manufacturing Engineering Research*, Lecture Notes in Mechanical Engineering,
https://doi.org/10.1007/978-3-032-09858-0_11

1 Introduction

The duplex microstructure was first identified in 1927 by Bain and Griffiths (Still 1994) and began to be commercially produced in 1930. Duplex stainless steel (DSS) is composed of Fe–Cr–N–Mo and features a biphasic microstructure, combining a ferritic matrix and an austenitic phase. The proportion of the ferritic phase ranges between 45 and 55% of the total volume, depending on the chemical composition and the thermal treatment applied to the material (Leite 2009; Fonseca et al. 2008). Its crystal structure is mixed, containing ferrite (body-centered cubic) and austenite (face-centered cubic) phases (Leite 2009). The refinement of austenitic grains gives the material high mechanical strength (Oliveira 2013).

Duplex stainless steels are widely used in structures, piping, and components across industries such as oil and gas, chemical, petrochemical, energy generation, pulp and paper, marine, offshore, and segments of the agribusiness sector. Their application in these areas is due to their excellent resistance in highly aggressive environments, where strict quality and surface finish requirements are imposed.

The continuous development of these steels has led to increasingly complex chemical compositions. Each added element aims to enhance specific properties of the material, although some additions may also introduce disadvantages, most notably, microstructural instability. During processing or service, DSS undergoes several phase transformations, ranging from the liquid state to extremely low temperatures. In most cases, these transformations are undesirable, requiring strict control to minimize or prevent their occurrence (Escriba et al. 2010).

Machining duplex stainless steel presents challenges due to its poor machinability. The production cost of super duplex components is high, as the process requires high productivity and high-performance tools to maintain economic viability. This type of steel exhibits a "pasty" cutting behavior, forming long, flexible chips that can adhere to the tool or create built-up edges. Major machining issues include rapid tool wear, poor surface finish, difficulty in chip breaking, and limitations on cutting speeds (Machado 2003).

Furthermore, duplex and super duplex stainless steels possess high mechanical strength and yield strength, approximately twice that of austenitic stainless steels. Other challenges include a high work hardening rate, low thermal conductivity, high fracture toughness, and a strong tendency for built-up edge (BUE) formation. The high contents of austenite and nitrogen also contribute to the material's difficult machinability, which is estimated to be about 30% lower than that of ferritic stainless steels (Koyee et al. 2014).

Tool wear is a critical issue in this process. The material's high hardness requires intense cutting forces, resulting in elevated temperatures in the cutting zone and, consequently, accelerated tool degradation. Understanding these forces is essential for machine tool design and has a direct impact on tool life, thus affecting the economic feasibility of the operation (Machado et al. 2009). Notch wear is the most common type of wear when machining this material, typically occurring in the lateral regions of the chip flow (Machado et al. 2009). This type of wear is frequently observed in the machining of heat-resistant materials such as nickel and titanium alloys, and stainless steel (Gamarra 2017).

In light of this scenario—characterized by low machinability, high cutting temperatures, and intense tool wear—cryogenic cooling emerges as an efficient alternative. The heat generated during machining is directly influenced by the cutting parameters, impacting both productivity and tool life (Dacas and Costa 2020). Cooling plays a key role in machining by lowering the temperature in the cutting zone, providing lubrication, and aiding in chip evacuation (Machado et al. 2009).

The use of conventional cutting fluids presents environmental and occupational health challenges, such as water pollution, soil contamination from improper disposal, dermatological risks, and smoke emissions caused by evaporation (Yildiz and Nalbant 2008). Since the 1950s, liquid nitrogen has been explored as a coolant for machining processes, offering an environmentally safe alternative. Additionally, it significantly improves process performance by increasing tool life, enhancing chip breakability, and providing superior surface finishes. Compared to flood emulsified jet cooling, cryogenic cooling can improve the machining efficiency of AISI 304 stainless steel by up to 21.36%, depending on the applied cutting parameters (Yildiz and Nalbant 2008).

Figure 1 illustrates a comparison between costs and productivity when using cryogenic and flood emulsified cooling techniques. It can be observed that, although cryogenic cooling entails higher initial costs, it also results in a significant increase in process productivity.

As with any technological application, cryogenic cooling presents both advantages and disadvantages. The main drawback is the need to maintain a highly pressurized tank. The storage and handling of highly pressurized cryogenic liquids can pose

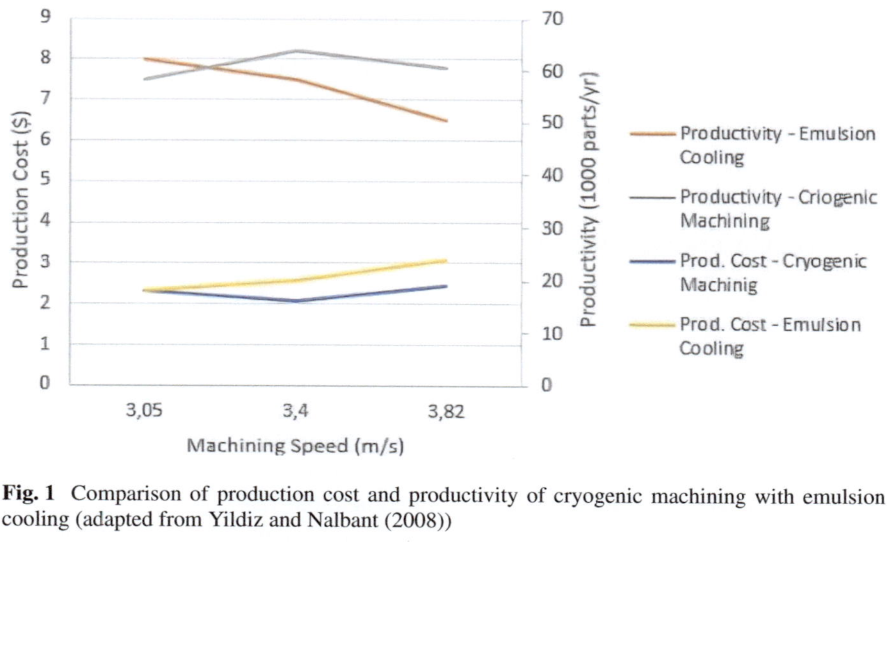

Fig. 1 Comparison of production cost and productivity of cryogenic machining with emulsion cooling (adapted from Yildiz and Nalbant (2008))

serious hazards, including the potential for explosions in certain situations. Additionally, the cost of cryogenic coolants is significantly higher when compared to conventional cutting fluids (Jadhav 2019).

On the other hand, achieving a high-quality surface finish becomes particularly challenging when machining hard-to-cut materials. Surface roughness is characterized by grooves and marks left by the cutting tool on the machined surface, which can range from 10^{-10} to 10^{-5} mm depending on the manufacturing process precision and the surface finish quality. Surface roughness plays a critical role in mechanical components, influencing properties such as sliding performance, wear resistance, press-fit capability, fluid and lubricant flow resistance, surface adhesion to protective coatings, corrosion and fatigue resistance, sealing capability, and visual appearance (Kalpakjian and Schmid 2001).

The surface roughness of a machined part is affected by numerous machining parameters, including tool geometry, part geometry, machine tool stiffness, workpiece material, cutting conditions, lubrication, cooling, and tool material (Gamarra 2017). Surface finish tends to improve when machining deflections—divided into three components: feed force (F_n), cutting force (F_c), and penetration force (F_p)—or vibrations are minimized, when the cutting edge radius is large, when both tool and workpiece are correctly aligned and centered, when the workpiece material is free of impurities or defects, when the machine spindle is aligned correctly and the guideways are unworn, and when the cutting process occurs without the formation of built-up edge. Surface finish is an essential feature of any machining process and the main requirement of many manufacturing, automotive, and aerospace applications (Branco et al. 2018).

This key parameter is also strongly correlated with several performance variables, including mechanical strength, fatigue resistance, fluid flow, lubrication, and heat transfer (de Amorim 2002).

The objective of this study is to investigate the influence of cryogenic cooling on surface roughness and cutting forces during the face milling of ASTM A890 GR 6A super duplex stainless steel, comparing cryogenic and flood cooling conditions, and to establish correlations and insights regarding these effects.

2 Experimental Procedure

A complete factorial experimental design was employed to define the machining trials. The same method, adapted to this study as described by Paschoalinoto et al. 2020. The machining operations were carried out using a Romi Discovery 560 machining center at the Maua Institute of Technology. The milling cutter used was a CoroMill 245, 45°—R245-063Q22-12M, equipped with Sandvik grade 2030 (M15–M30) inserts, code R245-12 T3 K-MM 2030. The workpiece material was ASTM A890 GR 6A. The input variables are presented in Table 1.

The test specimens were prepared as 20 × 20 × 20 mm cubes. To measure the machining forces, a 3D piezoelectric load cell (Kistler model 9367C) was mounted

Table 1 Input variables

Cooling method	Feed (mm/tooth)	Machining speed (m/min)	Cutting depth (mm)
Flooded	0.10	155	0.6
Cryogenic	0.05	125	0.3

on the machining center's worktable (Fig. 2). Data acquisition and storage were performed using the HBM Quantum system. The test specimen was fixed in a vise mounted on the load cell, and the experiments were conducted accordingly.

For conventional abundant lubrication machining, a mineral cutting fluid (Boreal BL from YPF) was used in a 1:19 ratio with water. In the cryogenic trials, liquid nitrogen was employed, stored in a Sempercrio SC-35 T tank, and delivered through a pressurized air system.

Cutting forces were analyzed using HBM's Catman software. A frequency analysis was performed via Fast Fourier Transform (FFT), and the signals collected from the experiments were subjected to a Butterworth low-pass filter. This filtering process allowed only signals below the dominant frequency to pass through, effectively removing unwanted noise from the actual signal representing cutting force measurements.

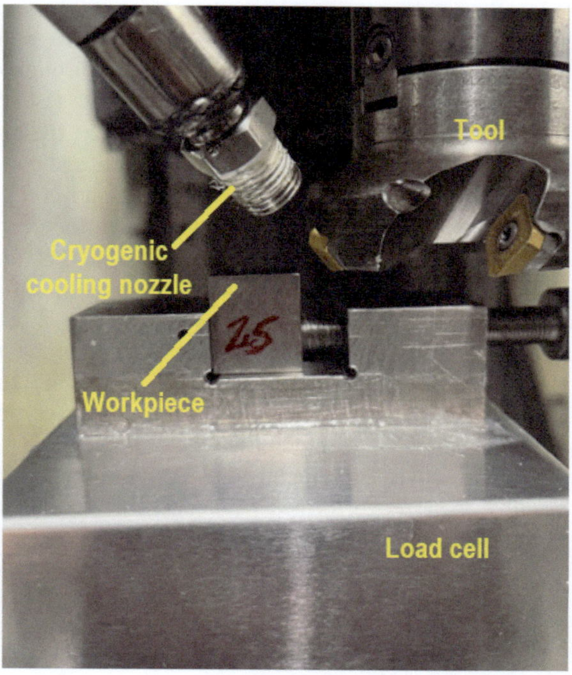

Fig. 2 Workpiece fixed on the load cell

Fig. 3 Roughness measurement

Surface roughness measurements were performed using a Mitutoyo SJ-301 profilometer (Fig. 3). For each roughness parameter, the reported value was the average of three measurements taken at the center of the machined surface.

3 Results and Discussions

Regarding cutting forces, both the cutting force and feed force exhibited higher average values when liquid nitrogen cooling was applied, by 40.8% and 17%, respectively. Table 2 presents the results obtained through data analysis and processing, while Fig. 4 shows a representative signal from test 15.

From Table 2, it can be observed that, under the lowest cutting parameter conditions—$v_c = 125$ mm/min, $f = 0.05$ mm/tooth, and $a_p = 0.30$ mm—the cutting forces were nearly the same for both emulsified and cryogenic cooling. The previously mentioned increase in forces is evident under the most severe cutting conditions: $v_c = 155$ mm/min, $f = 0.10$ mm/tooth, and $a_p = 0.60$ mm.

From Figs. 5 and 6, it is evident that the cutting force is higher when cryogenic cooling is applied. This increase can be explained by the fact that, at low temperatures, the material undergoes compression, becoming more resistant (Zhao and Hong 1992). This occurs because matter contracts and the atomic lattice spacing decreases. Nalbant and Yildiz (2011) observed that, under cryogenic cooling, the hardness of AISI 304 stainless steel increased by 21 BHN.

In contrast, surface roughness showed the opposite behavior. As shown in Figs. 7, 8, and 9, the tests performed with cryogenic cooling achieved, on average, a 26.5% reduction in R_a values, 25.3% in R_z, and 21.8% in R_p. This improvement is attributed to the lower temperature in the cutting zone and reduced adhesion at the tool–workpiece interface. Consequently, there is no chip adhesion at the cutting edge nor edge buildup, which results in less tool wear (Dhananchezian et al. 2018; Dhananchezian and Kumar, 2011). Agrawal et al. (2021) stated that this reduction in surface roughness is due to the cryogenic fluid's ability to dissipate the heat generated by friction in the machining region, thereby minimizing heat-induced damage to both the cutting

Table 2 Force results and cutting data

F_{fn} (N)	F_c (N)	F_p (N)	Cooling method	Feed (mm/rev)	Machining speed (m/min)	Cutting depth (mm)
Mean	Mean	Mean				
– 16.33	9.73	14.51	Flooded	0.05	125	0.3
– 27.90	19.04	19.62	Flooded	0.05	125	0.6
– 15.76	9.56	12.74	Flooded	0.05	155	0.3
– 24.40	18.60	18.19	Flooded	0.05	155	0.6
– 27.15	16.57	19.75	Flooded	0.10	125	0.3
– 56.22	42.58	37.74	Flooded	0.10	125	0.6
– 26.25	17.10	18.99	Flooded	0.10	155	0.3
– 61.43	47.04	42.76	Flooded	0.10	155	0.6
– 15.87	9.15	11.89	Cryogenic	0.05	125	0.3
– 34.40	25.15	28.12	Cryogenic	0.05	125	0.6
– 14.59	9.21	10.57	Cryogenic	0.05	155	0.3
– 30.67	25.15	25.17	Cryogenic	0.05	155	0.6
– 18.39	7.22	15.06	Cryogenic	0.10	125	0.3
– 66.41	58.94	60.22	Cryogenic	0.10	125	0.6
– 44.14	35.71	43.31	Cryogenic	0.10	155	0.3
– 74.30	83.24	100.30	Cryogenic	0.10	155	0.6

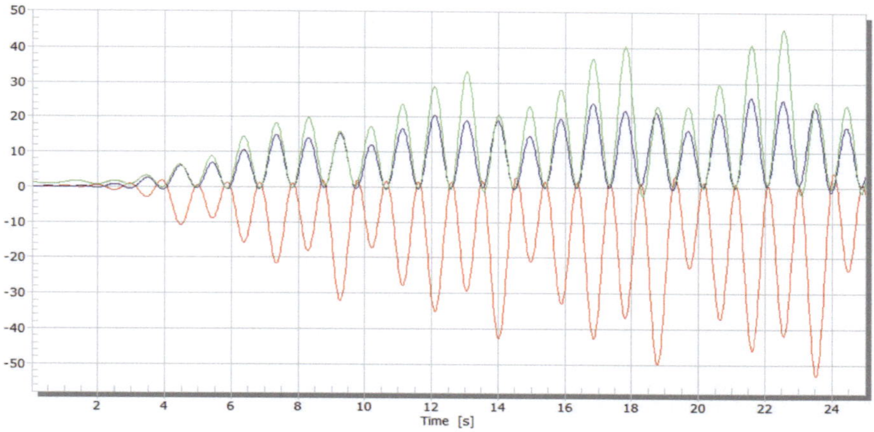

Fig. 4 Graph of filtered cutting force data. Test 15

tool and the machined surface. Better roughness outcomes were also reported by Baig et al. (2022) when using cryogenic cooling. These results can be supported by the findings of Souflas et al. (2021), who concluded that reduced tool wear significantly influences the final surface roughness of the workpiece.

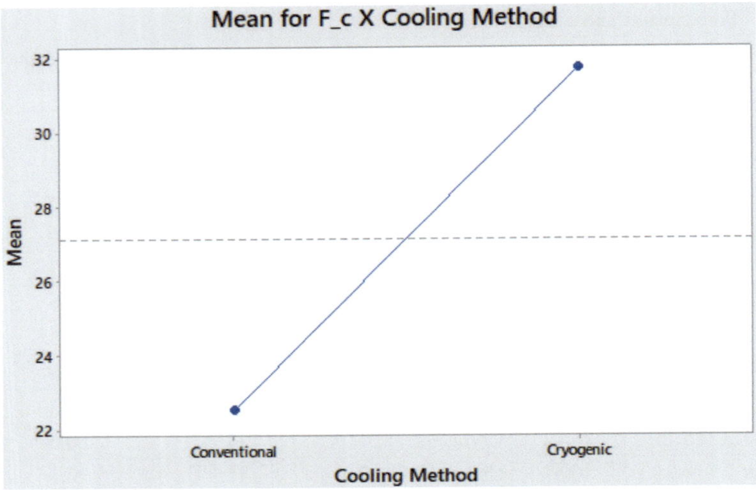

Fig. 5 Main effects for F_c

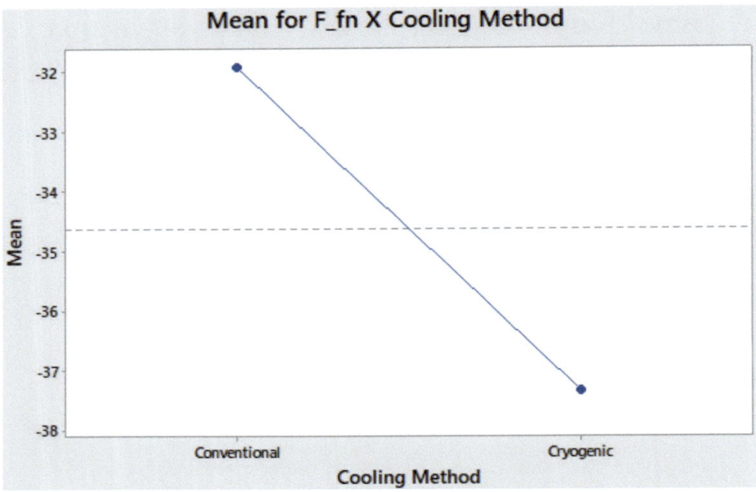

Fig. 6 Main effects for F_{fn}

4 Conclusions

Considering the detailed examination and interpretation of the experimental results obtained throughout the study, it is possible to derive and present the following set of conclusions, which summarize the main findings and highlight the most relevant trends and implications observed during the investigation.

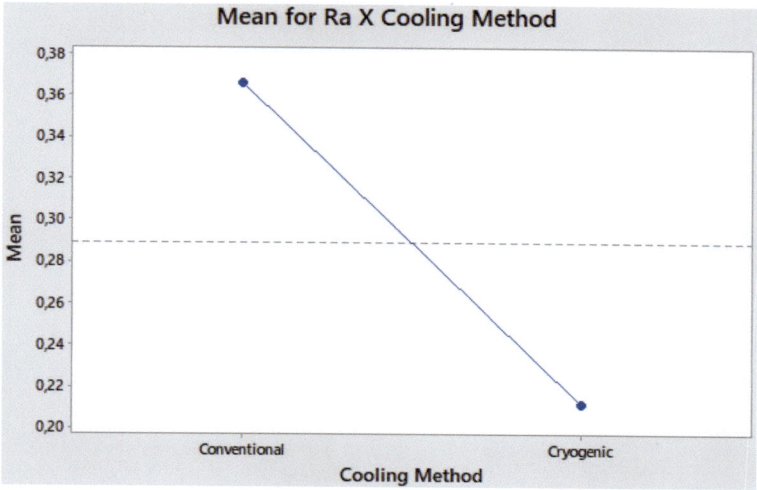

Fig. 7 Main effects for R_a

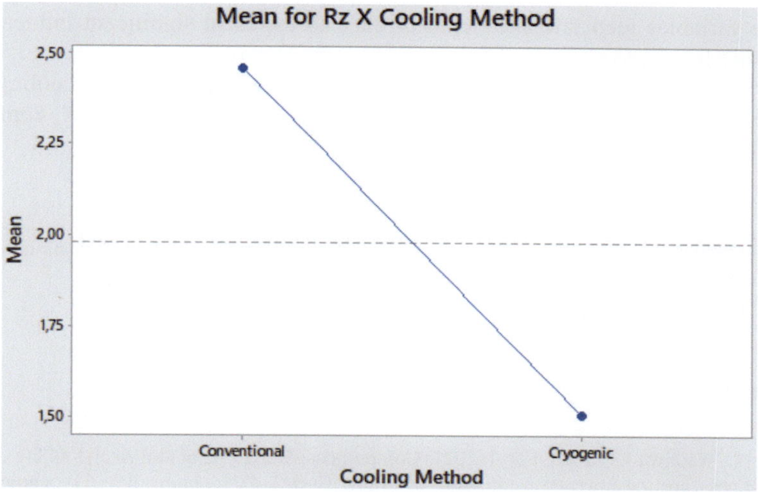

Fig. 8 Main effects for R_z

- For all influence graphs, the steep slope of the lines indicates that cryogenic cooling has a significant impact on the studied variables.
- A clear order of magnitude was observed among the analyzed force components: $F_n > F_p > F_c$.
- Cutting forces were higher under cryogenic cooling conditions. Both the cutting force and the feed force showed higher average values when liquid nitrogen was used, with increases of 40.8% and 17%, respectively.

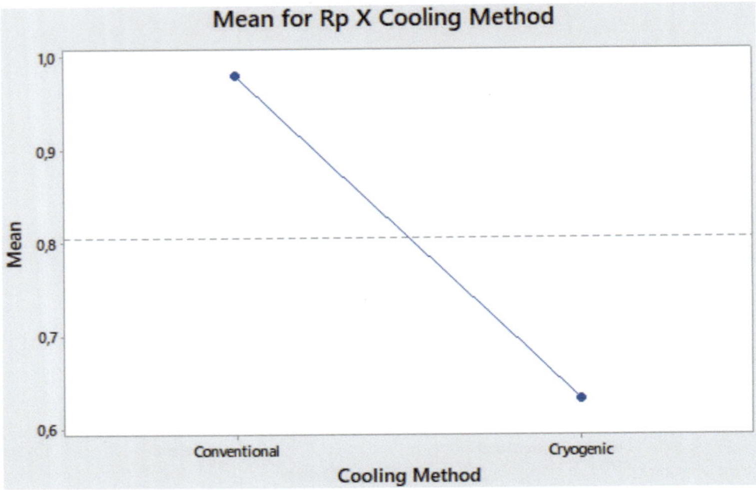

Fig. 9 Main effects for R_p

- The variables feed rate and depth of cut had the most significant influence on cutting forces, regardless of the cooling method used.
- Surface roughness was consistently improved when cryogenic cooling was applied, with reductions of 26.5% in R_a, 25.3% in R_z, and 21.8% in R_p compared to flooded cooling.

Responsibility for Information The authors are solely responsible for the information included in this work.

References

Agrawal C, Wadhwa J, Pitroda A et al (2021) Comprehensive analysis of tool wear, tool life, surface roughness, costing and carbon emissions in turning Ti–6Al–4V titanium alloy: Cryogenic versus Wet Machining. Tribol Int 153:106597. https://doi.org/10.1016/j.triboint.2020.106597

Baig A, Jaffery SHI, Khan MYA, Alruqi M (2022) Statistical analysis of surface roughness, burr formation and tool wear in high speed micro milling of inconel 600 alloy under cryogenic, wet and dry conditions. Micromachines 14:13. https://doi.org/10.3390/mi14010013

Branco FK, Delijaicov S, Bordinassi ÉC, Bortolussi R (2018) Surface integrity analysis in the hard turning of cemented Steel AISI 4317. Mater Res. https://doi.org/10.1590/1980-5373-mr-2017-1032

Dacas FP, Costa R (2020) Avaliação da Refrigeração de Corte Criogênica na Usinagem CNC do Aço Inoxidável Martensítico AISI 420. 2:35–47. https://doi.org/10.21166/CPITT.V2I1.2031

De Amorim HJ (2002) Estudo da Relação entre Velocidade de Corte, Desgaste de Ferramenta, Rugosidade e Forças de Usinagem em Torneamento com Ferramenta de Metal Duro. Master's thesis, Universidade do Rio Grande do Sul

Dhananchezian M, Pradeep Kumar M (2011) Cryogenic turning of the Ti–6Al–4V alloy with modified cutting tool inserts. Cryogenics 51:34–40. https://doi.org/10.1016/j.cryogenics.2010.10.011

Dhananchezian M, Priyan MR, Rajashekar G, Narayanan SS (2018) Study the effect of cryogenic cooling on machinability characteristics during turning duplex stainless steel 2205. Mater Today: Proc 5:12062–12070. https://doi.org/10.1016/J.MATPR.2018.02.181

Escriba DM, Materna-Morris E, Plaut RL, Padilha AF (2010) Intermetallic phase precipitation in duplex stainless steels during high temperature exposure. Mater Sci Forum 636–637:478–484. https://doi.org/10.4028/www.scientific.net/msf.636-637.478

Fonseca MC, Barbosa MF, Bartholo RN, Marques Jr. A, Pardal JM, Araujo AC (2008) Estudos das Tensões Residuais Geradas na Usinagem por Fresamento de Amostras de Aço Superduplex. Paper presented at the V Congresso Nacional de Engenharia Mecânica, Salvador, Brazil, 25–28 Aug 2008

Gamarra JR (2017). Estratégias de Usinagem no Torneamento de Aço Inoxidável Super Duplex (UNS S32750). Master's thesis, Universidade de Campinas.

Jadhav S (2019) Cryogenic machining: a sustainable solution. Int J Res Appl Sci Eng Technol 7:465–468. https://doi.org/10.22214/ijraset.2019.12077

Koyee RD, Heisel U, Schmauder S, Eisseler R (2014) Experimental investigation and multiobjective optimization of turning duplex stainless steels. Int J Manuf Eng 2014:1–13. https://doi.org/10.1155/2014/921081

Leite RA (2009) Avaliação da Tenacidade à Fratura de uma Junta Soldada de um Aço Inoxidável Super Duplex com a Utilização de Proteção Catódica. Master's thesis, Universidade Federal do Rio Grande do Sul

Machado AR, Abrão A, Coelho R, da Silva MB (2009) Teoria da Usinagem dos Materiais. Blucher

Machado IF (2003) An investigation of the effect of machining parameters on strain-induced martensite formation during turning in an austenitic stainless steel. Paper presented at the International Congress of Mechanical Engineering, São Paulo, Brazil, 10–14 Nov 2003

Nalbant M, Yildiz Y (2011) Effect of cryogenic cooling in the milling process of AISI 304 stainless steel. Trans Nonferrous Metals Soc China 21:72–79. https://doi.org/10.1016/s1003-6326(11)60680-8

Oliveira CA (2013) Torneamento do aço inoxidável super duplex UNS S32750 e influência na resistência à corrosão. Master's thesis, Universidade Estadual de Campinas

Paschoalinoto NW, Batalha GF, Bordinassi EC et al (2020) MQL strategies applied in Ti-6Al-4V alloy milling—comparative analysis between experimental design and artificial neural networks. Materials 13:3828. https://doi.org/10.3390/ma13173828

Schmid SR (2001) Manufacturing engineering and technology: international edition. Prentice Hall

Souflas T, Bikas H, Ghassempouri M et al (2021) A comparative study of dry and cryogenic milling for directed energy deposited IN718 components: effect on process and part quality. Int J Adv Manuf Technol 119:745–758. https://doi.org/10.1007/s00170-021-08313-7

Still JR (1994) The use of duplex stainless steel in the offshore oil and gas industry. Weld Met Fabr 62:293–298

Yildiz Y, Nalbant M (2008) A review of cryogenic cooling in machining processes. Int J Mach Tools Manuf 48:947–964. https://doi.org/10.1016/j.ijmachtools.2008.01.008

Zhao Z, Hong SY (1992) Cooling strategies for cryogenic machining from a materials viewpoint. J Mater Eng Perform 1:669–678. https://doi.org/10.1007/bf02649248

Performance of Different Dielectrics in Carbide Removal by WEDM

Victória Gasperin, Matheus Fenerich Sgarbi, Cassiano Rodrigues Moura, Igor dos Santos Roik, Rodrigo Blödorn, and Giovani Conrado Carlini

Abstract Carbide consists of tungsten particles in a non-oxidized ceramic phase (WC) bound by a metallic phase (Co), resulting in high hardness and mechanical resistance at elevated temperatures. However, it is low in machinability, and few machining processes can manufacture it. The thermal effect of successive electrical discharges without tribological mechanisms enhances removal. There is little research in the literature on the removal process by wire electrical discharge machining (WEDM) with reciprocating molybdenum wire in WC-Co and its performance under different conditions. Thus, this work aims to evaluate the effects of varying discharge energy in deionized water and hydrocarbons as dielectrics. Five lateral increments (Δ_y) [μm] at two discharge energy levels [mJ] were used for the two dielectrics, keeping other variables constant. The experiments showed that increasing energy discharge contributes to a higher material removal rate. The hydrocarbon dielectric presented the highest wire advance rate [mm/min]. The surface texture exhibited a reduction in average roughness (*Sa* [μm]) at the lowest energy density, indicating better stability and reduced morphological distortion in deionized water compared to the hydrocarbon.

Keywords Tungsten carbide · WEDM · Dielectric fluid · Three-dimensional roughness · Technological performance

1 Introduction

Cemented carbide is defined as a composite material composed of carbide particles bonded with a metallic phase. Its basic structure consists of tungsten carbide (WC) combined with a relatively ductile metal binder, such as cobalt (Co), resulting in a

V. Gasperin · M. F. Sgarbi · C. R. Moura · G. C. Carlini (✉)
Instituto Federal de Santa Catarina—IFSC, Jaraguá do Sul, Santa Catarina, Brazil
e-mail: giovani.carlini@ifsc.edu.br

I. S. Roik · R. Blödorn
Centro Universitário de Brusque—UNIFEBE, Brusque, Santa Catarina, Brazil

V. E. Beal and A. S. Ribeiro Junior (eds.), *Advances in Manufacturing Engineering Research*, Lecture Notes in Mechanical Engineering,
https://doi.org/10.1007/978-3-032-09858-0_12

non-oxide ceramic composite. According to Amorim et al. (2010), powder metallurgy is extensively employed in the production of tungsten carbide–cobalt (WC-Co) components. The performance of tungsten carbide is primarily characterized by its hardness, transverse rupture strength, and fracture toughness. About its applications, parameters such as the cobalt content in the metallic binder phase and the grain size of the ceramic particles are critical for optimizing these properties. In WC-based composites, grain sizes typically range from less than 0.2 μm to several micrometers (Carlini et al. 2020).

The precision metalworking industry often deals with materials exhibiting high hardness and mechanical strength, such as WC-Co. These materials pose significant challenges for conventional machining processes involving defined cutting geometries due to their mechanical properties. Consequently, non-conventional manufacturing methods, such as electrical discharge machining (EDM) and wire electrical discharge machining (WEDM), become more viable alternatives. These processes employ thermal energy generated by successive electrical discharges to remove material, thereby eliminating the occurrence of tribo-mechanical interactions at the tool–workpiece interface (Poroś 2021).

In the context of WEDM, two wire-feeding configurations are commonly used: single-pass and reciprocating systems. In single-pass systems, the wire is fed at speeds ranging from 2 to 10 m/min during machining and is subsequently discarded. Typically, brass or zinc-coated copper wires are employed, with tensile strengths ranging from 500 to 900 N/mm^2 (Reolon et al. 2019). In contrast, reciprocating systems utilize a closed-loop mechanism in which the wire is repeatedly wound and unwound from a spool.

When sensors detect the wire reaching its winding limit, the discharge energy is temporarily deactivated to allow for reversal of the wire's motion. Once the reversal is complete, electrical pulses are resumed. In this reciprocating mode, the wire repeatedly traverses the machining gap until it is excessively elongated and the loss of its circular cross-section necessitates replacement. The wire used in reciprocating systems is composed of molybdenum and is typically tensioned to approximately 1200 N/mm^2 (Carlini et al. 2021).

This process offers significant technological and productivity advantages, as it eliminates tribological mechanisms and mechanical deformations typically associated with contact-based cutting tools. As such, the primary input variables of the process include discharge current $\bar{\imath}_e$ (A), average discharge voltage $\bar{u}_e$ (V), and discharge duration t_e (μs). The product of these variables yields the discharge energy W_e (mJ), as described by Eq. (1), presented by Klocke and König (2007).

$$W_e = \int_0^{t_e} u_e(t) \cdot i_e(t) \cdot \mathrm{d}t \tag{1}$$

A common element across all electrical discharge machining (EDM) systems is the dielectric medium, which is typically used in liquid form. According to Klocke and König (2007), the functions of the dielectric fluid include: enabling ionization of the medium; facilitating the formation of the discharge channel by concentrating the energy density; flushing debris generated within the machining gap; promoting cooling and thermal stabilization of the discharge zone; and ensuring electrical insulation between the wire electrode and the workpiece.

The primary dielectric fluids reported in the literature for the WEDM process include hydrocarbons, which may be used in pure form or as emulsions (Carlini et al. 2018), and deionized water. Deionized water can be obtained through distillation or via a deionization process involving submicron filtration and exposure to mixed-bed ion exchange resins (both cationic and anionic) to remove free electrons from the medium. Niamat et al. (2017) conducted experiments comparing the performance of hydrocarbon-based dielectrics with that of deionized water. The results indicated higher electrode wear and greater material removal rates when pure hydrocarbons were used, while deionized water led to increased surface roughness.

Studies by Kunieda and Kitamura (2018) demonstrated that the material removal rate during successive electrical discharges varied significantly with increased discharge frequency, depending on the dielectric employed. Discharge distributions showed a higher probability of energy concentration in hydrocarbon media under short pulse intervals. However, the lower viscosity of deionized water provides enhanced efficiency in debris removal from the machining gap and superior cooling capability.

The present study aims to evaluate the effects of dielectric variation in the reciprocating WEDM process under two levels of discharge energy and five levels of lateral offset increments during the machining of cemented carbide. The investigation focuses on process performance by analyzing the influence of these parameters on wire feed rate (mm/min) and surface texture, with quantitative assessment of average surface roughness *Sa* (μm).

2 Experimental Procedures

The methodological framework employed in this research was structured systematically and organized into sequential stages, as follows: definition of the independent input parameters for the machining process; preparation and assembly of the experimental setup for conducting the trials; sample preparation and machining using the WEDM technique; characterization and quantification of the technological performance of the process; measurement of surface texture following machining; and, finally, evaluation and presentation of the obtained results, culminating in the development of the study's conclusions.

Figure 1 illustrates a stage of the WEDM process applied to a cemented carbide sample, performed without full immersion in the dielectric medium. The figure also highlights the clamping system used, providing a clear view of the experimental configuration adopted in this study.

Fig. 1 WEDM machining process of tungsten carbide without workpiece immersion, using deionized water as the dielectric fluid

2.1 Materials

The equipment used to perform experiments was manufactured by GFMS, model FW1U, and is equipped with a five-axis numerical control system. Its reciprocating spool mechanism distinguishes this model. For the generation of electrical discharges, a molybdenum wire with a nominal diameter (d_w) of 0.18 mm was employed, tensioned to approximately 1200 N/mm^2 using a counterweight-based loading system. The pulse generator was configured to operate under isoenergetic conditions, characterized by precise control of both the discharge duration (t_e) and the inter-pulse interval (t_o), using rectangular pulses with identical periods. Table 1 presents the independent input variables that were held constant throughout the experimental trials.

Deionized water produced by reverse osmosis was employed as the dielectric medium, with continuous treatment via ion exchange resins located in the system's reservoir. The deionized water exhibited an electrical conductivity of approximately 15.0 μS/cm, as reported by Carlini et al. (2021). The hydrocarbon-based dielectric consisted of an 8% solution in deionized water, resulting in an electrical conductivity of approximately 3.1 μS/cm (Table 2). Continuous filtration of the dielectric medium was performed using a pleated cartridge filter capable of retaining particles as small as 5 μm, ensuring high fluid purity throughout the machining process.

To achieve efficient material removal in the WEDM process, an appropriate combination of electrical and non-electrical parameters is essential. As reported by

Table 1 Independent input variables used in the WEDM equipment

Variable	Symbol	Quantity	Unit
Lateral infeed	Δ_y	10; 25; 50; 75 and 100	%
Open-circuit voltage	$\hat{u}_i$	– 70.0	V
Dielectric pressure	P_{in}	0.25	MPa
Wire tensile strength	–	1200	N/mm^2
Wire diameter	d_w	0.18	Mm
Wire feed rate	W_s	8.0	m/s

Table 2 Physical properties of the hydrocarbon-based and deionized water dielectrics

Property	Deionized water	Hydrocarbon 8 (%)	Unit
Viscosity at 40 °C	1.0×10^{-3}	2.4×10^{-3}	Pa s
Density at 20 °C	1.00	1.15	g/cm^3
Electrical conductivity	15.0	3.1	μS/cm
Specific heat capacity	4.2	1.9	J/g K
Heat of evaporation	2260	210	J/g

Adapted from Reolon et al. (2019), Kunieda and Kitamura (2018), Carlini et al. (2019), Klocke (2009)

Johny and Thiagarajan (2020), electrical parameters have a significant impact on the overall process performance. Although it is common practice to describe a specific WEDM regime in terms of the discharge energy applied across the working gap, it is equally important to consider the influence of the intrinsic properties of the machined material, which can directly affect both the process effectiveness and the quality of the final product.

The raw material used in this study consisted of a solid cylindrical tungsten carbide bar with submicron grain size (0.5–0.8 μm), classified as grade K30 and supplied by Ceratizit©. The specimens prepared for the experiments were machined to final dimensions of 22.5 × 2.5 mm in the rectangular cross-section, which was previously ground, and a thickness of 8 mm, resulting in a total machining perimeter of 50 mm. Additionally, a sample measuring 8 × 8 × 2.5 mm was separated and subjected to grinding and polishing processes to prepare it for subsequent characterization via scanning electron microscopy (SEM) and energy-dispersive spectroscopy (EDS). Figure 2 shows the spectral points and the weight percentage distribution (%) of the elements present in the analyzed tungsten carbide.

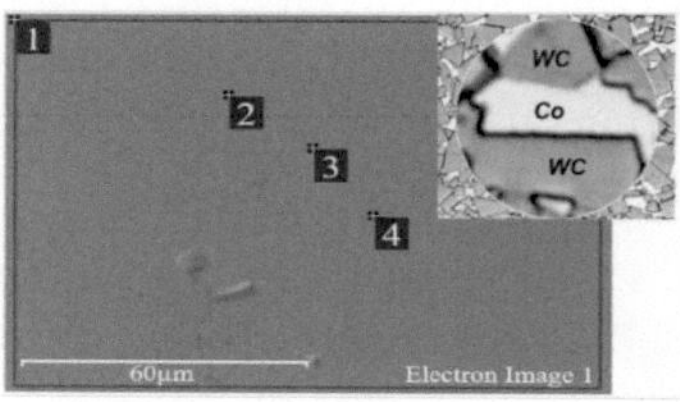

Espectra (wt%)	C	Co	W	Others	Total
Region 1	10.59	8.73	80.29	0.39	100.00
Point 2	10.30	9.11	80.13	0.46	100.00
Point 3	10.88	6.57	82.18	0.36	100.00
Point 4	10.93	8.88	79.68	0.51	100.00
Mean	10.68	8.32	80.57	0.43	100.00
Standard deviation	0.29	1.18	1.10	0.07	

Fig. 2 SEM and EDS characterization showing the weight percentage (wt%) of elements in the submicron-grained K30 grade tungsten carbide

2.2 Methods

The machining duration and the wire feed rate, V_f [mm/min], for each operation were rigorously controlled using the software and interface provided by the WEDM equipment. The study encompassed five variations by adjusting the percentage of lateral increment (Δ_y), calculated based on the wire diameter ($d_w = 0.18$ mm). These variations were directly applied by modifying the frontal working gap, with Δ_y values set at 100, 75, 50, 25, and 10%. Accordingly, the width of the electrical discharges relative to the wire diameter resulted in spacings of $\Delta_y = 180$ μm, 135 μm, 90 μm, 45 μm, and 18 μm, respectively. Figure 3 schematically illustrates the lateral increment arrangement (Δ_y), featuring a front view and a top view to facilitate understanding of the geometric distribution applied to the process.

After the machining process, the samples underwent a thorough cleaning procedure to ensure efficient removal of impurities and surface residues. Initially, the surfaces were manually brushed using polyester bristles, followed by washing and immersion in acetone (C_3H_6O) to eliminate residues resulting from the machining process. Subsequently, the samples were treated in an ultrasonic cleaning system operating at a frequency of 40 kHz for 3 min, using anhydrous isopropyl alcohol (C_3H_8O) as the cleaning medium. The strategic use of these chemical components aimed to minimize the presence of impurities that could compromise the accuracy of subsequent measurements of surface topography and texture. Furthermore, the application of ultrasonic cleaning contributes to the effective removal of microscopic particles that might interfere with the characterization of roughness patterns.

Surface texture analysis began with a detailed examination of the three-dimensional surface morphology using focus variation microscopy (BrukerAlicona® G5) equipped with a 20× optical lens, covering an area of 5.0 mm × 5.0 mm. The evaluated data employed the Gaussian filter according to ISO 16610-71, first order (ISO 2014). A vertical resolution of 50 nm and a lateral resolution of 1.5 μm were selected. Following ISO 25178-2 (ISO 2021), a cut-off sampling length (λ_c) of 250 μm with a profile length of 1.25 mm was established. Alicona MeasureSuite 5.3.1® software was chosen to process the parameter data, ensuring precise and comprehensive analysis of surface texture and relevant volumetric measurements.

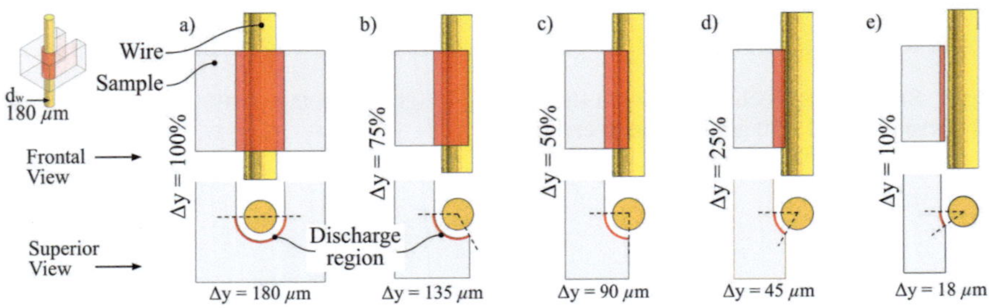

Fig. 3 Area of electrical discharge application when varying the lateral infeed (Δ_y)

The amplitude parameter analyzed was *Sa*, which specifically represents the arithmetic mean of the absolute ordinate $Z(x, y)$ within the selected evaluation area. Although the chosen *Sa* parameter does not provide a detailed representation of the topographic features of the machined tungsten carbide surface, it allows for a comparative assessment of the obtained texture, facilitating correlation with amplitude parameters commonly discussed in the literature and widely used in industry.

3 Results

This research aims to establish guidelines that aid in understanding wire electrical discharge machining (WEDM) using molybdenum wire on K30 grade tungsten carbide, by comparing the performance of deionized water and hydrocarbon as dielectric media. To achieve this objective, the experiments were conducted in two distinct stages. The first stage focused on evaluating the process efficiency in terms of material removal rate as a function of lateral infeed, applied at five different levels. This approach enabled a detailed analysis of the process productivity.

The second stage concentrated on the three-dimensional analysis of the surface texture of the samples, investigating the effects of discharge energy variations in both dielectric fluids used. This procedure aimed to correlate energy parameters with the resulting surface characteristics, providing a comprehensive understanding of the influence of dielectric media on the surface finish and quality of the machined workpiece.

3.1 Three-Dimensional Surface Texture Evaluation

The operational parameters and the conditions of the dielectric medium used directly influence the characteristics of surface texture in the electrical discharge material removal process. The greater the volume of material melted and ejected with each discharge pulse, the deeper the craters formed on the surface, resulting in increased roughness of the machined surface. Thus, prolonged discharge duration leads to an intensification of peaks and protrusions generated by the electrical discharges, significantly impacting the final morphology of the surface texture.

The analysis presented in Fig. 4 qualitatively highlights the differences in the obtained texture morphologies, as well as providing a metric comparison of the influence of the variation in the type of dielectric medium used during the machining process.

According to Kunieda and Kitamura (2018), the plasma channel support points during the electrical discharge process are not randomly defined. When the interval between pulses is reduced, but still sufficiently long to allow the complete cycle of plasma formation and extinction, the craters formed exhibit protrusions of lesser intensity. Even under stable machining conditions, the discharge locations do not

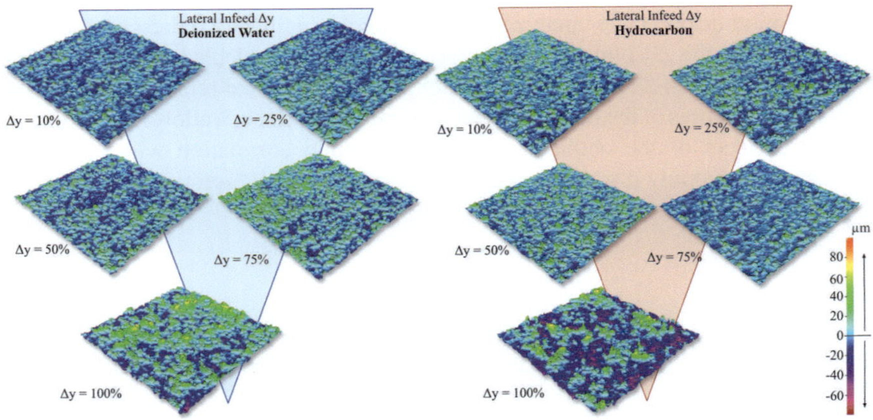

Fig. 4 Resulting textures for different dielectrics and lateral infeed (Δ_y)

distribute uniformly along the wire and the length of the cut. As described by Kitamura et al. (2015), subsequent discharges tend to occur at the periphery of the bubble formed by the implosion of the previous discharge, resulting in a heterogeneous distribution of discharge events along the machining path.

This dynamic implies that more stable discharges contribute to lower surface roughness, even if the absolute roughness values remain relatively high. Furthermore, when process variables favor stability at higher frequencies of discharges per unit time, a positive synergy emerges between process efficiency and the quality of the resulting surface texture. However, when the amount of by-products becomes excessive, these by-products act as current-carrying bridges that hinder plasma channel formation or reduce its intensity, leading to decreased efficiency and increased texture amplitude.

The increase in discharge energy, combined with the specific properties of different dielectrics, exerts a direct influence on the quality of the machined surface. It is observed that the highest roughness magnitudes were obtained under conditions with higher lateral infeed increments (Δ_y). This increase favored greater material removal, which is beneficial from a process efficiency standpoint.

However, it also intensified the accumulation volume of by-products in the working gap, which remains narrow due to limited discharge energy. This condition contributed to a higher incidence of short circuits and abnormal discharges, resulting in more pronounced distortions on the processed surfaces (Carlini et al. 2021). Additionally, the interaction between dielectric variables and discharge parameters directly affects process stability, influencing the repeatability of discharges and the predictability of electrode-workpiece wear. Thus, it underscores the importance of optimizing operational parameters to mitigate undesirable effects such as premature tool degradation and increased dispersion of errors during mechanical manufacturing.

3.2 Technological Performance

The material removal rate in the electrical discharge machining process is influenced not only by the amount of material removed per discharge pulse but also by the stability of the process. During machining, an interval of several microseconds is required for the plasma channel, formed due to high temperatures, to be extinguished entirely (Kunieda et al. 2005). When the interval between discharges is reduced, the subsequent discharge may occur near the previous one, taking advantage of the residual temperature gradient at the discharge site. Additionally, a shortened interval may limit the time available for the dielectric fluid to clean the working gap properly.

As highlighted by Amorim and Weingaertner (2005), contamination of the working gap occurs during the electrical discharge machining process due to the presence of electro-eroded particles and by-products resulting from dielectric breakdown. These residues can have both positive and negative impacts on the technological performance, affecting the material removal rate, electrode wear, surface finish, and other factors. The discharge energy [$W_e \approx u_e \cdot i_e \cdot t_e \quad (\mathrm{mJ})$] induced in the working gap is considered the primary factor determining the removal performance in WEDM (Reolon et al. 2019). Therefore, material removal rates expressed in mm/min were quantified to evaluate the technological performance. Figure 5 presents the process performance behavior for the use of deionized water and hydrocarbon as dielectric fluids.

A qualitative analysis of Fig. 5 demonstrates that, under identical energy density and discharge duration conditions, operations using hydrocarbon-based dielectric fluid exhibited lower average surface roughness. As reported by Davim (2010), surface roughness tends to decrease with the reduction of discharge duration and, consequently, the applied energy density. This phenomenon can be explained by the fact that higher energy densities produce a denser plasma, intensifying the Joule effect. Such intensification increases the depth of protrusions and accentuates the peaks formed by the resolidified, non-ejected layer at the discharge periphery.

One of the functions of the dielectric fluid is to maintain the plasma channel formed during the electrical discharge. In this context, hydrocarbon demonstrated superior performance, as highlighted by Kunieda and Kitamura (2018). The authors observed that the plasma channel support locations are not randomly distributed. Even under stable machining conditions, discharges tend to occur at the periphery of bubbles generated by the implosion of previous discharges (Kitamura et al. 2015). When the pulse interval is sufficiently short to extinguish the plasma generated by the previous discharge, but long enough to avoid overlap, hydrocarbon provides more favorable conditions for technological performance.

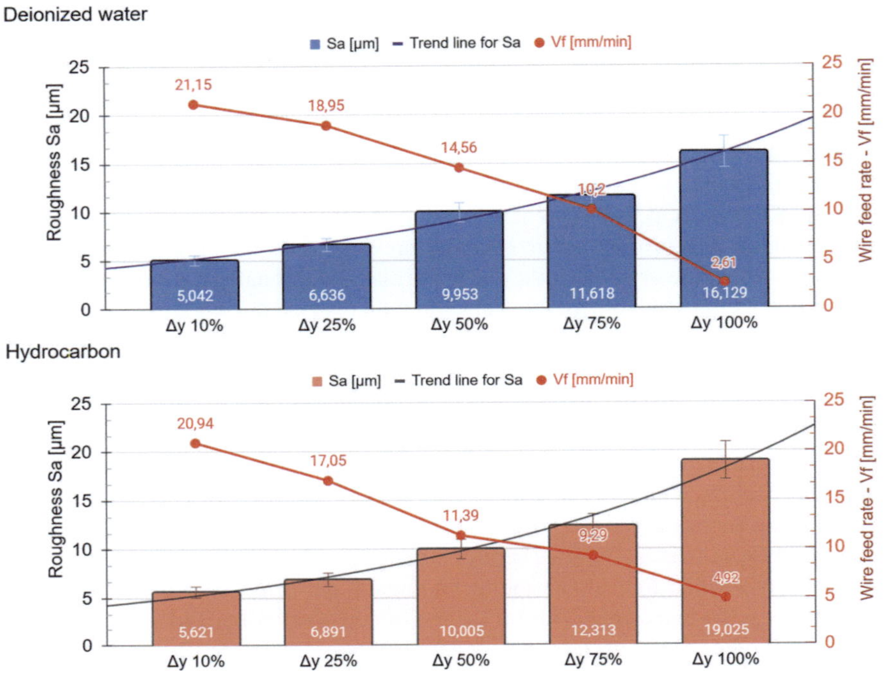

Fig. 5 Process technological performance versus. texture behavior for each dielectric fluid as a function of lateral infeed variation (Δ_y)

4 Conclusions

This study evaluated the effect of two dielectric fluids and their influence on the technological performance of the process and the surface texture during tungsten carbide removal by WEDM using molybdenum wire, with deionized water as the dielectric medium at five levels of lateral infeed. Historically, there are few publications on WEDM using molybdenum wire, and its technological behavior is not yet fully understood. Based on the results presented, the following conclusions can be drawn: (i) Increasing the discharge energy promoted a higher wire feed rate, resulting in greater productivity during machining. However, when using the lateral infeed of $\Delta_y = 100\%$, a significant reduction occurred due to difficulties in cleaning the working gap. (ii) The hydrocarbon-based dielectric exhibited better process performance across all tested lateral infeed levels. (iii) The condition using deionized water resulted in lower surface roughness. The low viscosity of deionized water aids in cleaning the working gap and reduces the incidence of abnormal discharges. (iv) Intermediate conditions with Δ_y between 25 and 75% resulted in a nonlinear increment in which the lower lateral infeed increases the V_f rate and reduces roughness.

Therefore, whenever possible, it is recommended to perform a semi-finishing operation with lateral infeed within the studied range after primary removal, followed by a final finishing pass with lower discharge energy.

Acknowledgments This work was supported by the Instituto Federal de Santa Catarina—IFSC, under Grant 02/2024/PROPPI, and by UNIFEBE for their collaboration in the research and development of surface characterizations.

Responsibility for Information The authors are solely responsible for the information included in this work.

References

Amorim FL, Weingaertner WL (2005) The influence of generator actuation mode and process parameters on the performance of finish EDM of a tool steel. J Mater Process Technol 166(3):411–416. https://doi.org/10.1016/j.jmatprotec.2004.08.026

Amorim FL, Weingaertner WL, Bassani IA (2010) Aspects on the optimization of die-sinking EDM of tungsten carbide-cobalt. J Braz Soc Mech Sci Eng 32:496–502. https://doi.org/10.1590/s1678-58782010000500009

Carlini GC, Amorim FL, Weingaertner WL (2019) Influence of different grades of CuW electrodes when die sinking ED-machining of cemented carbide. Int J Adv Manuf Technol. https://doi.org/10.1007/s00170-019-03982-x

Carlini GC, da Silva C, Torres RD, Soares P, Weingaertner WL, Amorim FL (2021) WED-machining performance by reciprocating molybdenum wire on Inconel 718 with water or hydrocarbon dielectrics. Int J Adv Manuf Technol 119:1853–1866. https://doi.org/10.1007/s00170-021-08386-4

Carlini GC, Moura CR, Amorim FL, Weingaertner WL (2020) On geometrical characteristics of WC-Co round cavities after ED-machining with different grades of CuW electrodes. Mater Res 23(2). https://doi.org/10.1590/1980-5373-mr-2020-0041

Carlini GC, Xavier FA, Weingaertner WL, Di Maiochi H, Fodi E (2018) Análise de parâmetros na remoção de metal duro utilizando WEDM. In: Anais do X Congresso Nacional de Engenharia Mecânica—CONEM, Rio Grande, Brasil. https://doi.org/10.26678/ABCM.CONEM2018.CON18-1266

Davim JP (2010) Surface integrity in machining. Springer, London. https://doi.org/10.1007/978-1-84882-874-2

ISO (2014) ISO 16610-71: geometrical product specifications (GPS)—filtration—Part 71: robust areal filters. International Organization for Standardization, Geneva. Available at: Accessed 6 Aug 2025. https://www.iso.org/standard/60159.html

ISO (2021) ISO 25178-2: geometrical product specifications (GPS)—surface texture: areal—Part 2: terms, definitions and surface texture parameters. International Organization for Standardization, Geneva. Available at: Accessed 8 Aug 2025

Johny A, Thiagarajan C (2020) Investigation of surface integrity and its optimization on pure titanium using molybdenum wire by reciprocated travelling WEDM—a review. Mater Today Proc 33:2581–2584. https://doi.org/10.1016/j.matpr.2019.12.251

Kitamura T, Kunieda M, Abe K (2015) Observation of relationship between bubbles and discharge locations in EDM using transparent electrodes. Precis Eng 40:26–32. https://doi.org/10.1016/j.precisioneng.2014.09.009

Klocke F (2009) Manufacturing processes 2, RWTH edn. Springer, Berlin, Heidelberg. https://doi.org/10.1007/978-3-540-92259-9

Klocke F, König W (2007) Fertigungsverfahren 3. Abtragen, Generieren und Lasermaterialbearbeitung, 5th edn. Springer, Berlin, Heidelberg. https://doi.org/10.1007/978-3-540-48954-2

Kunieda M, Kitamura T (2018) Observation of EDM gap phenomena difference in water and oil using transparent electrode. Procedia CIRP 68:342–346. https://doi.org/10.1016/j.procir.2017.12.065

Kunieda M, Lauwers B, Rajurkar KP, Schumacher BM (2005) Advancing EDM through fundamental insight into the process. CIRP Ann 54(2):64–87. https://doi.org/10.1016/s0007-8506(07)60020-1

Niamat M, Sarfraz S, Aziz H, Jahanzain M, Shehab E, Ahmad W, Hussain S (2017) Effect of different dielectrics on material removal rate, electrode wear rate and microstructures in EDM. Procedia CIRP 60:2–7. https://doi.org/10.1016/j.procir.2017.02.023

Poroś D (2021) Comparative analysis of different WEDM strategies applied to cut WC-Co cemented carbides. Adv Sci Technol Res J 15(4):126–135. https://doi.org/10.12913/22998624/142561

Reolon LW, Laurindo CAH, Torres RD, Amorim FL (2019) WEDM performance and surface integrity of Inconel alloy IN718 with coated and uncoated wires. Int J Adv Manuf Technol 100(5–8):1981–1991. https://doi.org/10.1007/s00170-018-2828-6

Surface Characteristics of ER70S-6 Micro Slots Produced by Additive Manufacturing After Micro Milling

André Luis Silva da Costa, Francisco Menezes Guimarães, Maksym Ziberov, and Déborah de Oliveira

Abstract Micro machining is increasingly adopted for manufacturing miniaturized components with high functional integration and stringent surface quality requirements. Due to the comparable scale of tool edge radius and material grain size, chip formation occurs on a grain-by-grain basis, challenging the direct application of conventional machining principles. This study investigates the micro milling of ER70S-6 steel produced via Wire and Arc Additive Manufacturing (WAAM) using Cold Metal Transfer (CMT). The work employed 400 μm (Ti, Al)N-coated tungsten carbide tools at a cutting speed of 6.2 m/min, feed per tooth of 0.50 μm and 4.00 μm, and depth of cut values of 20 μm and 40 μm. Surface characterization was performed through confocal microscopy at magnifications of 5 ×, 10 ×, and 50 ×. Results indicate that cutting depth had negligible influence on surface quality, whereas feed rate significantly affected machining marks. The feed of 4.0 μm/tooth provided the most stable cutting conditions and best surface finish. The inherent microstructural and topographic features of WAAM-fabricated material were also found to be a dominant factor influencing the final surface integrity.

Keywords Micro machining · Surface quality · Additive manufacturing · WAAM · ER70S-6

1 Introduction

Additive manufacturing (AM) is a process that enables the fabrication of objects through successive layer-by-layer printing guided by a 3D model. This process allows the production of complex geometries, reduction of material waste, and customized parts (Vafadar et al. 2021). Accordingly, the ISO/ASTM 52,900 standard (ISO/ASTM 2015) defines seven categories of AM used for component fabrication: Binder Jetting (BJ), Powder Bed Fusion (PBF), Sheet Lamination (SL), Directed

A. L. S. da Costa · F. M. Guimarães · M. Ziberov · D. de Oliveira (✉)
University of Brasilia, Brasilia, DF, Brazil
e-mail: oliveira.deborah@unb.br

V. E. Beal and A. S. Ribeiro Junior (eds.), *Advances in Manufacturing Engineering Research*, Lecture Notes in Mechanical Engineering,
https://doi.org/10.1007/978-3-032-09858-0_13

Energy Deposition (DED), Material Extrusion (ME), Material Jetting (MJ), and Vat Photopolymerization (VP).

Metal AM processes are classified based on the feedstock form and heat source. Material supply can be via metal powder or wire, which are fused using various heat sources such as electron beams, lasers, and electric arcs (Lewandowski and Seifi 2016). Examples of powder deposition include PBF, which employs an electron beam directed at powder to fuse it into the desired shape, and LENS, which uses a laser to melt powder layer by layer to build the final geometry. Wire-feed methods utilize an electric arc to deposit one layer at a time, building a three-dimensional structure sequentially (Frazier 2014).

More specifically, in Wire and Arc Additive Manufacturing (WAAM), an electric arc serves as the heat source to melt and deposit material layer by layer, forming beads (Norrish et al. 2021). The significant heat input inherent to this technique necessitates careful control of deposition parameters, material composition, and heat treatments to manage the part's thermal history and achieve the desired microstructure (Jin et al. 2020). Improper parameter selection can induce defects that compromise mechanical strength and reduce fatigue life, critical for parts subjected to cyclic stresses (Barroqueiro et al. 2019; Silva et al. 2024).

Although wire feed offers advantages such as high deposition rates and the ability to build large-volume parts, it presents challenges, including a greater need for machining to achieve the desired surface quality compared to powder-based methods. As highlighted by Lin et al. (2024), surface quality directly impacts critical factors such as fatigue resistance, tribological properties, and the aesthetic appearance of the part. And the machining of WAAM parts might possess different challenges than the similar wrought material (Oliveira et al. 2025a).

In this context, micro machining emerges as an effective solution to overcome surface finishing limitations in parts produced by Wire Arc Additive Manufacturing (WAAM), especially when high precision at the microscale is a central requirement. Widely used in medical, aerospace, and microelectronics fields, micro machining allows rigorous control of the machined surface, an essential factor for achieving high quality that directly affects the performance and functionality of micro milled components (Assis et al. 2025).

Therefore, this study aims to evaluate surfaces obtained by micro milling with 400 μm tungsten carbide tools coated with (Ti, Al)N, using the following cutting parameters: cutting speed of 6.2 m/min, feed per tooth of 0.50, 1.00, 2.50, and 4.00 μm, and cutting depths of 20, 30, and 40 μm. The micro machined sample was previously deposited by WAAM, specifically the Cold Metal Transfer (CMT) process, using ER70S-6 wire.

2 Methodology

The sample was fabricated using the WAAM process, employing an automated system with a MAXR23-S42-H42-C42 Cartesian robot from Schneider Electric, which offers a movement precision of 0.1 mm. The configured process on the power source was synergic Cold Metal Transfer (CMT), with a bidirectional deposition strategy and no interpass temperature control. Table 1 presents the deposition parameters used for the sample, which were based on the work of Da Costa et al. (2025).

The additive material used was ER70S-6 wire, with a diameter of 1.2 mm, whose chemical composition is detailed in Table 2. For shielding during the process, an Ar + 18% CO_2 gas mixture was used, with the gas flow rate varying from 15 to 20 L/min to ensure arc stability and deposition quality.

The deposited wall, with a total length of 350 mm, was built up in 30 layers, requiring a total fabrication time of 22 min. Figure 1 provides a visual representation highlighting the leveling and uniformity of the deposited wall.

The micro milling of the samples fabricated with ER70S-6 wire was performed on a VEGA XH7132 vertical machining center. This machine features a worktable

Table 1 Deposition parameters

Parameter	Value
Step over [mm]	1.2
Current [A]	74
Tension [V]	13.5
Horizontal torch travel speed [mm/s]	8.0
Vertical torch travel speed [mm/s]	24.0
Direction	Bidirectional

Table 2 Chemical composition and mechanical properties of ER70S-6 wire (ESAB 2025)

Chemical composition	%	σy [MPa]	σu [MPa]	E [GPa]	v
Carbon [C]	0.070	420	500	207	0.29
Manganese [Mn]	1.500				
Silicon [Si]	0.900				
Sulfur [S]	0.010				
Phosphorus [P]	0.013				

Fig. 1 WAAM deposited sample

measuring 920 × 320 mm, a spindle speed ranging from 60 to 6000 rpm, and a feed rate adjustable between 2.5 and 3000 mm/min. The main motor delivers 3.7 kW of power and can support tools weighing up to 8 kg. In this experiment, a spindle speed of 5000 rpm was used. The cutting tool employed was the MS2MSD0040 model from Mitsubishi, with a diameter of 400 μm, a cutting length of 800 μm, two cutting edges with a radius of 1 ± 0.1 μm, and a maximum wear tolerance of 5%. Figure 2 contains the cutting tool with its main dimensions.

The micro machining was performed using the BIO100e cutting fluid, characterized by good durability, high lubricity, and effective cooling and anti-corrosive properties. The process parameters were defined based on de Paiva Silva et al. (2023) and Silva et al. (2025), considering variations in feed per tooth (f_z), cutting speed, and depth of cut (a_p). Accordingly, the feed per tooth was set to 0.5, 1.0, 2.5, and 4.0 μm, while the depth of cut started at 40 μm and was reduced by 10 μm every four micro milled slots. The cutting speed was kept constant at 6.2 m/min throughout the process. Table 3 presents the parameters applied during the micro milling of the ER70S-6 sample.

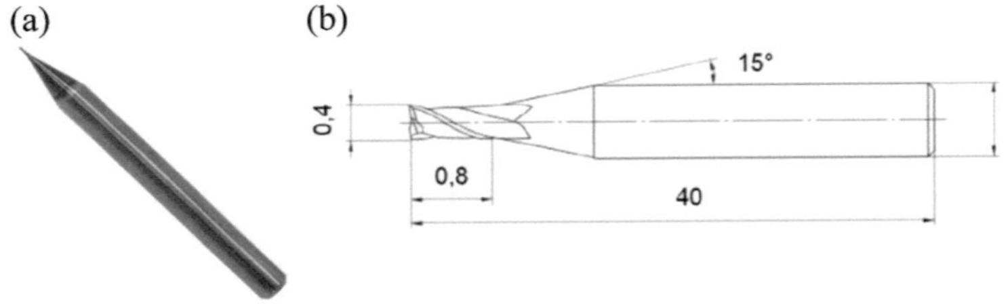

Fig. 2 Microtool. **a** MS2MSD0040 micro end mill and **b** micro end mill dimensions (Mitsubishi 2024)

Table 3 Cutting parameter for micro milling WAAM ER70S-6

Slot	f_z [μm]	a_p [μm]	Lubrication [l/min]	n [rpm]
1	0.5	20	5.88	5000
2	1.0			
3	2.5			
4	4.0			
5	0.5	30		
6	1.0			
7	2.5			
8	4.0			
9	0.5	40		
10	1.0			
11	2.5			
12	4.0			

For the purpose of analyzing the surface quality of the face milling on the ER70S-6 sample produced by the WAAM process using CMT, a qualitative surface evaluation was conducted. This analysis employed an Olympus LEXT OLS4100 laser scanning microscope, following the standard recommendations and applying a Gaussian filter. The surface images were captured at 50 × magnification, allowing detailed observation of the micro machined surface characteristics.

3 Results and Discussions

3.1 *Surface Quality as a Function of Depth of Cut (a_p)*

The surface quality analysis was performed on slots 4, 8, and 12, where the cutting depth was varied. Figure 3 presents confocal microscope images alongside 2D surface maps generated in MATLAB. It can be observed that the surface texture changes with increasing cutting depth. In slot 4 ($a_p = 20$ µm), the interaction between the cutting tool and the material occurs efficiently, resulting in continuous material removal, as indicated by well-defined circular marks. Furthermore, the length indicated by the yellow arrow corresponds to the mark length, which is similar to the slot length, as shown in Fig. 3a.

With increasing depth, in slot 8 ($a_p = 30$ µm), a slight reduction in the regularity of the circular marks is noticeable, represented by the white arrow in Fig. 3b, suggesting an initial instability in the cutting process. Finally, in slot 12 ($a_p = 40$ µm), Fig. 3c

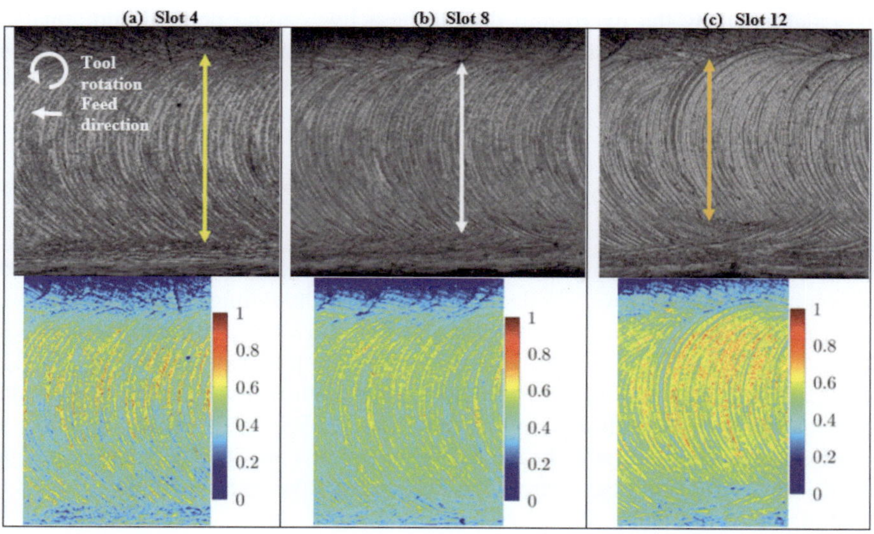

Fig. 3 Machined surface **a** $a_p = 20\mu m$, **b** $a_p = 30\mu m$ and **c** $a_p = 40\mu m$

highlights, with an orange arrow, that the circular marks become more compact and smaller compared to the cutting tool diameter. This pattern indicates periods of inactivity during micro milling, compromising the continuity of material removal and resulting in a surface with more evident irregularities.

Additionally, the surface maps provide a detailed view of the topographical features, allowing identification of texture and surface quality variations as a function of cutting depth. For the map of slot 4, a uniform topographic distribution is observed, reflecting an active micro milling cycle. As the cutting depth increases, as in slot 12, the map reveals accumulation of uncut material resulting from inactivity periods during micro milling. These irregularities are characterized by zones with lower amplitudes and abrupt transitions, indicating interruptions in effective contact between the tool and the material surface.

Similar circular patterns were also reported by Ziberov et al. (2020) in a study on micro milling of Ti-6Al-4 V, where the characteristic surface marks were associated with repetitive interactions between the micro mill and the material. These marks reflect the cyclic trajectory of the tool during the process, which, as in this work, is influenced by tool geometry and cutting dynamics. Therefore, regardless of the studied material, circular patterns emerge as a characteristic phenomenon of micro milling, directly related to the cyclic contact between the tool and the workpiece. This emphasizes how cutting depth and the cyclic motion of the micro endmill affect both the regularity of surface marks and the final surface quality.

Not only does the cutting depth directly influence the marks left on the machined surface, but also the interaction between the tool and the material. In this context, De Oliveira et al. (2025b), while studying the micro milling of In718, observed characteristic surface marks associated with repetitive interactions between the tool and the material. These circular marks, similar to those observed in this work, reflect the pattern left by the micro mill during the active cutting period. Additionally, adhered particles on the surface, resulting from the formation of built-up edges (BUE), contribute to the irregularity of the surface texture (Ziberov et al. 2016).

3.2 *Surface Quality as a Function of Feed Per Tooth (f_z)*

For this surface quality analysis, slots 1, 2, 3, and 4 were selected, varying the feed per tooth. The procedure was similar to that applied in the cutting depth investigation, using a confocal microscope for image acquisition and MATLAB for topography, as illustrated in Fig. 4. As a result, for feed values of 0.5 μm (slot 1), 1.0 μm (slot 2), and 2.5 μm (slot 3), a defect in image capture was identified: a line running through the center of the slot, indicated by the red arrows in Fig. 4a, b, and c. This defect may be associated either with the micro milling process or with errors during image acquisition by the confocal microscope, which failed to scan this region properly, resulting in the formation of a band along the slot.

On the other hand, for the feed of 4.0 μm (slot 4), a more uniform surface texture was observed, without visible defects. The circular marks showed a regular pattern,

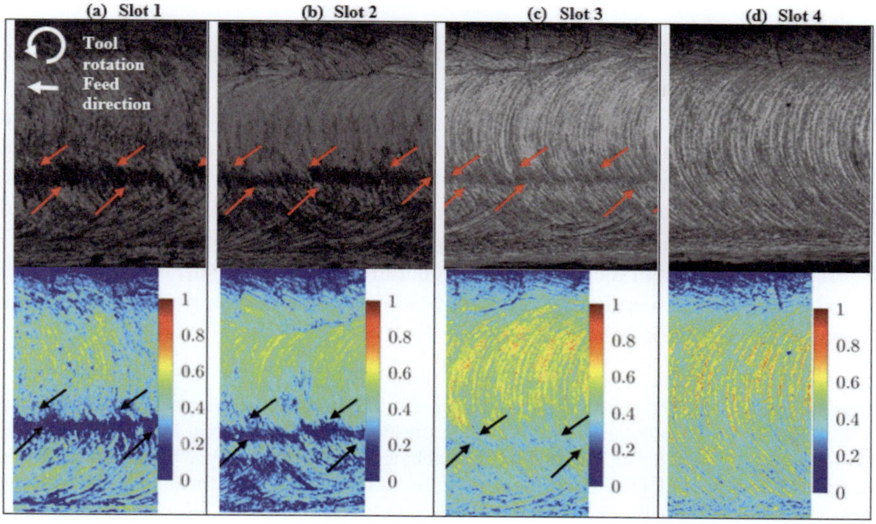

Fig. 4 Machined surface **a** $f_z = 0, 5\mu m$, **b** $f_z = 1, 0\mu m$, **c** $f_z = 2, 5\mu m$ and **d** $f_z = 4, 0\mu m$

with dimensions equivalent to the tool diameter or slot width, indicating more stable and efficient material removal at this feed rate.

In the topography analysis, the results obtained corroborate the findings of Assis et al. (2025), who investigated the influence of feed per tooth and cutting speed on the surface quality of low carbon steel. The authors observed that an increase in feed per tooth promoted more pronounced formation of craters and burrs, resulting in irregular and discontinuous surfaces, features that compromise the final quality of the micro machined part. Finally, Assis et al. (2015) state that when the feed per tooth (f_z) is less than or equal to the tool's cutting edge radius, the interaction between the material and the micro mill results in a more uniform surface free of defects. Conversely, when f_z exceeds the cutting edge radius, intense shear deformations occur, promoting negative rake angles, and increasing burr formation (Oliveira et al. 2025c).

4 Conclusions

After conducting micro milling tests on the ER70S-6 sample produced by WAAM and analyzing the surface quality of the slots, the following conclusions were drawn:

- Well-defined and uniform circular marks predominated at lower depths of cut (a_p = 20 μm), reflecting a continuous and efficient material removal.

- Greater depths of cut (a_p = 40 μm) resulted in more compact and irregular patterns, indicating periods of inactivity during micro milling, which compromised surface uniformity.
- Feed rates of 0.5 μm, 1.0 μm, and 2.5 μm exhibited image capture defects, with a line appearing at the center of the slot. Conversely, a feed rate of 4.0 μm resulted in more stable material removal.
- The results emphasize the need for precise control of the depth of cut (a_p) and feed per tooth (f_z) parameters to ensure surface quality.
- Proper adjustment of these parameters guarantees machined surfaces with greater uniformity and fewer defects, especially in applications requiring high precision and strict geometric tolerances.

Acknowledgements The authors acknowledge the Department of Mechanical Engineering (ENM), Faculty of Technology (FT), University of Brasília (UnB), the Graduate Program in Mechanical Sciences (PCMEC), the Undergraduate Teaching Dean's Office (DEG), and the Graduate Studies Dean's Office (DPG). They also gratefully acknowledge CNPq for funding through the projects 405499/2022-1, 402730/2023-2, 443148/2023-6, and 200018/2025-6.

Competing Interests There are no competing interests to declare.

Responsibility for Information The authors are solely rete sponsible for the information included in this work.

References

Assis CLF, Jasinevicius RG, Rodrigues AR (2015) Micro end-milling of channels using ultrafine-grained low-carbon steel. Int J Adv Manuf Technol 77:1155–1165

Assis CF, Mecelis GR, Coelho TR, Rodrigues AR (2025) Enhancing surface quality in micro end-milling of ultrafine-grained low carbon steel micromolds through strategic toolpath selection. Int J Adv Manuf Technol 137:3077–3093. https://doi.org/10.1007/s00170-025-15312-5

Barroqueiro B, Andrade-Campos A, Valente RAF, Neto V (2019) Metal additive manufacturing cycle in aerospace industry: a comprehensive review. J Manuf Mater Process 3:52

da Costa ALS, de Paiva RL, de Oliveira D, Ziberov M (2025) Influence of interlayer temperature and deposition method on the wall geometry and Vickers microhardness profile of ER70S-6 parts manufactured by additive manufacturing using CMT. J Manuf Mater Process 9:93. https://doi.org/10.3390/jmmp9030093

da Silva CP, Silva GP, Santos MC, Ziberov M, Malcher L (2024) Hybrid search methodology for mechanical characterization of material produced via WAAM assuming Gurson porous material. J Braz Soc Mech Sci Eng 46:1–16

de Oliveira D, Goncalves MV, Ribeiro GM, da Costa ALS, Regueiras L, Silva T, de Jesus A, Malcher L, Ziberov M (2025a) On the use of compressed air and synthetic biodegradable cutting fluid to enhance the surface quality of WAAM–CMT manufactured low-alloy steel parts during post-processing milling with different cooling–lubrication strategies. J Manuf Mater Process 9:193. https://doi.org/10.3390/jmmp9060193

de Oliveira D, Ziberov M, de Paiva RL, da Silva MB (2025b) An experimental evaluation of cutting parameters influence on the surface integrity and tool wear mechanisms on the dry micromilling of austenitic alloy Inconel 718. Wear. https://doi.org/10.1016/j.wear.2025.205789

de Oliveira D, Gomes MC, Ziberov M, Montagnana MB, Jackson MJ, da Silva MB (2025c) On the use of extremely high feed per tooth as an alternative to reduce size effects during the micro milling of critically low machinability alloy Inconel 718. Int J Adv Manuf Technol 139:3891–3908. https://doi.org/10.1007/s00170-025-16152-z

de Paiva SG, de Oliveira D, Malcher L (2023) Numerical study of the minimum uncut chip thickness in micro-machining of Inconel 718 based on Johnson-Cook isothermal model. Int J Adv Manuf Technol 127:2707–2721. https://doi.org/10.1007/s00170-023-11573-0

ESAB (2025) Weld 70S-6 copper-coated ER70S-6 solid MIG wire. ESAB United States. Available at: https://esab.com/us/nam_en/products-solutions/product/filler-metals/mild-steel/mig-wires-tig-rods-gmaw-gtaw/weld-70s-6/. Accessed on August 13, 2025

Frazier WE (2014) Metal additive manufacturing: a review. J Mater Eng Perform 23:1917–1928

ISO/ASTM 52900 (2015) Additive manufacturing, general principles, terminology. International Standard, 1st edn, pp 1–19

Jin W, Zhang C, Jin S, Tian Y, Wellmann D, Liu W (2020) Wire arc additive manufacturing of stainless steels: a review. Appl Sci 10:1563

Lewandowski JJ, Seifi M (2016) Metal additive manufacturing: a review of mechanical properties. Annu Rev Mater Res 46:151–186

Lin YT, Tsai MY, Yen SY, Lung GH, Yei JT, Hsu KJ, Chen KJ (2024) Comparing the performance of rolled steel and 3D-printed 316L stainless steel. Micromachines 15:353

Mitsubishi (2024) Solid end mills. https://www.mitsubishicarbide.net/mht/enuk/solid_end_mills/10000222/20060775. Accessed 5 Dec 2024

Norrish J, Polden J, Richardson I (2021) A review of wire arc additive manufacturing: development, principles, process physics, implementation and current status. J Phys D Appl Phys 54:473001

Silva GP, Gomes MC, da Silva MB, Malcher L, de Oliveira D (2025) Experimental validation of cutting forces modeling in micromilling of Inconel 718 considering material heterogeneity and wear-induced tool edge rounding. J Manuf Process 146:303–319. https://doi.org/10.1016/j.jmapro.2025.04.101

Vafadar A, Guzzomi F, Rassau A, Hayward K (2021) Advances in metal additive manufacturing: a review of common processes, industrial applications, and current challenges. Appl Sci 11:1213

Ziberov M, da Silva MB, Jackson M, Hung WNP (2016) Effect of cutting fluid on micromilling of Ti-6Al-4V titanium alloy. Proced Manuf 5:332–347

Ziberov M, de Oliveira D, da Silva MB, Hung WNP (2020) Wear of TiAlN and DLC coated microtools in micromilling of Ti-6Al-4V alloy. J Manuf Process 56:337–349

Effects of the Volume-to-Surface Area Ratio in Cylindrical Plunge Grinding with Cryogenic Pre-cooling

Ézio Carvalho de Santana, Eraldo Jannone da Silva, Daniel Weber, Benjamin Kirsch, and Jan Christian Aurich

Abstract Currently, metal working fluids (MWFs) are the standard coolants applied in grinding processes, avoiding the formation of surface cracks, burning, and residual tensile stresses. Due to sustainability-related issues, alternatives to MWF have been extensively researched. A few of these alternatives include minimum quantity lubrication (MQL), dry grinding, and the application of cryogenic fluids. Regarding cryogenic coolants, the usual approach consists of directing a cryogenic jet at the cutting zone, whereas pre-cooling has been a less explored strategy. The pre-cooling strategy aims to reduce the workpiece's temperature below cryogenic temperatures beforehand, followed by dry machining, resulting in lower peak temperature values compared to regular dry machining. The application of cryogenic pre-cooling in cylindrical plunge grinding has been tested, and it has been claimed that its effectiveness is tied to the ratio between the volume and surface area of the workpiece. As a follow-up study, this work aims to evaluate both experimentally and through thermal simulations how the volume-to-surface ratio influences the thermal behavior of the workpiece as well as its dimensional accuracy. More specifically, two cylindrical AISI 4140H workpieces with different volume-to-surface ratios were ground, temperature measurements were performed at depths of 2 mm and 4 mm below the external surface, electric power consumption during grinding was also monitored and the initial and final diameter of the workpieces were measured. Furthermore, thermal simulations were also conducted to evaluate the temperature at the surface during the process, assuming ideal conditions. It was found that larger workpieces are able to more easily conserve cryogenic temperature given their higher heat capacity at the expense of worse dimensional accuracy due to higher thermal contraction. Workpieces with smaller volume-to-surface, on the other hand, present lower heat capacity and higher dimensional accuracy. Therefore, larger workpieces demand more technical know-how in order to account for the contraction and expansion of the workpiece during grinding, while smaller workpieces require better thermal

É. C. de Santana (✉) · E. J. da Silva
University of São Paulo, São Carlos, SP, Brazil
e-mail: eziocds@usp.br

D. Weber · B. Kirsch · J. C. . Aurich
Institute for Manufacturing Technology and Production Systems, RPTU, Kaiserslautern, Germany

V. E. Beal and A. S. Ribeiro Junior (eds.), *Advances in Manufacturing Engineering Research*, Lecture Notes in Mechanical Engineering,
https://doi.org/10.1007/978-3-032-09858-0_14

insulation to reduce heat transfer between the workpiece and environment, keeping cryogenic temperatures for a longer period.

Keywords Cylindrical plunge grinding · Cryogeny · Pre-cooling · Thermal simulation · Sustainability

1 Introduction

A large amount of heat is generated during grinding processes, which can adversely affect the surface integrity and dimensional accuracy of workpieces. Traditionally, metalworking fluids (MWFs) are employed to mitigate heat generation and enhance the grinding process. While effective, the use of MWFs poses environmental challenges, as their disposal can lead to soil and water contamination, presenting both ecological and regulatory concerns (Najiha et al. 2016). In contrast, cryogenic cooling offers a cleaner alternative, eliminating the need for complex disposal procedures associated with conventional coolants, thereby reducing environmental impact and improving workplace safety. In Fig. 1, the trend of different forms of lubrication and heat removal methods is depicted, presenting lubrification-free grinding as the ultimate goal from a sustainability standpoint. In grinding processes, most studies on cryogenic cooling have focused on supplying the cryogenic fluid during the operation, while the pre-cooling method remains largely underexplored. Furthermore, experiments in the field of cryogenic grinding have predominantly been conducted using surface grinding machines. Thus, there is a lack of research on the application of cryogenic techniques for cylindrical plunge grinding processes (Abedrabbo et al. 2021). The feasibility of performing cylindrical plunge grinding in dry conditions with pre-cooling of the workpiece was tested both experimentally and with the aid of numerical simulations by Oliveira et al. (2015). In that study, it was proposed that the viability of the method may depend on the dimensions of the workpiece, more specifically, the volume-to-surface area ratio (V/A).

Thus, the present study aims to ascertain more in-depth how the size of the workpiece affects the efficiency and performance of the pre-cooling strategy. To assess this, two grinding experiments were performed with different workpiece sizes. Additionally, two thermal simulations were conducted to evaluate the difference in thermal behavior of the workpieces under ideal conditions.

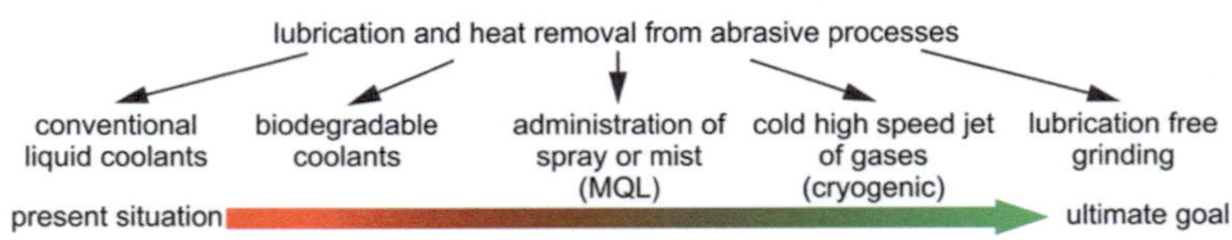

Fig. 1 Different forms of lubrication and heat removal methods. Inasaki et al. 1993 Adapted from Inasaki et al. ()

2 Methodology

Given that both experimental and numerical approaches were employed, the methodology of this work is divided into Subsect. 2.1 Materials and experimental procedure and 2.2 Thermal simulations.

2.1 *Materials and Experimental Procedure*

The workpieces were machined from two different alloy steel AISI 4140 billets. Table 1 depicts the chemical composition of both billets.

To evaluate the volume-to-surface ratio effect on the pre-cooling strategy, two different geometries were designed. The first geometry, denoted as D30, was manufactured from billet #1 and presents a nominal diameter of 30 mm, representing the workpiece with a smaller volume-to-surface ratio. The second geometry, denoted as D100, presents a nominal diameter of 100 mm, representing the workpiece with higher volume-to-surface ratio. The technical drawings of workpieces D30 and D100 are shown in Fig. 2.

Only section B (sec. B) from the workpieces was machined (grinding length b = 40 mm), while section A was used to position the driving carrier on the workpieces. Additionally, cavities were drilled through electro discharge machining in both workpieces for thermocouple insertion during the experiments, with T1 and T2 being 2 and 4 mm below the external surface.

Table 1 Chemical composition of the billets

Billet	C	Mn	Si	P	S	Cr	Ni	Mo	W	V	Cu
#1	0.39	0.94	0.25	0.01	0.006	1.04	0.18	0.23	0	0.01	0.01
#2	0.40	0.87	0.26	0.01	0.007	1.03	0.03	0.22	0	0.00	0.09

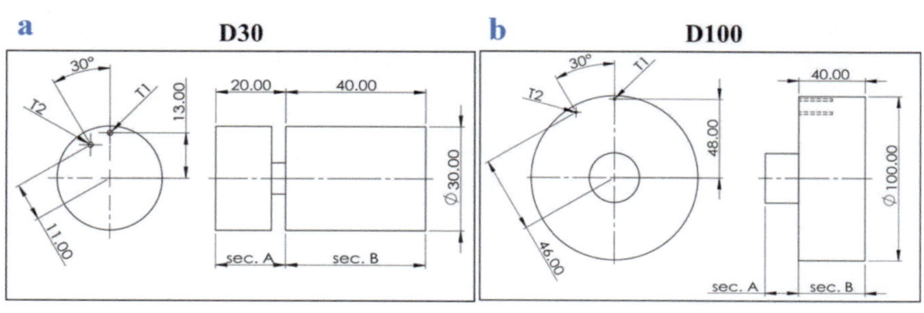

Fig. 2 Technical drawing of workpieces D30 **a** and D100 **b**

The workpieces were subject to quenching, being kept in a salt bath at a temperature of 860 °C for 30 min, followed by quick cooling to 170 °C and subsequent tempering, resulting in a predominantly martensite microstructure and average surface hardness of 58 HRC.

A total of two grinding experiments were conducted on the grinder Zema Numerika G-800—HS, one experiment for each workpiece geometry (D30 and D100). An Aluminum oxide (Al_2O_3) grinding wheel with 38 A 80 J VHB specifications and 500 mm of diameter was employed in the experiments. Before grinding, 3 dressing steps with a diamond stationary fliesen tool were performed, adopting an overlap ratio value of $U_d = 7$, effective contact width $b_d = 0{,}66$ mm and dressing depth of cut $a_p = 0{,}02$ mm.

The grinding parameters are described in Table 2 and were computed such that the specific material removal rate was kept at a value of $Q'_w = 0, 83$ mm^3/(mm.s), cutting speed $v_s = 30$ m/s and a diameter stock removal value (Δd) per cycle of 0,3 mm. The initial diameter of the workpieces was measured at ambient temperature, which was then inputted into the CNC program, therefore, no adjustments to account for the material's thermal contraction were implemented in the tests conducted.

A diagram was elaborated, as shown in Fig. 3, to illustrate the procedure adopted during the experiments.

Replace the current table values with the correct ones which are placed on the right side of the wrong values.

The temperature-measuring system consisted of a slip-ring device that rotates in conjunction with the workpiece during grinding. The temperature signals are then transmitted to an external receptor device connected to a NI-USB-6341 via Bluetooth. The raw data is then retrieved from a standard Labview application from a notebook also connected to the NI-USB-6341 device. The power consumption measurements, on the other hand, were acquired from the SINAMICS G120—CU320 control module present in the machine. Both measurement systems have been previously employed to validate thermal simulations of grinding with pre-cooling and more details are available at Santana et al. (2023).

After the experiments, a moving average filter was applied to both temperature and power measurements to reduce the noise present in the data collected. The size of the filters applied was of 61 points for temperature signals (sampling rate: 1000 Hz) and 11 points for power consumption signals (sampling rate: 100 Hz).

Table 2 Workpiece and grinding parameters of experiments

Work-piece	V/A (mm)	Q'_w (mm^2/s)	v_s (m/s)	Δd (mm)	Number of cycles	Spark-in (s)	Spark-out (s)	Infeed rate (mm/min)
D30	5.5	0.83	0.30	0.3	2	17.0	4.0	0.53
D100	11.1	0.83	0.30	0.3	2	56.8	4.0	0.16

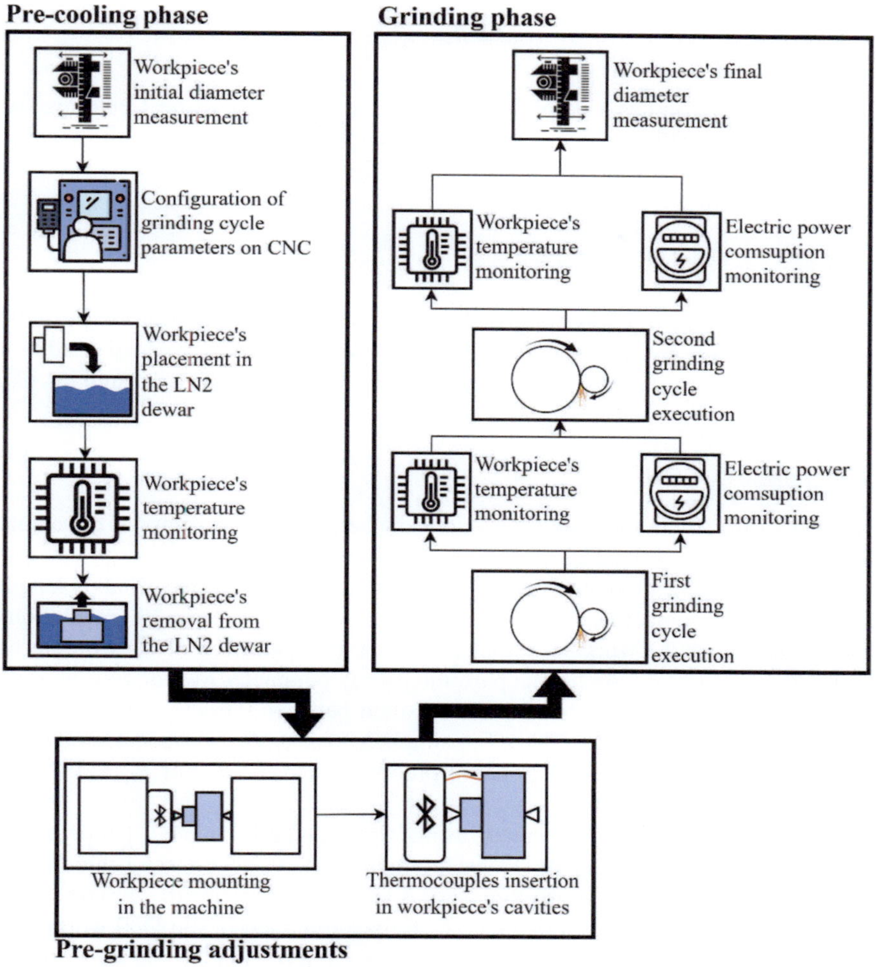

Fig. 3 Experimental procedure flow chart

2.2 Thermal Simulations

Two simulations were executed to evaluate the thermal behavior of the workpieces considering both present equivalent initial temperature ($T_o = -196$ °C) and constant magnitude of the moving heat flux over time $q_w(t) = q$, in order to evaluate how the volume-to-surface ratio of the workpieces impacts the pre-cooling strategy under ideal grinding conditions. The constant simulated heat flux magnitude q was determined through Eq. (1).

$$q = R_w \frac{u \cdot Q'_w}{l_c} \tag{1}$$

where R_w (dimensionless) is the energy partition parameter of heat entering the workpiece, l_c (mm) is the geometric contact length and u (J/mm^3) is the specific grinding energy, which was determined by Eq. (2).

$$u = \int \frac{P_{\text{net}}}{V_r} \mathrm{d}t \tag{2}$$

The $\int P_{\text{net}} \mathrm{d}t$ quantity was computed through numerical integration (trapezoidal rule) of the power consumption measurements over time minus no-load condition power consumption. The material volume removal V_r is determined through Eq. (3).

$$V_r = \frac{\pi b}{4} \left(d_w^2 - [d_w - \Delta d_a]^2 \right) \tag{3}$$

where d_w (mm) is the workpiece's initial diameter and Δd_a (mm) is the actual diameter stock removal value.

The area of application of the moving heat flux is delimited by the grinding length b and geometric contact length l_c. This moving heat flux performs 100 revolutions per minute around the workpiece, matching the rotational speed of the workpieces during the experiments. As for the partition energy parameter R_w, for conventional grinding with vitrified wheels, its value is often between 0.6 and 0.9 (Marinescu et al. 2016; Guo and Malkin 2000). For simplicity's sake, R_w was considered to be the same for both workpieces' grinding processes, and a value of $R_w = 0.7$ was adopted for both simulations.

The simulations were executed using the standard coupled temperature-displacement module of Abaqus. Transient implicit analysis with step solutions via the Full Newton algorithm were employed. A temporal discretization of 0.001 s per step was adopted for both simulations, with the simulation of the D30 workpiece concluding after 17,000 steps and of the D100 workpiece after 56,800 steps, corresponding to the spark-in duration of one cycle in number of steps. The mesh of the D30 workpiece was comprised of 2580 C3D8T and 540 C3D6T linear elements, while the D100 workpiece presented 5200 C3D8T and 720 C3D6T linear elements. The center supports, on the other hand, contain 3691 C3D4T-type elements, as shown in the cross-sectional views of each assembly meshes in Fig. 4.

Although finer mesh sizes and lower temporal discretization would be necessary to more accurately evaluate the workpiece's temperature throughout the process, they still suffice for relative comparisons to assess the volume-to-surface ratio's influence on the workpiece's overall thermal behavior.

Regarding the thermal modeling, convection between the workpiece and air, as well as thermal contact conductance between the workpiece and center supports were also included in order to better characterize the thermal behavior of the pre-cooled workpieces. More details about these boundary conditions implementation

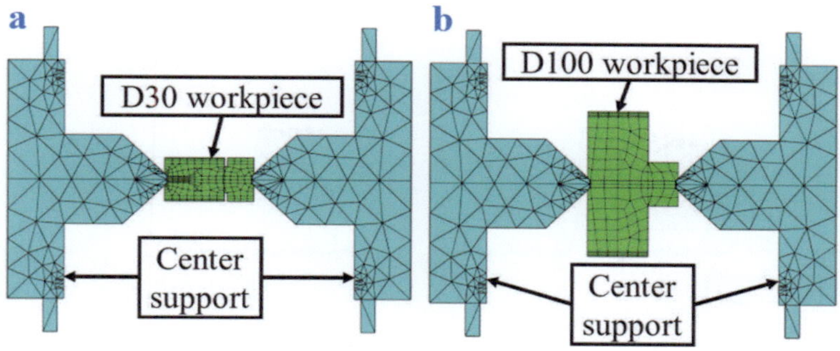

Fig. 4 Cross-sectional renderings of workpieces D30 **a** and D100 **b** depicting the internal mesh configurations

and validation of the thermal models for pre-cooling applications can be found at Santana et al. (2023).

3 Results

In Subsect. 3.1. "Experimental results," the technical challenges and experimental aspects regarding the pre-cooling strategy as a function of the workpiece size can be observed. The results of the thermal simulations are contained in Subsect. 3.2. "Numerical results" which aims to evaluate the pre-cooling strategy under ideal conditions.

3.1 Experimental Results

The power consumption measurements of the experiments are plotted in Fig. 5.

The time intervals delimited in Fig. 6 are defined as follows:

- Δt_1: First grinding cycle duration (spark-in + spark-out);
- δt_1: Time interval between the start of the first grinding cycle (CNC control reaches the inputted workpiece diameter) and first contact between grinding wheel and workpiece;
- Δt_2: Time interval between the end of the first grinding cycle and start of the second grinding cycle;
- Δt_3: Second grinding cycle duration (spark-in + spark-out);
- δt_2: Time interval between the start of the second grinding cycle (CNC control reaches the inputted workpiece diameter) and first contact between grinding wheel and workpiece;

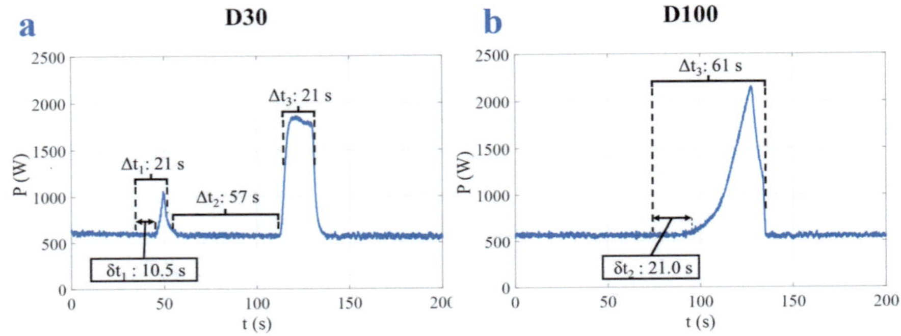

Fig. 5 Power measurements over time during the grinding experiments on workpieces D30 **a** and D100 **b**

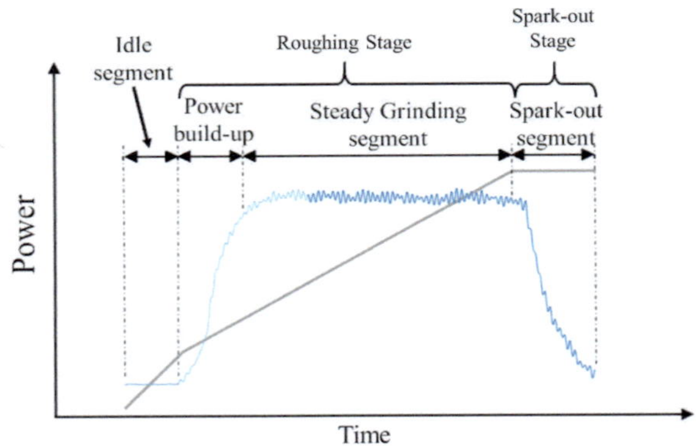

Fig. 6 Standard power cycle of a grinding process with roughing and spark-out stages. The blue curve represents the power consumption, and the gray curve represents the infeed. 2022 Adapted from Ahmer et al. ()

As described in Sect. 2.1, two grinding cycles were performed for each experiment. For the D30 workpiece's first grinding cycle, 10.5 s (δt_1) had elapsed before the first contact between the grinding wheel and workpiece after the cycle start was verified, comprising half of the total grinding cycle duration, including spark-out (10.5 out of 21.0 s). A more standard power curve behavior for a single grinding cycle comprised of a roughing stage (spark-in) and spark-out stage is depicted in Fig. 6. Comparing the more standard power curve from Fig. 6 with the Δt_1 segment from Fig. 5a, it becomes clear that for the first grinding cycle, no steady grinding segment is present. Instead, only power build-up, followed by the spark-out stage in quick succession are verified, producing a peak-shaped curve. The second grinding cycle (Δt_3 segment), on the other hand, presented much more uniform contact between the grinding wheel and workpiece throughout the process, which is reflected in the more standard behavior observed.

For the power curve of the D100 workpiece on Fig. 5b, despite two grinding cycles also been executed, no actual contact between grinding wheel and workpiece occurred in the first grinding cycle, therefore no Δt_1, Δt_2 and δt_2 were specified in Fig. 5b. In the second grinding cycle (Δt_3 segment), the contact was uneven throughout the whole process, producing a power curve without a steady grinding segment being verified.

The predominant pattern of uneven contact verified in the power cycle curves can be attributed to the thermal contraction of the workpieces due to cryogenic pre-cooling. Equation (4) can be used to compute the variation of the workpiece's diameter as a function of its temperature.

$$\Delta d_t = \alpha \cdot d_w \cdot \left(T_{\text{amb}} - T_{\text{cryo}}\right) \quad (4)$$

where Δd_t (mm) is the thermally induced diameter variation, α (°C− 1) is the thermal expansion coefficient, T_{amb} (°C) is the room temperature and T_{cryo} (°C) is the workpiece's temperature after the pre-cooling step.

The initial diameter at room temperature, maximum thermal contraction and actual stock removed from the D30 and D100 workpieces, as well as the volume removed and specific grinding energy are depicted in Table 3.

The maximum diameter contraction expected of the workpieces was computed assuming $\alpha = 12.2 \times 10^{-6}$ °C − 1, $T_{\text{amb}} = 25$ °C and $T_{\text{cryo}} = -\ 196$ °C. Since two cycles were executed, each with a preset stock removal value of 0, 3 mm, a total stock removal of 0, 6 mm was expected. Therefore, the stock removal relative error for each workpiece was 39% (D30) and 77% (D100). The higher error for the D100 workpiece is attributed to the higher thermal contraction due to the cryogenic pre-cooling strategy, which is a direct consequence of the higher volume-to-surface ratio of the workpiece. As such, larger workpieces require more know-how of the process to account for the diameter variation at the beginning and along the grinding cycle.

The graphs in Fig. 7 depict both T1 and T2 curves, which represent the temperature measurements at positions 2 mm and 4 mm below the external surface of the workpieces, respectively. Figure 7a depicts these results for the D30 workpiece, while Fig. 7b depicts the D100 workpiece's registered temperature measurements.

Comparing the peak temperatures of workpieces D30 (T1 = 177.6 °C and T2 = 146.7 °C) and D100 (T1 = − 39.3 °C and T2 = − 40.7 °C) shows that the larger workpiece (D100) was better suited to the pre-cooling strategy, despite the worse dimensional accuracy verified previously. The reason for that is due to the difference in heat capacity of the workpieces. The heat capacity is the proportionality constant between the heat that is necessary to result in a unit change of an object's temperature

Table 3 D30 and D100 workpiece's dimensional information and specific grinding energy values

Workpiece	Initial diameter (mm)	Diameter contraction (mm)	Actual diameter stock removal (mm)	Volume removed (mm^3)	Specific grinding energy (J/mm^3)
D30	30.000	0.081	0.366	685.7	27.4
D100	100.000	0.270	0.136	1023.3	22.5

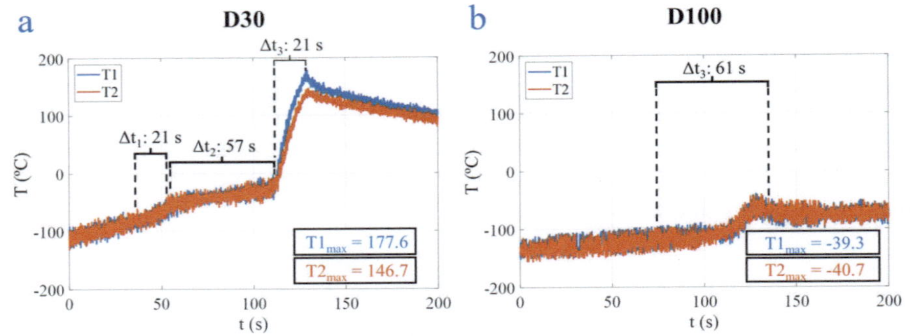

Fig. 7 Temperature measurements over time during grinding experiments of the D30 **a** and D100 **b** workpieces at 2 mm (T1) and 4 mm (T2) below the workpieces' external surface

(Halliday et al. 2013). This is an extensive property, therefore, proportional to the workpiece's mass, and its effects are twofold:

- Initial temperature: During the time interval between the end of pre-cooling and the start of the grinding process, the workpiece continuously exchanges heat with the ambient. Larger workpieces (higher volume-to-surface ratio) are more prone to keeping cryogenic temperatures for longer, given their higher heat capacity;
- Heat accumulation: Larger workpieces are able to absorb more heat generated during the grinding process as a result of their higher heat capacity, reducing overall temperature increase;

Another implication that concerns the volume-to-surface ratio of workpieces in plunge cylindrical grinding subjected to pre-cooling is related to the rate and amount of material to be removed during the process. Given a grinding process with fixed Δd (diameter stock removal) and Q'_w (specific material removal rate), higher diameters (higher volume-to-surface ratio) will require higher cycle durations, as depicted in Table 2. The total grinding cycle duration of the D30 workpiece was 21.0 s, while the duration for the D100 workpiece was 60.9 s. Longer grinding cycles mean that the effects of natural heat transfer phenomena are able to accumulate for longer and more significantly affect the workpiece's thermal behavior, reinforcing the necessity of more know-how for larger workpieces.

3.2 Numerical Results

Given the thermal contraction that occurred during the experiments, causing highly irregular grinding cycles (particularly the first grinding cycle of both workpieces), and the different pre-cooling conditions of the workpieces, where the D30 (smaller volume-to-surface ratio) workpiece registered temperature values near room temperature at the beginning of the second grinding cycle (see Fig. 7a), thermal simulations

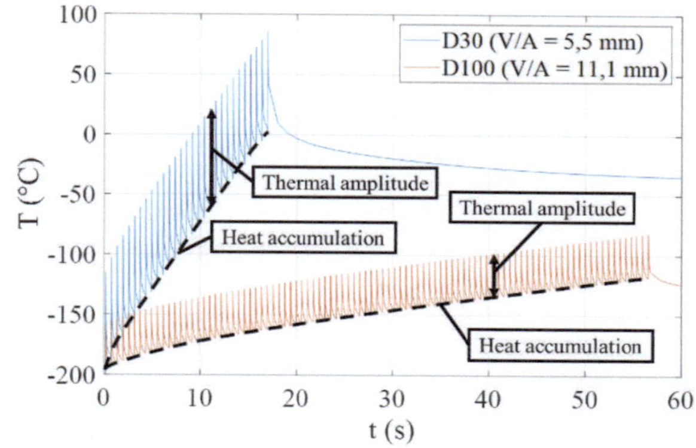

Fig. 8 Simulated temperature values at the surface for the D30 and D100 simulations

were conducted to better examine the influence of the volume-to-surface ratio of the workpieces on the pre-cooling strategy efficacy under more ideal conditions.

Two thermal simulations of a single grinding cycle were conducted, one for each workpiece geometry. Uniform contact between the grinding wheel and workpiece throughout the cycle was assumed and an initial temperature value of − 196 °C was defined for all nodes of both workpieces. The temperature curves of two correspondent nodes positioned at the workpiece's external surface are plotted in Fig. 8, one node from each simulation (D30 and D100 workpieces).

The difference in thermal behavior of the workpieces as a function of the volume-to-surface ratio was classified into two categories: instantaneous and cumulative. The instantaneous thermal effect is reflected by the thermal amplitudes highlighted in Fig. 8. The difference in thermal amplitude is a result of the difference in depth of cut (a_e) and also of different peripherical speed values of the workpieces, which results in different geometric contact length (l_c) values and in how fast the moving surface heat flux will travel a certain distance in the simulations. Equations (5) and (6) were used to compute a_e and l_c (Malkin and Guo 2008), resulting in $a_e = 5.28$ μm and $l_c = 0.39$ mm for the D30 workpiece and $a_e = 1.59$ μm and $l_c = 0.36$ mm for the D100 workpiece. This difference in geometric contact length impacts the peak temperatures reached since heat will flow across different surface area values.

$$a_e = \frac{Q'_w}{\pi d_w n_w} \tag{5}$$

$$l_c = \sqrt{\frac{a_e d_s}{1 + d_s/d_w}} \tag{6}$$

The workpiece's peripherical speed, which is proportional to the workpiece's diameter ($v = \omega \ldots d_w/2$) will be lower for the smaller workpiece since the rotational speed of both workpieces are equal ($\omega = 100$ rpm), therefore the moving heat flux will take more time to travel a certain region, effectively increasing the interaction time between the region and the moving surface heat flux.

The cumulative effect, on the other hand, is reflected by the heat accumulation trend also highlighted in Fig. 8. The difference in thermal behavior for different-sized workpieces further diverges due to heat capacity differences. Heat accumulation in the workpiece's volume is more critical for smaller workpieces, which can be observed by its steeper lower temperature trending curve depicted in Fig. 8.

4 Conclusions

In plunge cylindrical grinding, this study has shown that the pre-cooling strategy is impacted by the volume-to-surface ratio of the workpiece through the following:

- Dimensional accuracy: Larger workpieces (higher volume-to-surface ratio) present inferior dimensional accuracy due to thermal contraction;
- Cryogenic temperature retention: Smaller workpieces (lower volume-to-surface ratio) present less heat capacity and, therefore, pre-cooling temperatures are preserved for a shorter duration while heat generated during grinding also accumulates faster.

Based on these conclusions, larger workpieces require more thermal analysis regarding contraction and expansion of the workpiece to achieve better dimensional accuracy. On the other hand, smaller workpieces demand better insulation to preserve cryogenic temperature at the start of grinding. Even then, diameter stock removal values Δd are narrower for smaller workpieces due to less capability to maintain cryogenic temperatures.

Acknowledgements The authors would like to thank the Conselho Nacional de Desenvolvimento Científico e Tecnológico (CNPq) and the Fundação de Amparo à Pesquisa do Estado de São Paulo (FAPESP) for their financial support of the project "Expanding the Possibilities of Cryogenic Assisted Grinding" (FAPESP Grant No. 2018/16416-5) and DeutscheForschungsgemeinschaft DFG with DFG-project number 440394762.

Competing Interests The authors declare that they have no known competing financial interests or personal relationships that could have appeared to influence the work reported in this paper.

Responsibility for Information The authors are solely responsible for the information included in this work.

References

Abedrabbo F, Soriano D, Madariaga A, Fernández R, Abolghasem S, Arrazola PJ (2021) Experimental evaluation and surface integrity analysis of cryogenic coolants approaches in the cylindrical plunge grinding. Sci Rep 11(1):20952. https://doi.org/10.1038/s41598-021-00225-6

Ahmer M, Sandin F, Marklund P, Gustafsson M, Berglund K (2022) Using multivariate quality statistic for maintenance decision support in a bearing ring grinder. Machines 10(9):794. https://doi.org/10.3390/machines10090794

Guo C, Malkin S (2000) Energy partition and cooling during grinding. J Manuf Process 2(3):151–157. https://doi.org/10.1016/S1526-6125(00)70116-2

Halliday D, Resnick R, Walker J (2013) Fundamentals of physics. John Wiley & Sons, 10a ed., p 524

Inasaki I, Tönshoff HK, Howes TD (1993) Abrasive machining in the future. CIRP Ann 42(2):723–732. https://doi.org/10.1016/S0007-8506(07)62535-9

Malkin S, Guo C (2008) Grinding geometry and kinematics. In: Malkin S, Guo C, Grinding technology: theory and applications of machining with abrasives. Industrial Press, 2a ed., p 43–47

Marinescu ID, Hitchiner MP, Uhlmann E, Rowe WB, Inasaki I (2016) Handbook of machining with grinding wheels. CRC Press, 2nd ed., p 25. https://doi.org/10.1201/b19462

Najiha MS, Rahman MM, Yusoff AR (2016) Environmental impacts and hazards associated with metal workingfluids and recent advances in the sustainable systems: a review. Renew Sustain Energy Rev 60:1008–1031. https://doi.org/10.1016/j.rser.2016.01.065

Oliveira JFG, Silva EJ, Coelho RT, Brozek L, Bottene AC, Marcos GP (2015) Dry grinding process with workpiece precooling. CIRP Ann Manuf Technol 64(1):329–332. https://doi.org/10.1016/j.cirp.2015.04.098

Santana EC, Souza GU, Silva EJ, Weber D, Kirsch B, Aurich JC (2023) Modeling and simulation of thermal effects in cylindrical plunge dry grinding with precooling. In: Proceedings of the 27th international congress of mechanical engineering—COBEM2023. Florianópolis, Brazil. https://doi.org/10.26678/ABCM.COBEM2023.COB2023-2374

Study of the Residual Stresses Induced by End Milling and Grinding of 350 Grade Maraging Steel in Solution-Treated and Aged Conditions

Hugo Mikael Santos Amâncio, Tatiane de Campos Chuvas, Túlio Salek Quintas, Paola Rodrigues Pereira, and Maria Cindra Fonseca

Abstract The surface integrity plays an important role in the service life of machined components and is characterized by the residual stresses and surface roughness generated by the machining processes. In this context, the use of different cutting parameters can substantially impact machining operations and, consequently, influence the characteristics of the machined surface. The present work aims to study the machining of 350 grade maraging steel in solution-treated and aged conditions by end milling, with cemented carbide inserts, and grinding, with oxide aluminum (Al_2O_3) wheel, correlating the cutting conditions with the behavior of residual stresses and the surface roughness. All the operations were carried out with the application of flood coolant. The effects of cutting parameters were evaluated by Analysis of Variance (ANOVA). The residual stresses were measured by X-ray diffraction technique, using the sin2ψ method. The results indicated that both milling and grinding processes induced compressive residual stresses in the solution-treated and aged steel, however, a better surface finish was obtained by milling operation.

Keywords Residual stresses · End milling · Grinding · 350 grade maraging steel · X-ray diffraction

H. M. S. Amâncio · T. S. Quintas · P. R. Pereira · M. C. Fonseca (✉)
Universidade Federal Fluminense, Niterói, RJ, Brazil
e-mail: mariacindra@id.uff.br

T. de C. Chuvas
Centro Federal de Educação Tecnológica Celso Suckow da Fonseca, Rio de Janeiro, RJ, Brazil

V. E. Beal and A. S. Ribeiro Junior (eds.), *Advances in Manufacturing Engineering Research*, Lecture Notes in Mechanical Engineering,
https://doi.org/10.1007/978-3-032-09858-0_15

1 Introduction

Residual stresses (RS) are the stresses that remain in a component after manufacturing processes, in the absence of thermomechanical loadings (Tabatabaeian et al. 2022). According to Macherauch and Kloos (1987), these stresses are self-equilibrating, meaning that the resultant of the forces and moments they produce is zero. They can be classified in three groups: macrostresses (type I), which are distributed throughout the bulk material, microstresses (type II), distributed in the scale of a few grains and grain boundaries, and type III, sub-microstresses, are heterogeneous in submicroscopic areas of a material, at atomic distances within a grain.

The machining-induced residual stresses play an important role in the service life of machined components and can be either beneficial or detrimental, depending on their nature and magnitude. Tensile residual stresses can accelerate crack nucleation and propagation, while compressive residual stresses produce the opposite effect, extending the service life of components (Masmiati et al. 2016).

Maraging steels are high-alloy ultra-high-strength steels formed by a Fe–Ni alloy matrix with low carbon content. According to Fonseca et al. (2021) these steels are hardened by the precipitation of intermetallic compounds when age treated. The high-strength-to-weight ratio of these steels, compared with conventional steels, makes them ideal for manufacturing molds and injection dies, as well as high-performance components used in the nuclear and automotive industries (Abbas et al. 2023).

Machining of high-strength hardened steels, such as maraging steel, involves high cutting forces, leading to increased energy consumption, excessive heat generation in the cutting zone and rougher surface finish, which affect the residual stresses formation in the process (Abbas et al. 2023). The study of residual stresses focuses on cutting conditions as the dominant factors in their formation, since during machining, high strain rates and temperatures act as the mechanisms that generate these stresses, influencing the service life of machined components (Saptaji et al. 2019).

During machining, most of the energy consumed is converted into heat and transferred to the chip, workpiece and tool, with the largest amount of heat being dissipated by the chip. Heating the workpiece can help reduce the shear stress, but may cause dimensional distortions due to the thermal expansion and increase tool wear (Sales et al. 2020).

As in all machining processes, residual stresses induced by milling result from the interaction between mechanical and thermal effects, therefore the selection of cutting parameters plays a key role in the behavior of these stresses. Among these parameters, feed rate and cutting speed stand out as the most important factors that influence the residual stresses in milling (Jiang et al. 2016).

Grinding is usually the final step in the manufacturing of machined components, especially in critical applications in the automotive and aerospace industries. Inadequate selection of grinding parameters can increase the risk of thermal damage of the material and the generation of tensile residual stresses. Proper selection of cutting parameters is essential for obtaining compressive or less tensile residual stresses, without compromising the surface quality of the component (Shen et al. 2019).

The present work aims to study the influence of cutting parameters in the behavior of residual stresses induced by end milling and grinding of 350 grade maraging steel, in solution-treated and aged conditions.

2 Materials and Methods

The material used in this work was a 350 grade maraging steel in the conditions of solution-treated, at 910 °C for 1 h and aged at 510 °C for 6 h. The mechanical properties of the material in both conditions are presented in Table 1 and the chemical composition is presented in Table 2.

The end milling (EM) and grinding (G) experiments with solution-treated (ST) and aged (AG) samples were carried out with the cutting parameters presented in Tables 3 and 4, respectively. Flood lubrication method was used in all experiments, with a natural-based, biodegradable cutting fluid (Hydria EP), diluted in water at a ratio of 1:20.

Table 1 Mechanical properties of 350 grade maraging steel

Condition	Yield stress (MPa)	Ultimate tensile stress (MPa)	Hardness (HRC)
Solution-treated	951	1097	37
Aged	2195	2297	59

Source Manufacturer

Table 2 Chemical composition of 350 grade maraging steel (wt.%)

C	S	N	H	P	W	Co	Ni
0.0023	0.0006	7.03	0.81	0.005	0.026	11.65	17.65
Si	Mn	Cr	Mo	V	Ti	Cu	Al
0.039	0.021	0.052	4.69	0.1	1.44	0.009	0.065

Source Manufacturer

Table 3 Cutting parameters—end milling (EM)

Condition	Spindle speed, N (rev/min)	Cutting speed, V_c (m/min)	Feed per tooth, f_z (mm/tooth)	Feed rate, V_f (mm/min)	Depth of cut, a_p (mm)
EM-AG 1	1194	60	0.02	48	0.5
EM-AG 2			0.04	96	
EM-AG 3	1592	80	0.02	64	
EM-AG 4			0.04	127	
EM-ST 1	2500	126	0.02	100	1
EM-ST 2	3500	176	0.01	70	0.3

Table 4 Cutting parameters—grinding (G)

Condition	Wheel speed, V_c (rev/min)	Feed rate, V_f (m/min)	Depth of cut, a_p (mm)
G-AG 1	3500	11	0.03
G-AG 2		17	
G-ST 1		11	0.06
G-ST 2		17	

The end milling experiments were carried out on a VEKER MV-760ECO CNC milling machine using a Ø16 mm BAPZ300R162S16 mill cutter with two EDPT 10T308PDERHD KC725M carbide inserts, coated with titanium nitride (TiN) and a corner radius of 0.8 mm, manufactured by Kennametal, as shown in Fig. 1a and b. The grinding experiments were carried out on a VEKER RVK-4515 surface grinding machine with a Norton 1SG80-KUH aluminum oxide (Al_2O_3) wheel. The samples used for the milling and grinding experiments with the material in solution-treated condition (EM-ST1, EM-ST2, G-ST1, and G-ST2) were taken from the sample illustrated in Fig. 1c, while the samples used for experiments with the material in aged condition (EM-AG1, EM-AG2, EM-AG3, EM-AG4, G-AG1, and G-AG2) were as illustrated in Fig. 1d.

The machining forces in the milling experiments with the solution-treated material (EM-ST1 and EM-ST2) were measured using a three-component data acquisition system composed of a Kistler 9257-BA piezoelectric dynamometer, a signal

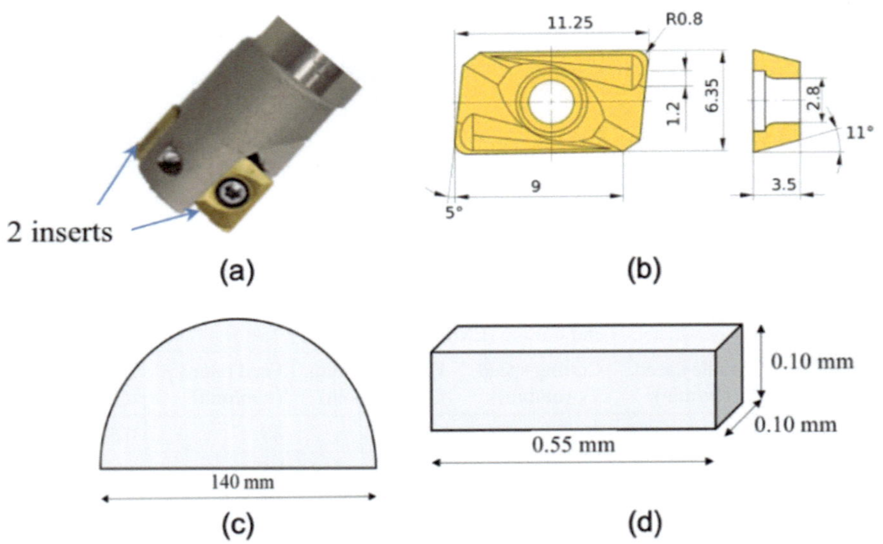

Fig. 1 Cutting tool used in the milling experiments: **a** Ø 16 mm mill cutter with two inserts, **b** carbide insert dimensions, **c** dimensions of solution-treated maraging steel sample, and **d** dimensions of aged maraging steel samples

Table 5 Parameter used in the residual stress analysis

Inclination angle ψ (°)	0, 20.7, 30, 37.7 and 45
Radiation	Crκα
Wavelength λ (Å)	2.29092
Exposure time (s)	25
X-ray tube voltage (kV)	30
X-ray tube current (mA)	6.7

amplifier, and an analog-to-digital converter. The force components were acquired in separate channels (F_x, F_y, and F_z), and the resultant cutting force was calculated using Eq. 1.

$$F_R = \sqrt{F_x^2 + F_y^2 + F_z^2} \tag{1}$$

The residual stresses were measured in the feed direction of the tool using the X-ray diffraction technique and the $\sin^2 \psi$ method, with the measurement parameters listed in Table 5. Measurements were taken at the center of each sample, avoiding the tool entry and exit regions, using the Xstress 3000 stress analyzer manufactured by Stresstech, as shown in Fig. 2.

The arithmetic mean surface roughness Ra was measured using a Taylor Hobson Surtronic 25 roughness tester, at four points spaced 10 mm apart from each other. The tester has a resolution of 0.01 μm and the evaluation length and cutoff were 4.00 mm and 0.80 mm, respectively.

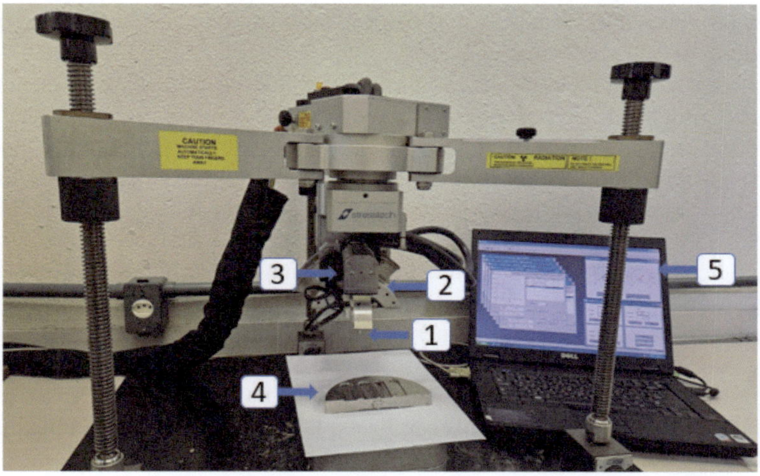

Fig. 2 Stress analyzer Xstress 3000: **a** collimator, **b** goniometer, **c** X-ray tube, **d** sample, **e** software

The parametric statistic tool Analysis of Variance (ANOVA) was applied to investigate the influence of the factors (cutting parameters) on the response variable residual stress. The normality of residuals and homocedasticity were verified with a significance level (α) of 0.05. The Shapiro–Wilk test was used to evaluate the normality of residuals, and the Cochran-Bartlett test was used to evaluate the homocedasticity of variances.

The analysis for the milling experiments with solution-treated maraging steel considered two levels of cutting conditions: low-level condition ($V_c = 126$ m/min; $a_p = 1$ mm and $f_z = 0.02$ mm/tooth) and high-level condition ($V_c = 176$ m/min; $f_z = 0.01$ mm/tooth and $a_p = 0.3$ mm), with 2 replicates, totaling 6 experiments. A full factorial design 22 with one replicate was applied for the aged maraging steel, totaling 8 experiments. The factors and levels considered were V_c (60 m/min and 80 m/min) and f_z (0.02 mm/tooth and 0.04 mm/tooth). One-way ANOVA was used for the grinding experiments with one replicate, totaling 4 experiments for each condition of the material, with feed rates V_f of 11 m/min and 17 m/min.

3 Results and Discussion

The results of residual stresses generated in the milling experiments are presented in Fig. 3. It is possible to observe that all cutting conditions resulted in compressive residual stresses. However, with an increase in cutting speed, compressive residual stress with higher magnitude was obtained in the solution-treated condition of the material, probably due to the use of flood cutting fluid, which reduced the thermal effects, making the plastic deformation effects dominant. These results are consistent with those reported by Tomaz et al. (2019), in their study on the surface integrity in end milling of 300 grade maraging steel.

Compressive residual stresses were also obtained in milling experiments with aged maraging steel. Considering that the aged steel exhibits higher hardness and mechanical strength, it requires higher machining forces for cutting to occur, which may explain why there was no significant difference in residual stresses when varying the cutting parameters within the employed range, as shown by ANOVA in Table 6. However, the smaller the *p*-value (< 0.05), the greater the influence of the analyzed factor, indicating that the feed per tooth (f_z) may have a greater effect, compared to cutting speed.

The results of residual stresses generated in the grinding experiments are presented in Fig. 4. Compressive residual stresses were obtained for the material in both solution-treated and aged conditions and increasing the table feed rate led to residual stresses of higher magnitudes. This can be explained by the fact that a higher feed rate increases the material removal rate, so a greater amount of heat is dissipated by the chip, thus reducing the thermal effects. As a result, mechanical effects became predominant, leading to more compressive residual stresses (Shen et al. 2019).

The results of machining force measurements in the milling experiments with the material in solution-treated condition are presented in Fig. 5. The complete signal

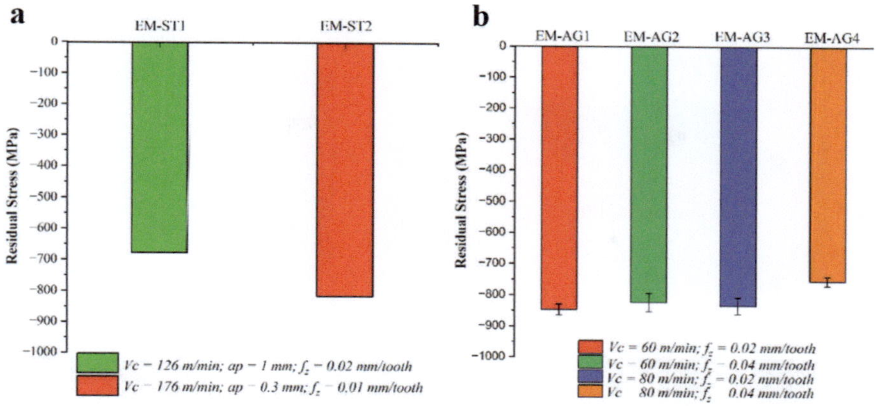

Fig. 3 Residual stresses induced by milling of 350 maraging steel in **a** solution-treated condition and **b** aged condition

Table 6 ANOVA for residual stresses in milling experiments

Factor	Degrees of freedom	Sum of squares	F	p-value
V_c	1	3321	0.2743	0.6281
f_z	1	5253	0.4339	0.5460
$V_c \times f_z$	1	1540	0.1272	0.7393
Residuals	4	48.424		

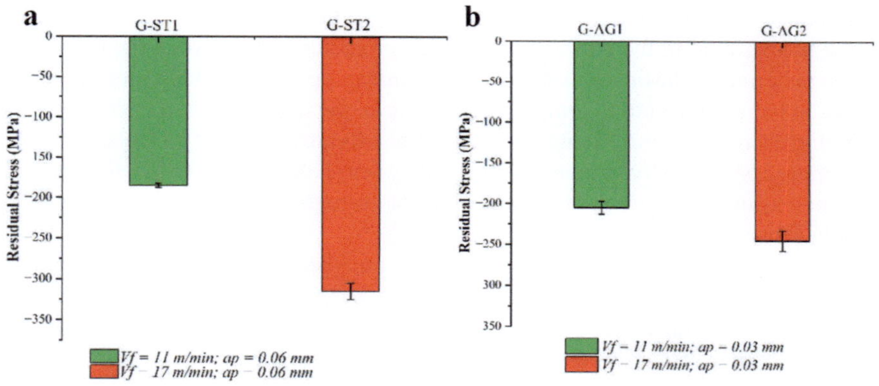

Fig. 4 Residual stresses in grinding experiments with 350 maraging steel in **a** solution-treated condition and **b** aged condition

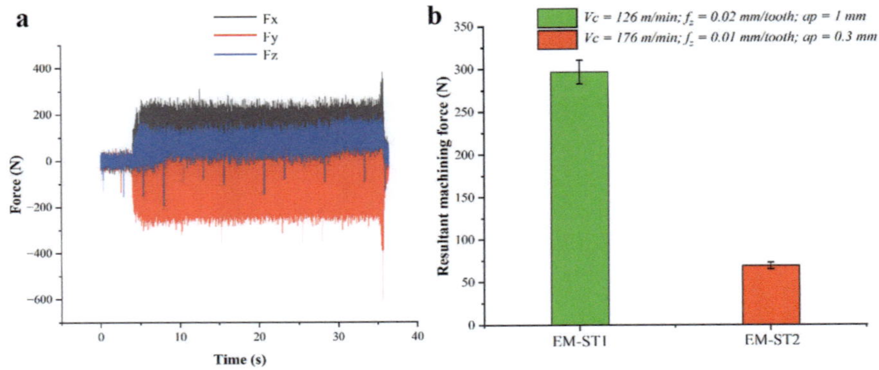

Fig. 5 Machining force: **a** complete signal for one sample, **b** resultant force

obtained for one sample is presented in Fig. 5a, showing the three components (F_x, F_y and F_z) and the resultant force is presented in Fig. 5b.

As shown in Fig. 5a, the force component F_z, which corresponds to the thrust force, is collinear with the tool axis and normal to the workpiece surface, contributing only to the resultant force.

From Fig. 5b, it can be observed that with higher feed, lower cutting speed and higher depth of cut there is an increase in material removal rate and in the resultant force. The opposite behavior is observed when cutting speed was increased simultaneously with a decrease in feed and depth of cut, resulting in a decrease in the resultant force of approximately 78%. Increasing cutting speed results in an increase of the temperature in the cutting zone, which promotes a decrease in the resistance of the material to plastic deformation. Since maraging steel has high thermal resistance, an increase in cutting temperature can make the material easier to cut, thereby reducing the required machining force. These results are consistent with results found in literature, specifically with those obtained by Varghese et al. (2019), who reported that increasing the cutting speed in milling of maraging steel led to a decrease in cutting forces.

When decreasing depth of cut, the contact area between tool and workpiece is reduced, resulting in a decrease of material removal rate and cutting forces. Similarly, feed per tooth influences this effect: increasing it causes each tooth to remove a greater volume of material per tool revolution.

The correlation between residual stresses and machining force is shown in Fig. 6.

From Fig. 6, it can be observed that a compressive residual stress of higher magnitude was generated when the resultant machining force was lower, which occurred with a higher cutting speed. It is well known that increasing the cutting speed results in a higher strain rate and higher cutting temperature, facilitating material removal. These results are consistent with those reported by Jiang et al. (2016), who investigated the effect of machining force and cutting temperature on residual stresses induced by milling. The authors calculated a correlation coefficient and observed

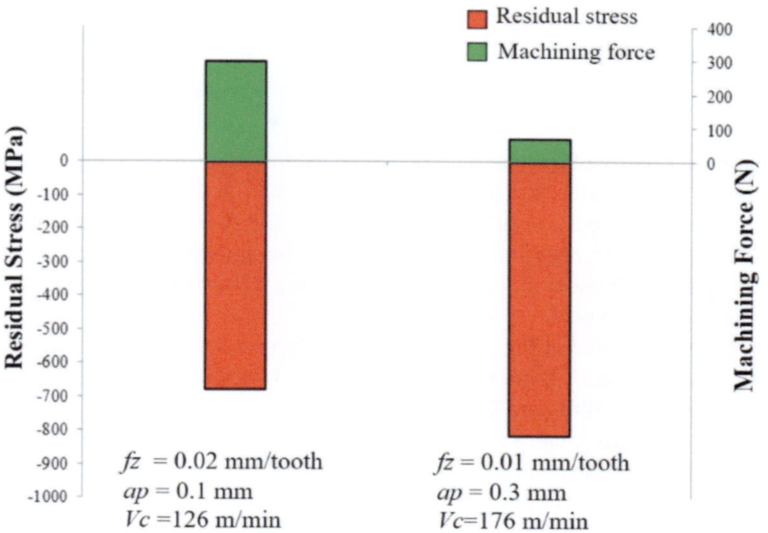

Fig. 6 Correlation between residual stress and machining force

that cutting force was predominant over thermal effects in the generation of residual stresses.

The mean value and standard deviation of surface roughness (Ra) obtained in each milling condition are presented in Fig. 7.

The results shown in Fig. 7a indicate that higher cutting speed, combined with lower feed rate and depth of cut, results in lower surface roughness, which characterizes a finer surface finish. This behavior can be attributed to the use of a new cutting edge in each operation, avoiding the effect of wear, as well as to the performance of the tool coating, which helps prevent the formation of a built-up edge (BUE),

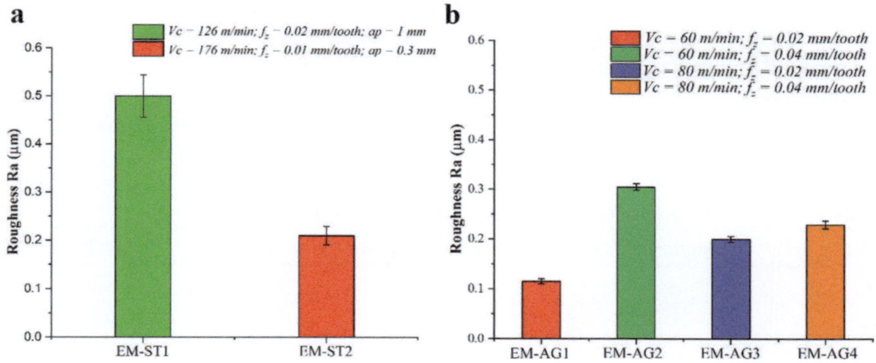

Fig. 7 Surface roughness (Ra) after milling of maraging steel in **a** solution-treated and **b** aged conditions

Table 7 Normality of residuals and homocedasticity tests

Normality of residuals	p-value	Homocedasticity	p-value
Shapiro–Wilks	0.5985	Cochran bartlett	0.8156

improving the quality of the machined surface. Additionally, decreasing depth of cut can improve process stability, as the tool interacts with a thinner layer of material, resulting in higher precision of the surface finish. The results of surface roughness obtained in the milling experiment with 350 grade maraging steel are similar to those reported by Padmalatha et al. (2017) when grinding 250 grade maraging steel with an aluminum oxide wheel.

The results presented in Fig. 7b, indicate that the best surface finish was achieved with the combination of higher cutting speed and lower feed rate. To verify these results, the Shapiro–Wilk and Cochran–Bartlett tests were performed, as shown in Table 6, and p-values greater than 0.05 qualified the data for parametric statistical analysis by ANOVA, as presented in Table 7.

Analyzing the results in Table 8, it can be stated that the variation of a parameter is significant if the p-value is lower than 0.05. Therefore, the feed per tooth (f_z) is the factor with the greatest impact on roughness during milling experiments, followed by its interaction with cutting speed.

The main effects of the factors for surface roughness are shown in Fig. 8, where the slope of the line represents the influence of the variation in factor levels. Figure 8a confirms the greater influence of f_z, indicating a better surface finish (lower roughness) with lower feed rates, while Fig. 8b shows the interaction of f_z with V_c, indicating that increasing both parameters simultaneously has a negative effect on roughness.

The average values of surface roughness Ra measured after the grinding experiments are shown in Fig. 9. The results show that the best surface finish (lower Ra value) was obtained with lower feed rate when grinding the material in both solution-treated and aged conditions. Although a greater difference can be observed for the aged condition, the statistical tests confirmed that there was no significant difference between the two levels of feed rate, as shown in Table 9.

In general, increasing the cutting speed and decreasing the feed rate and depth of cut resulted in lower machining forces and surface roughness, as well as compressive residual stresses of higher magnitudes. In the context of manufacturing, these results are desirable, as the decrease in the resultant machining force results in lower energy

Table 8 ANOVA for surface roughness in milling experiments

Factor	Degrees of freedom	Sum of squares	F	p-value
V_c	1	0.00005	0.0690	0.8058
f_z	1	0.24200	33.3793	0.0044
$V_c \times f_z$	1	0.01280	17.6552	0.01368
Residuals	4	0.00290		

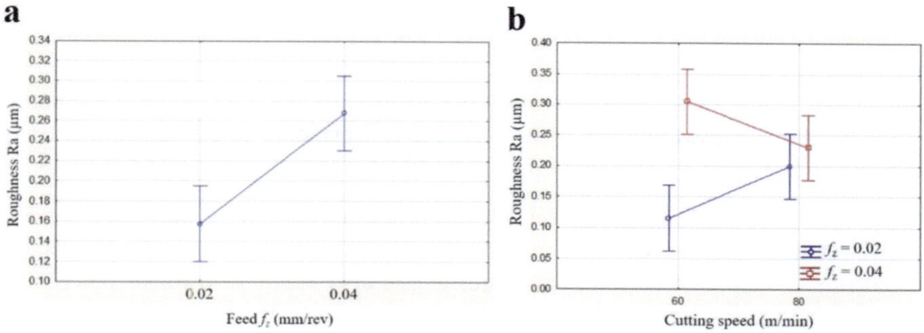

Fig. 8 Main effects: **a** effect of f_z **b** interaction effect of f_z and V_c

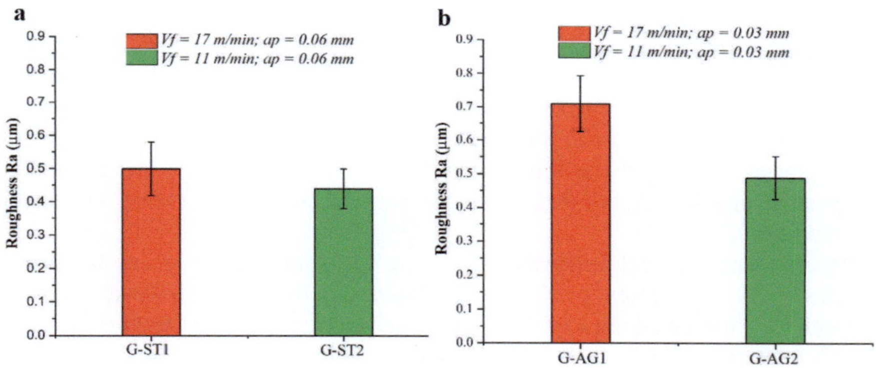

Fig. 9 Surface roughness (Ra) after grinding of maraging steel in **a** solution-treated and **b** aged conditions

Table 9 ANOVA for surface roughness in grinding experiments

	Factor	Degrees of freedom	Sum of squares	F	p-value
Solution-treated	V_f	1	0.0049	0.3643	0.607
	Residuals	2	0.0269		
Aged	V_f	1	0.44100	4.3663	0.1718
	Residuals	2	0.02020		

consumption, while surface compressive residual stresses contribute to extending service life by retarding or even preventing crack nucleation and propagation (Sutanto and Madl 2018).

4 Conclusions

This study, which experimentally investigated the influence of cutting parameters on the residual stresses induced by end milling and grinding of 350 grade maraging steel in solution-treated and aged conditions, allows the following conclusions:

- End milling of 350 maraging steel in solution-treated conditions induced compressive residual stresses of higher magnitudes when combining higher cutting speed with lower depth of cut and feed rate, which also resulted in lower machining forces. Compressive residual stresses were also generated in the milling of maraging steel in aged condition, however, no significant variation in the magnitude of the stresses was observed for the different cutting conditions employed. Grinding of the material in both solution-treated and aged conditions induced compressive residual stresses and the stresses with higher magnitudes were obtained when using higher feed rate.
- For the maraging steel in solution-treated condition, milling with lower feed rate and depth of cut combined with higher cutting speed resulted in lower surface roughness. In the aged condition, each parameter combination resulted in different roughness values, with the best result obtained for the lowest feed rate and cutting speed. In grinding experiments, lower surface roughness was obtained with lower feed rate.
- Statistical analysis confirmed that there were no significant differences in residual stresses for the different cutting conditions used in this work. However, feed per tooth (f_z) presented the lowest p-values, indicating a tendency for this factor to influence more significantly the behavior of residual stresses when varied over a wider range. The influence of cutting speed (V_c) as a second-order factor in interaction with f_z was more significant than its influence as a main factor, indicating that it affects the behavior of the residual stresses more expressively when it is varied simultaneously with feed per tooth.

Acknowledgements This study was financed in part by the Coordenação de Aperfeiçoamento de Pessoal de Nível Superior—Brasil (CAPES)—Finance Code 001. The authors would also like to thank the Brazilian research agencies CNPq (Process 304327/2021-2) and FAPERJ, for the financial support.

Responsibility Notice The authors are the only responsible for the material included in this work.

References

Abbas AT, Helmy MO, Al-Abduljabbar AA, Soliman MS, Hasan AS, Elkaseer A (2023) Precision face milling of maraging steel 350: an experimental investigation and optimization using different machine learning techniques. Machines. https://doi.org/10.3390/machines11111001

Fonseca DPMD, Feitosa ALM, Carvalho LGD, Plaut RL, Padilha AF (2021) A short review on ultra-high-strength maraging steels and future perspectives. Mater Res 24(1):e20200470. https://doi.org/10.1590/1980-5373-mr-2020-0470

Jiang X, Li B, Wang L, Wang Z, Li H (2016) An approach to evaluate the effect of cutting force and temperature on the residual stress generation during milling. Int J Adv Manuf Technol 87(5–8):2305–2317. https://doi.org/10.1007/s00170-016-8605-5

Macherauch E, Kloos KH (1987) Origin, measurements and evaluation of residual stress. Sci Technol

Masmiati N, Sarhan AAD, Hassan MAN, Hamdi M (2016) Optimization of cutting conditions for minimum residual stress, cutting force and surface roughness in end milling of S50C medium carbon steel. Measurement 86:253–265. https://doi.org/10.1016/j.measurement.2016.02.049

Padmalatha T, Jalli M, Konapalli S (2017) Influence of process parameters of surface grinding on maraging steel. Int J Mater Sci 12:127–135

Sales WF, Schoop J, da Silva LRR, Machado AR, Jawahir IS (2020) A review of surface integrity in machining of hardened steels. J Manuf Process 58:136–162

Saptaji K, Afiqah SN, Ramdan RD (2019) A review on measurement methods for machining induced residual stress. IJoCED 1(2):106. https://doi.org/10.35806/ijoced.v1i2.64

Shen S, Li B, Guo W (2019) Residual stresses distributions in grinding of 3J33 maraging steel with miniature electroplated CBN wheel. MATEC Web Conf 256:01002

Sutanto H, Madl J (2018) Residual stress development in hard machining—a review. In: IOP Conference series: materials science and engineering, vol 420, p 012031. https://doi.org/10.1088/1757-899X/420/1/012031

Tabatabaeian A, Ghasemi AR, Shokrieh MM, Marzbanrad B, Baraheni M, Fotouhi M (2022) Residual stress in engineering materials: a review. Adv Eng Mater 24(3):2100786. https://doi.org/10.1002/adem.202100786

Tomaz ÍV, Pardal JM, Cindra Fonseca M (2019) Influence of minimum quantity lubrication in the surface quality of milled maraging steel. Int J Adv Manuf Technol 104(9–12):4301–4311. https://doi.org/10.1007/s00170-019-04262-4

Varghese V, Akhil K, Ramesh MR, Chakradhar D (2019) Investigation on the performance of AlCrN and AlTiN coated cemented carbide inserts during end milling of maraging steel under dry, wet and cryogenic environments. J Manuf Process 43:136–144. https://doi.org/10.1016/j.jmapro.2019.05.021

Removal Mechanisms in Sintered Alumina by Flat Lapping with Cycloidal Kinematics

Marcelo Alem Foloni, Arthur Alves Fiocchi, Thiago Ferreira Querino, Thiago Valle França, Luiz Eduardo de Ângelo Sanchez, and Carlos Alberto Fortulan

Abstract Hardness and brittleness of advanced sintered ceramics pose significant machining challenges, typically performed by abrasive processes. Widely used in industry, flat lapping reconciles shape correction and finishing of difficult-to-machine materials. Traditionally, lapping employs oils as a lapping medium that are harmful to the environment when not disposed of properly, which increases production costs. On the other hand, there is growing demand for sintered ceramic parts and components for the mechanical, electronic, and photonic industries that require sustainable processes to remove material without introducing microstructural defects. Within this context, this study investigates the flat lapping of polycrystalline alumina disks using a CNC lap-grinding machine reconfigured for precision ceramic lapping. The alumina workpieces, produced via single-action uniaxial cold pressing at 80 MPa and sintered at 1600 °C for two hours, exhibited a polycrystalline microstructure composed of grain clusters (~ 100 μm), individual grains (~ 5 μm), and closed pores (~ 25 μm). Porosity negatively affected surface roughness, increasing Ra, Rq, Rz, Rt, and Rp values, while reducing Rv and Rsk and significantly increasing Rku. Lapping tests were conducted using an aluminum plate, single-crystalline diamond abrasives (54 μm), and distilled water as the lapping medium. The best surface finish was achieved at 25 kPa pressure, with Ra = 0.894 ± 0.166 μm and Rku = 13.866 ± 7.964, indicating plateau-like surfaces. Material removal mechanisms were predominantly brittle, governed by three-body abrasion, with limited ductile contributions from two-body abrasion. Higher pressures enhanced plastic deformation and widened removal tracks, improving surface quality without compromising finish. Material removal rates (MRRs) were up to three times higher at 25 kPa compared to 20 kPa. The use of distilled water proved effective and sustainable, eliminating

M. A. Foloni (✉) · T. V. França · L. E. de Ângelo Sanchez
São Paulo State University, Bauru, SP, Brazil
e-mail: marcelo.foloni@unesp.br

A. A. Fiocchi · T. F. Querino
Federal University of Uberlândia, Uberlândia, MG, Brazil

C. A. Fortulan
São Paulo University, São Carlos, SP, Brazil

V. E. Beal and A. S. Ribeiro Junior (eds.), *Advances in Manufacturing Engineering Research*, Lecture Notes in Mechanical Engineering,
https://doi.org/10.1007/978-3-032-09858-0_16

the need for solvent-based cleaning and reducing environmental impact. The Fiocchi CNC machine demonstrated high precision despite low structural rigidity, enabled by engineering solutions. The results confirm the feasibility of achieving deterministic, efficient, and sustainable ceramic machining.

Keywords Advanced ceramic · Removal mechanism · Ductile-to-brittle transition · Cycloidal flat lapping · Sustainable process

1 Introduction

The same intrinsic properties that make advanced ceramics highly desirable for a wide range of industrial applications—such as high hardness, thermal stability, and chemical inertness—also pose significant challenges to their processing. As a result, these materials are classified as difficult-to-machine (Rakshit and Das 2019; Bharathi et al. 2021), often requiring machine tools with exceptionally high structural rigidity to withstand elevated cutting forces and suppress vibrational instabilities during machining operations (Brinksmeier et al. 2010).

Abrasive machining plays a critical role in the production of ceramic components, particularly in achieving the dimensional precision and surface integrity required by design specifications following the sintering process (Pawar et al. 2015; Zhong 2003). However, to meet these stringent requirements, ceramics must be machined in a manner that avoids crack formation and minimizes subsurface damage. These challenges often render conventional machining techniques unsuitable, as they typically rely on brittle material removal mechanisms that compromise structural integrity (Alao 2024).

In the machining of brittle materials, material removal mechanisms are generally classified into two distinct modes: brittle and ductile. The brittle mode involves material removal through the initiation, propagation, and intersection of microcracks, which can compromise surface integrity and structural reliability. In contrast, the ductile mode is characterized by plastic deformation, where material is removed in the form of highly sheared chips, enabling smoother surfaces and reduced subsurface damage (Bifano et al. 1991).

Both brittle and ductile material removal mechanisms can occur during the lapping of brittle materials. Lapping is classified as a precision machining process in which abrasive grits are suspended in a fluid medium to form a slurry. These abrasives, distributed across the lapping plate, act as the cutting tools when the plate and abrasives encounter the surface of the workpiece. The effectiveness and outcome of the lapping process are influenced by several key input parameters, including: (a) the composition and concentration of the abrasive slurry, (b) the applied load on the workpiece surface, (c) the relative velocity between the workpiece and the lapping plate, (d) the kinematic motion pattern between the two surfaces, (e) the physical and mechanical properties of the materials involved, (f) the total lapping duration, and

(g) the functional requirements of the final component, which dictate the precision and surface finish needed (Chang et al. 2000).

The condition of the abrasives at the interface between the workpiece and the lapping plate determines the prevailing abrasion regime during the process. Two-body abrasion occurs when abrasives are fixed or embedded into the surface of the lapping plate, resulting in sliding contact with the workpiece. In contrast, three-body abrasion takes place when the abrasives are free to roll or move loosely between the workpiece and the lapping plate. The dominant abrasion mode is influenced by factors such as the ductility, malleability, and surface finish of the lapping plate, as well as the applied load or pressure during machining. Notably, both abrasion modes can simultaneously induce either brittle or ductile material removal mechanisms, depending on the specific process conditions (Chang et al. 2000; Touge and Matsuo 1996; Fiocchi 2014; Pyun et al. 2018; Kim et al. 2015).

The material removal rate (MRR), a key indicator of process productivity, tends to increase with higher applied loads and greater lapping plate speeds. However, MRR typically decreases over time as the process progresses. This behavior is observed in both two-body and three-body abrasion regimes, with two-body abrasion generally yielding higher MRR due to the fixed nature of the abrasives. Despite its lower MRR, three-body abrasion often results in inferior surface quality, leading to increased roughness on the machined workpiece (Kim et al. 2015).

Various abrasives and machining techniques have been developed and investigated over time to enable ductile-mode material removal in brittle and hard-to-machine materials. However, the application of cycloidal flat lapping using an aluminum lapping plate and distilled water as the lapping medium remains scarcely explored, particularly regarding the correlation between key process parameters and the material removal mechanisms in sintered alumina. Furthermore, there is a pressing need to optimize the cycloidal lapping parameters to minimize the introduction of microstructural defects in sintered alumina components.

Therefore, this study aims to investigate the influence of lapping pressure on the material removal mechanisms in sintered alumina during cycloidal flat lapping using an aluminum lapping plate. The objective is to identify process conditions that promote ductile-mode removal, enabling the generation of high-quality surfaces through an environmentally sustainable and deterministic approach, while minimizing the induction of microstructural defects that may compromise the performance of ceramic components under both static and dynamic loading conditions.

2 Materials and Methods

The Fiocchi lap-grinding machine tool (Fiocchi 2010, 2014; Fiocchi et al. 2015, 2016) was employed in exploratory flat lapping experiments on sintered alumina to evaluate the ductile-to-brittle transition behavior. These tests aim to support future optimization of the abrasive process, targeting predominantly ductile-mode material removal.

Aluminum 5052F was selected as the lapping plate material, measuring 340 mm in diameter and 25.4 mm in thickness, owing to its favorable mechanical properties for abrasive anchoring and its lower cost relative to tin and copper. Its hardness and malleability promote plastic deformation under compressive loads, enabling effective mechanical retention of abrasives and facilitating two-body abrasion with a predominance of ductile-mode material removal via micro-cutting.

Prior to testing, the flat surfaces of the lapping plates were precision-faced on a lathe to correct deviations in flatness and surface roughness. The lapping plate was mounted on the machine's plate holder over a 5 mm EVA disk to compensate for geometric misalignments in the mechanical interface and to dampen vibrations. A containment barrier was installed along the plate's perimeter to retain the slurry—composed of fluid, abrasives, and debris—preventing its ejection during rotation. Additionally, a bulkhead (deflector) was employed to continuously redirect the slurry from the outer edge back to the lapping interface, ensuring consistent abrasive action throughout the process.

A two-step abrasive procedure was adopted: (1) conditioning of the lapping plate, followed by (2) flat lapping of sintered alumina disks. In the first stage, the lapping plate was textured using #100 SiC sandpaper bonded to a flat glass disk, which served as the sandpaper holder. This assembly was mounted onto the spindle via a self-alignment vacuum chuck (Fiocchi 2014; Fiocchi et al. 2016). For the conditioning process, both the lapping plate (N_l) and the sandpaper holder (N_s) were rotated at 100 rpm under a load of 50 N for 30 s, with 10 mL of distilled water uniformly applied to the plate surface prior to texturing. The purpose of conditioning was to modify the surface texture of the lapping plate by introducing randomly oriented scratches with geometries conducive to the mechanical lodging and anchoring of abrasives. This anchoring occurs through plastic deformation of the aluminum plate under compressive lapping loads, thereby promoting a two-body abrasion mechanism and favoring ductile-mode material removal.

In the subsequent lapping experiments, three sintered alumina workpieces were mounted in a custom-designed, mechanically compliant fixture fabricated via additive manufacturing using ABS thermoplastic. This workpiece holder was bonded to a glass disk and secured onto the vacuum chuck of the lap-grinding machine prior to each test. The fixture has a diameter of 120 mm, a thickness of 3.0 mm, and accommodates up to three workpieces with diameters of 40 ± 0.5 mm. During lapping, both the workpiece holder (N_w) and the lapping plate (N_l) were rotated at 100 rpm. One workpiece was designated as sacrificial and used throughout all tests, while the remaining two were unique to each experimental condition and labeled as workpieces 1 through 4. The conditioning and lapping procedures are schematically illustrated in Fig. 1.

Two distinct lapping conditions were investigated by varying the applied pressure and processing time. The lapping plate was textured only once, prior to the initial test, and remained unchanged throughout the subsequent experiments. Under each

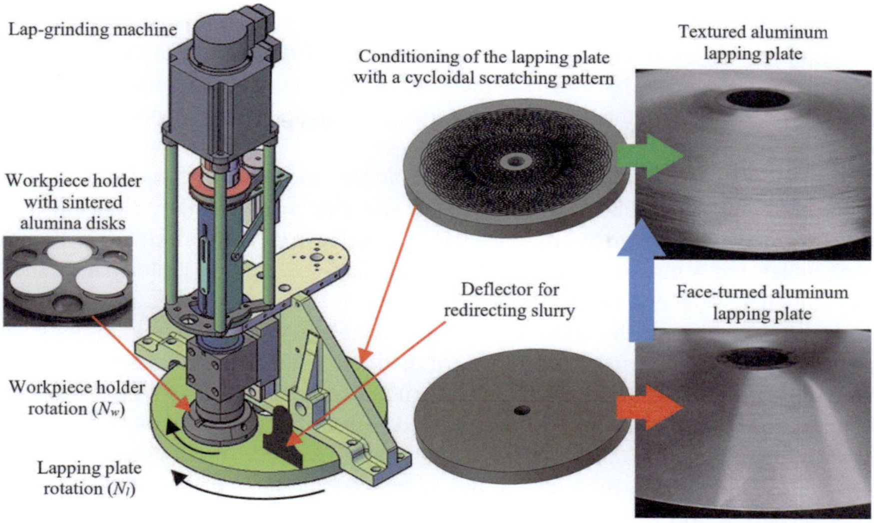

Fig. 1 Schematic representation of the main components of the Fiocchi CNC lap-grinding machine tool and the two-step process: (1) surface texturing of the aluminum lapping plate, and (2) flat-lapping of sintered alumina disks

condition, three sintered alumina workpieces were lapped for 30 min. A schematic cross-sectional representation of the lapping process is provided in Fig. 2.

The lapping process was periodically interrupted to measure the surface roughness and mass of the workpieces, enabling the calculation of the material removal rate (MRR) in grams per minute (g/min). Prior to and following the abrasive process, the workpieces were characterized using scanning electron microscopy (SEM) and profilometry by confocal microscopy (PCM). The surface roughness of the lapping

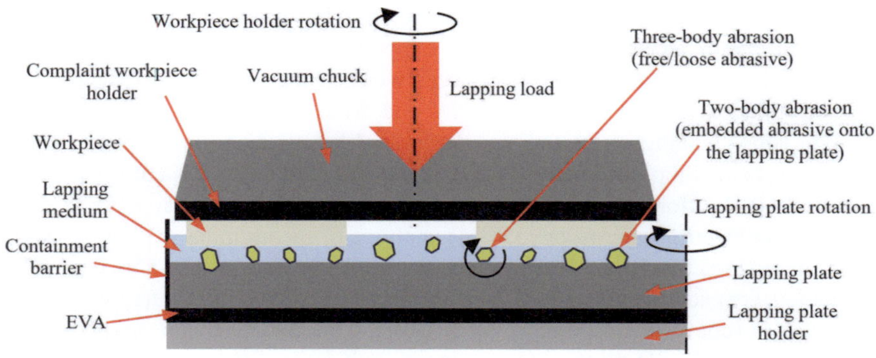

Fig. 2 Schematic cross-sectional representation of the flat lapping process with cycloidal kinematics, illustrating the aluminum lapping plate, abrasive grits, workpieces, workpiece holder, and the interaction mechanisms—two-body and three-body abrasion—acting on sintered alumina

plate was measured before and after the texturing procedure; however, post-lapping measurements were not performed to avoid potential damage to the contact-type roughness meter caused by abrasives anchored on the lapping plate surface. The input lapping parameters and the output characteristics evaluated on the lapped workpieces are summarized in Table 1.

Almatis A1000-SG calcined alumina workpieces were fabricated via single-action uniaxial cold pressing at 80 MPa, followed by sintering in atmospheric air at 1600 °C for two hours, as described by (Bukvic 2015), resulting in disks approximately 40 mm in diameter and 4 mm in thickness. After texturing, the lapping plate was cleaned using distilled water and vacuum suction to remove residual particles. Lapping preparation proceeded with the addition of 1 g of single-crystalline diamond abrasive (average grit size of 54 μm) and 25 mL of distilled water, applied with the lapping plate stationary. The three workpieces were mounted onto the vacuum chuck. Given that the lap-grinding machine tool is equipped with computerized numerical control (CNC), the application and removal of the lapping load via the workpiece holder feed were standardized. Both the lapping plate (N_l) and the workpiece holder (N_w) were rotated at 100 rpm, and the workpieces were fed toward the plate at a feed rate of 50 mm/min until the target lapping pressure was reached. Upon completion of the lapping cycle, the load was withdrawn and the workpieces retracted at the same feed rate, while rotation was maintained. Rotational motion ceased only after complete disengagement of the workpieces from the lapping plate. Considering the plate rotation, the containment barrier at its periphery, and the presence of the deflector, the abrasive slurry was uniformly distributed across the plate surface and continuously redirected toward the lapping interface, ensuring consistent abrasive action throughout the process.

The masses of the alumina disks and diamond abrasives were measured using a Shimadzu analytical balance, model ATX 224R, with a resolution of 0.0001 g. Surface roughness of both the workpieces and the lapping plate was assessed using a Taylor Hobson Surtronic 3+ roughness tester, equipped with a skidded diamond stylus featuring a tip radius of 2 μm. The roughness parameters Ra, Rq, Rz, Rsk, Rku, Rt, Rp, and Rv were measured following ISO 4288 (2008), employing a cut-off length of 0.8 mm, a Gaussian filter, and a sampling length of 4 mm.

Prior to characterization, the workpieces were cleaned in a Schuster ultrasonic cleaner, model L100, operating at 100 W and 50 kHz for 5 min using isopropyl alcohol. This cleaning procedure was performed before measurements of surface roughness, confocal profilometry (PCM), mass, and gold coating via physical vapor deposition (PVD) for scanning electron microscopy (SEM) analysis.

Scanning electron microscopy (SEM) analyses of the alumina workpieces were performed using Zeiss EVO MA10 and Tescan VEGA 3 microscopes. The acquisition parameters for each image are indicated within the respective micrographs.

Brightfield confocal profilometry (PCM) using white and blue light was performed with a Leica DCM 3D microscope equipped with 10 ×, 20 ×, and 100 × objectives, supported by Leica DCM3D software version 3.2.3 and TalyMap Gold version 5.0.3.4995.

Table 1 Experimental procedure for flat lapping with cycloidal kinematics, employing a malleable aluminum lapping plate for machining sintered alumina disks

Test		Input parameters		Output parameters
		Constant	Variable	
1	Workpiece 1 Workpiece 2	100 rpm for both the workpiece holder (N_w) and the lapping plate (N_l); 1 g single-crystalline diamond abrasive ($Ø = 54\ \mu m$); 25 mL distilled water; 30 min lapping time.	$P_1 = 20$ kPa	Roughness (Ra, Rq, Rz, Rsk, Rku, Rt, Rp, Rv); Alumina mass; Electrical average lapping power; Scanning electron microscopy (SEM); Profilometry by confocal microscopy (PCM).
2	Workpiece 3 Workpiece 4		$P_2 = 25$ kPa	

The average electrical power consumed by the three-phase induction motor during the 30-min lapping process—responsible for driving the lapping plate—was measured using a Ciber three-phase electrical multimeter, model CVM NGR-96, with an acquisition rate of 1 sample per minute.

All experiments were conducted under controlled environmental conditions of 20 ± 1 °C and relative humidity below 65 %, following ISO/IEC 17025:2017.

3 Results and Discussions

Representative scanning electron microscopy (SEM) micrographs of the sintered alumina workpieces at different magnifications are shown in Fig. 3. The microstructure of the polycrystalline alumina consists of grain agglomerates with heterogeneous dimensions and predominantly polygonal geometries, with an average agglomerate size of approximately 100 μm. Closed pores with diameters near 25 μm are homogeneously distributed along the grain boundaries and within the agglomerates. The average grain size is approximately 5 μm, with a predominantly equiaxed morphology.

The surface and subsurface pore density of the sintered alumina workpieces was considered high, adversely affecting the surface finish. This elevated porosity, along with the presence of grain agglomerates, was attributed to the compaction method employed. Single-action uniaxial pressing at 80 MPa was insufficient to eliminate inter- and intra-agglomerate porosity, thereby preserving the geometric signature of the green body agglomerates in the sintered microstructure and compromising densification during sintering. In addition to compaction-related defects, a density gradient was observed in the green body, which contributed to crack formation at the edges of some workpieces. These cracks typically originated at agglomerate boundaries during demolding, as illustrated in Fig. 3b. To improve green body uniformity and reduce defect formation, cold isostatic pressing is recommended as an alternative compaction technique.

SEM micrographs at four different magnifications for the four alumina workpieces lapped for 30 min under varying lapping pressures are presented in Fig. 4. The results indicate a predominance of brittle material removal mechanisms, primarily governed by three-body abrasion. Instances of plastic deformation—associated with micro-grooving, micro-ploughing, and micro-cutting—were less frequent, with micro-grooving being the most prevalent among them.

Micro-grooving, micro-ploughing, and micro-cutting are characteristic features of two-body abrasion, indicating that some abrasives were mechanically anchored—albeit temporarily—onto the lapping plate surface. This transient anchoring is suggested by the groove morphology, which typically exhibits lengths below 150 μm and non-linear trajectories, implying abrasive displacement during interaction. An increase in the normal force applied to the abrasives enhanced plastic deformation in the alumina, thereby increasing abrasive penetration. This deeper engagement of the abrasives promoted a shift in the dominant removal mechanisms: lower normal forces

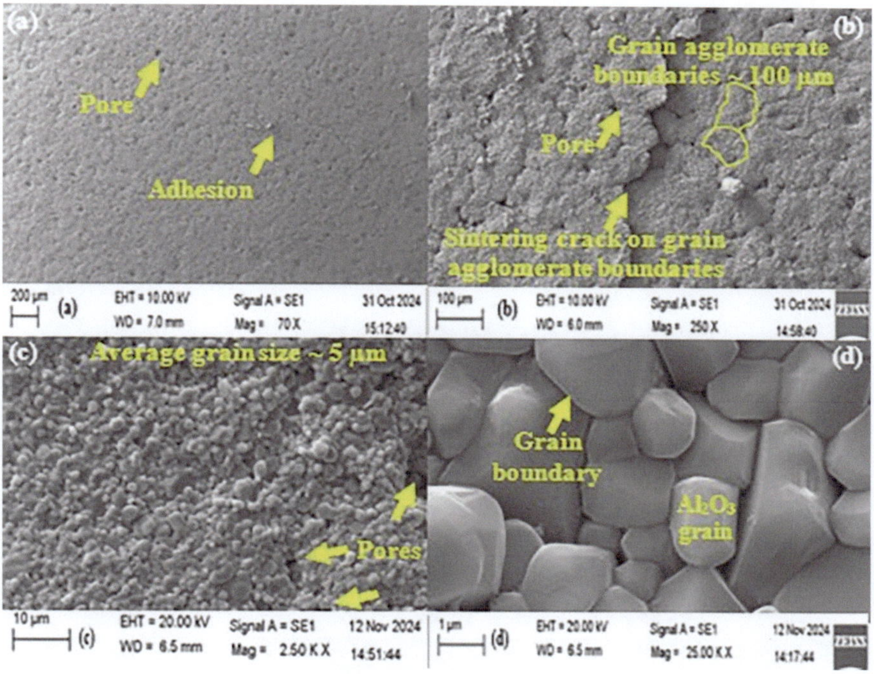

Fig. 3 Scanning electron microscopy (SEM) micrographs at different magnifications of polycrystalline alumina disks (Almatis A1000-SG calcined alumina), fabricated via single-action uniaxial cold pressing at 80 MPa and sintered in atmospheric air at 1600 °C for two hours

favored micro-grooving, whereas higher forces led to micro-ploughing and, subsequently, micro-cutting. The increased penetration resulting from elevated normal forces also induced the nucleation and propagation of intergranular and transgranular cracks, with intensified crack propagation observed during the unloading phase as the abrasives disengaged from the surface. Additionally, higher lapping pressures resulted in more pronounced plastic deformation of the alumina, evidenced by wider scratches across the polycrystalline microstructure.

The nature of lapping planetary kinematics, combined with a constant apparent lapping pressure (i.e., constant applied load without variation in the apparent contact area of the workpieces), promotes uniform surface abrasion in random directions, thereby producing an isotropic surface finish—independent of the roughness measurement direction. Successive localized plastic deformations introduce dislocations into the crystal lattice, increasing localized mechanical strength and, consequently, material brittleness. This microstructural evolution favors the predominance of brittle-mode material removal.

The mechanisms of micro-grooving and micro-ploughing do not necessarily result in immediate material removal in the form of chips. Instead, the accumulation

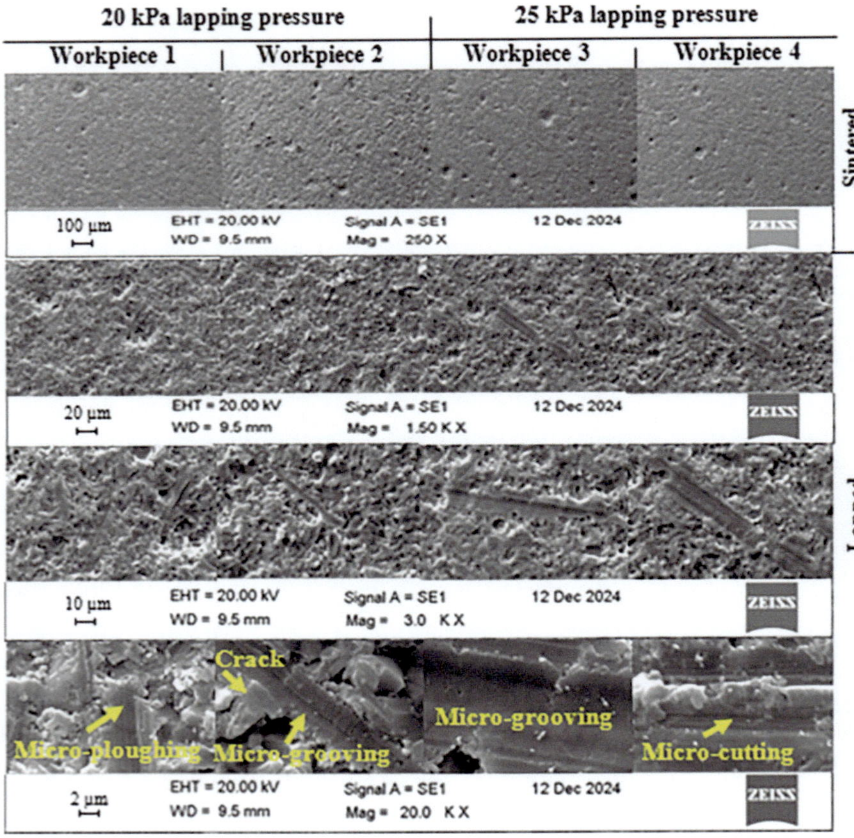

Fig. 4 Scanning electron microscopy (SEM) micrographs of polycrystalline alumina workpieces in the as-sintered condition (top row) and after 30 min of lapping on an aluminum 5052F lapping plate. The process employed 1 g of single-crystalline diamond abrasives (54 µm), rotational speeds of 100 rpm for both the workpiece holder (Nw) and the lapping plate (Nl), and 25 mL of distilled water as the lapping medium. Microstructural changes are shown as a function of lapping pressures of 20 kPa and 25 kPa

of localized deformation caused by repeated abrasive interactions leads to micro-cracking—characterized by the nucleation and propagation of cracks—which ultimately results in chip formation. This phenomenon, commonly referred to as material pulverization, is typical in abrasive machining of materials with high hardness and low fracture toughness, such as-sintered alumina. Micro-grooving induces localized stress that plastically deforms the grains of the alumina microstructure, producing shallow grooves with negligible edge elevation. In contrast, micro-ploughing involves more pronounced plastic deformation, generating elevated edges due to lateral material flow from the groove center—known as the ploughing effect. Micro-cutting occurs when the abrasive geometry and penetration depth exceed the critical threshold for chip formation, resulting in complete material shearing with minimal residual plastic deformation. The channels produced by micro-cutting are characterized by parallel striations and an absence of residual cracking on the newly formed surface.

To corroborate the ductile and brittle material removal mechanisms identified in the SEM micrographs presented in Fig. 4, additional SEM and PCM analyses are shown in Fig. 5 as a function of lapping pressure. These results highlight the prevalence of brittle mechanisms, including intergranular and transgranular fracture, as well as representative examples of ductile mechanisms such as micro-ploughing and micro-cutting.

The use of distilled water in the lapping process was justified by four fundamental roles: (1) as a lapping medium, it facilitates abrasive transport and, due to its lower viscosity compared to conventional lapping oils, reduces the hydrodynamic bearing effect on non-textured lapping plates. This enhances abrasive engagement with both the workpiece and the lapping plate by decreasing the thickness of the interface; (2) as a cutting fluid, in addition to aiding chip removal, it provides superior cooling performance relative to oils, albeit with limited lubricating capacity—which, in this study, was desirable to intensify abrasive action; (3) in alumina ceramics, distilled water promotes localized corrosion at crack's tips, accelerating crack propagation and thereby contributing to brittle-mode material removal; and (4) from a sustainability perspective, distilled water eliminates the need for oils and solvents during both lapping and post-process cleaning stages, reducing health risks in the workplace and minimizing environmental impact and production costs associated with improper disposal of chemical agents.

Average lapping powers of 160 W and 205 W were recorded for lapping pressures of 20 kPa and 25 kPa, respectively. Although acoustic emission (AE) was not instrumentally monitored, both lapping conditions produced an audible sound characteristic of three-body abrasion, with perceptibly greater acoustic energy at higher pressure. This sound is partially attributed to the release of acoustic energy associated with the rupture of strong chemical bonds in alumina. The crystal structure of alumina is composed of both ionic and covalent bonds: ionic bonds arise from electrostatic attraction between aluminum (Al^{3+}) and oxygen (O^{2-}) ions. In contrast, covalent bonds result from electron sharing between these atoms. Although ionic bonding predominates, the significant covalent character contributes to alumina's intrinsic properties, including high hardness, low fracture toughness, elevated melting point, corrosion resistance, chemical inertness, and electrical insulation.

Figure 6 presents the surface roughness profiles and corresponding parameters of the ceramic workpieces in three conditions: as-sintered (Fig. 6a), and after 30 min of lapping under applied pressures of 20 kPa (Fig. 6b) and 25 kPa (Fig. 6c). The roughness profiles shown correspond to the two workpieces that exhibited the lowest arithmetic mean roughness (Ra) values for each condition. The roughness parameters are reported as the arithmetic mean and standard deviation, calculated from five measurements taken on two distinct workpieces per lapping condition, resulting in a total of 20 measurements for each parameter. This methodology ensures statistical robustness and metrological reliability in the surface characterization of the lapped ceramics.

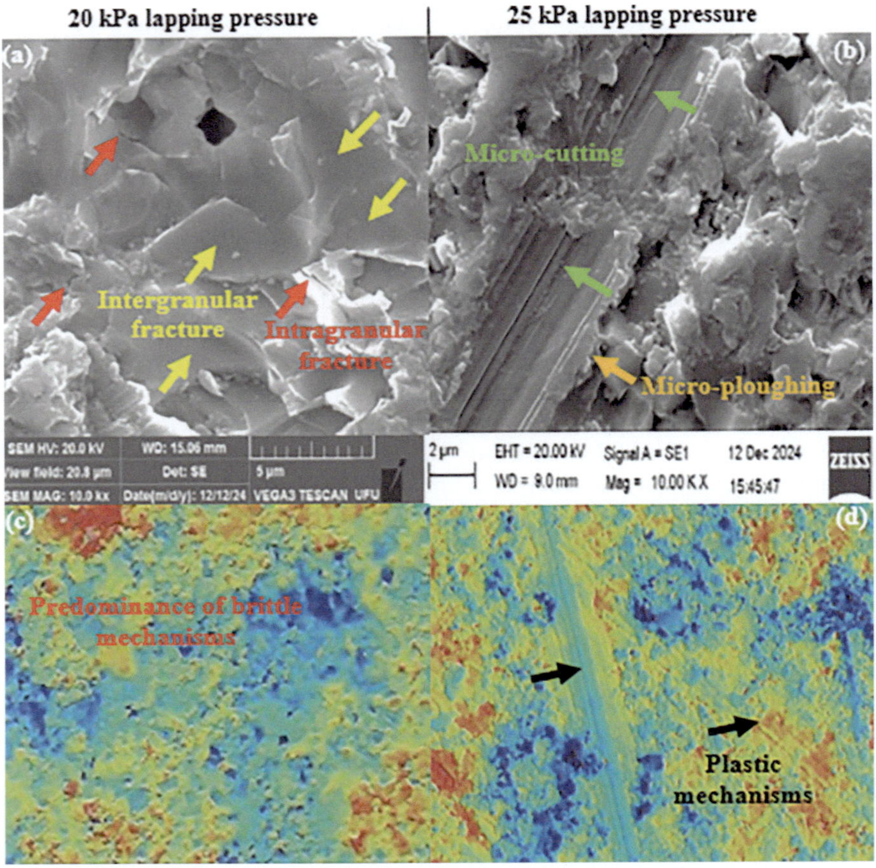

Fig. 5 Microstructural and topographical characterization of sintered polycrystalline alumina workpieces lapped for 30 min on an aluminum 5052F lapping plate using 1 g of single-crystalline diamond abrasives (54 μm), rotational speeds of 100 rpm for both the workpiece holder (N_w) and the lapping plate (N_l), and 25 mL of distilled water as the lapping medium. (**a**, **b**) Scanning electron microscopy (SEM) micrographs and (**c**, **d**) profilometry confocal microscopy (PCM) images acquired with a 100 × objective in bright field mode using (**c**) blue light and (**d**) white light, corresponding to lapping pressures of (**a**, **c**) 20 kPa and (**b**, **d**) 25 kPa

An increase in lapping pressure led to improved surface quality, as evidenced by the reduction in roughness parameters with increasing pressure, except for the kurtosis (Rku), as shown in Fig. 6b and c. The observed rise in Rku, along with the decrease in skewness (Rsk) to negative values, indicates the process's effectiveness in generating plateau-like surface topographies characterized by a predominance of valleys and the removal of surface peaks. This surface morphology is advantageous for tribological applications, as it promotes lubricant retention—an essential feature in the manufacturing of mechanical seals, bearings, and other sliding components.

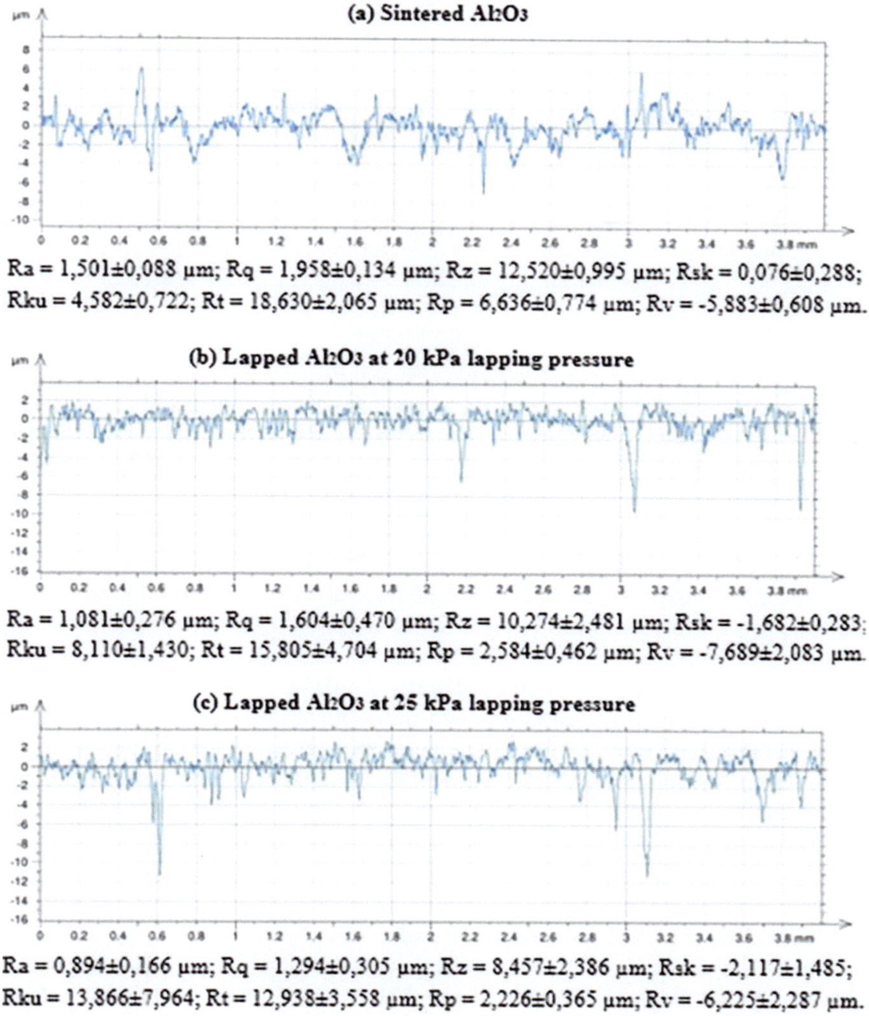

Fig. 6 Surface roughness profiles and corresponding roughness parameters of sintered polycrystalline alumina workpieces: **a** as-sintered, and **b**, **c** after 30 min of lapping on an aluminum 5052F lapping plate using 1 g of single-crystalline diamond abrasives (54 μm), rotational speeds of 100 rpm for both the workpiece holder (N_w) and the lapping plate (N_l), and 25 mL of distilled water as the lapping medium. Subfigures (**b**) and (**c**) correspond to lapping pressures of 20 kPa and 25 kPa, respectively

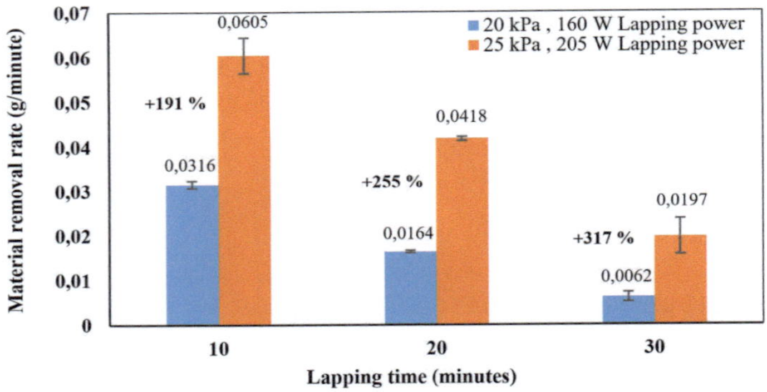

Fig. 7 Lapping power in Watts and material removal rate (MRR) in g/min of sintered polycrystalline alumina workpieces lapped for 30 min on an aluminum 5052F lapping plate using 1 g of single-crystalline diamond abrasives (54 μm), rotational speeds of 100 rpm for both the workpiece holder (N_w) and the lapping plate (N_l), and 25 mL of distilled water as the lapping medium. Lapping power and MRR are presented as a function of lapping time and applied pressures of 20 kPa and 25 kPa

Figure 7 illustrates the material removal rates (MRR) as a function of lapping time and applied lapping pressure. The tests conducted at 25 kPa exhibited higher MRR values compared to those performed at 20 kPa. For both pressure conditions, the MRR progressively decreased over the 30-min lapping duration. These findings are consistent with the results reported by (Kim et al. 2015), reinforcing the correlation between increased pressure and enhanced initial material removal, followed by a gradual decline due to surface stabilization and reduced abrasive efficiency.

The observed decline in material removal rate (MRR) over time can be attributed to the fact that abrasives and lapping medium were introduced only at the beginning of the lapping process and remained confined above the lapping plate throughout the operation. As lapping progressed, the slurry gradually accumulated alumina debris from the ceramic workpieces and aluminum particles or chips from the lapping plate itself. Although the aluminum plate experienced wear, it was less pronounced than that of the alumina due to the anchoring—or charging—effect of the abrasives. Additionally, the single-crystalline diamond abrasives used in the process exhibit lower friability compared to polycrystalline diamond abrasives, making them less prone to fracture and the formation of new cutting edges, which further contributes to the reduction in MRR.

It is noteworthy that the anchoring action of the abrasives effectively minimized wear on the aluminum lapping plate. No wear tracks or grooves were observed on the plate surface even after 90 minutes of continuous lapping. However, the slurry acquired a grayish coloration, confirming the presence of aluminum particles and indicating that material removal from the lapping plate did occur.

The Fiocchi lap-grinding machine tool, reconfigured for CNC-controlled lapping, demonstrated a notable capability for machining precision ceramic components, despite lacking the high structural rigidity typically associated with conventional grinding machines. This performance is primarily attributed to the combined stiffness of the workpieces themselves, which were consistently arranged in sets of three in a triangular configuration. The simultaneous lapping of three workpieces contributes to the stabilization of the abrasive removal process and effectively suppresses chatter, even in a machine with relatively low structural stiffness.

The integration of a self-aligning workpiece holder, a vacuum clamping system, and a compliant mechanism ensures additional process stability. These features collectively prevent elastic deformation of thin workpieces due to clamping forces and maintain consistent parallelism between the workpiece surfaces and the lapping plate. The degrees of freedom of the mechanical system allow dynamic self-adjustment, thereby improving surface uniformity.

As a result, the study demonstrated that deterministic, efficient, sustainable, and high-precision lapping outcomes can be achieved using domestically developed technologies. This contributes meaningfully to the advancement of Brazil's innovation ecosystem by aligning scientific and technological capabilities with strategic demands in the manufacturing of advanced materials.

4 Conclusions

The polycrystalline alumina workpieces analyzed in this study exhibited microstructures characterized by homogeneous grain agglomerates or grain clusters (~ 100 μm), individual grains (~ 5 μm), and closed pores (~ 25 μm), all of which are influenced by the ceramic processing route. The manufacturing process involved single-action uniaxial cold pressing at 80 MPa, followed by sintering in atmospheric air at 1600 °C for two hours.

The presence of porosity had a detrimental effect on surface roughness, leading to elevated average values of Ra, Rq, Rz, Rt, and Rp. Conversely, the parameters Rv and Rsk were reduced, while Rku showed a significant increase. These results suggest that the porous microstructure contributes to peak removal and valley formation, which is reflected in the statistical behavior of the roughness parameters.

The condition Rku > 3 combined with a reduction in Rsk < 0 demonstrates the effectiveness of the lapping process under the studied parameters in producing surfaces characterized by plateau-like topographies with a predominance of valleys and removal of surface peaks. This surface morphology is particularly advantageous for tribological applications, as it facilitates lubricant retention. Such characteristics are essential for components operating under sliding conditions, including hydrodynamic and hydrostatic bearings, where surface texture plays a critical role in minimizing friction and wear while enhancing operational efficiency.

The optimal surface roughness results were obtained under the following lapping conditions: highest applied pressure (25 kPa), 30 minutes of lapping, aluminum lapping plate, 25 mL of distilled water as the lapping medium, and single-crystalline diamond abrasives (54 μm). Under these parameters, the surface roughness values achieved were Ra = 0.894 ± 0.166 μm, Rq = 1.294 ± 0.305 μm, Rz = 8.457 ±

2.386 μm, Rsk = − 2.117 ± 1.485, Rku = 13.866 ± 7.964, Rt = 12.938 ± 3.558 μm, Rp = 2.226 ± 0.365 μm, and Rv = − 6.225 ± 2.287 μm.

Regarding the material removal mechanisms, the process was predominantly governed by brittle-mode removal through three-body abrasion, characterized by free-rolling abrasives. This mechanism facilitated both intergranular and transgranular microfracturing of the ceramic surface. In contrast, a lesser contribution was observed from ductile-mode mechanisms such as micro-grooving, micro-ploughing, and micro-cutting, which are typically associated with two-body abrasion involving abrasives anchored to the lapping plate. These findings highlight the influence of abrasive dynamics on the surface integrity of polycrystalline ceramics and reinforce the role of abrasive mobility in determining the dominant material removal mode during precision lapping.

Increasing the lapping pressure resulted in enhanced plastic deformation on both the surface and subsurface regions of the polycrystalline alumina workpieces. This led to the formation of wider material removal tracks within the microstructure, contributing to improved surface quality. These observations suggest that, under specific conditions, ductile-mode removal mechanisms—typically associated with higher pressures—can play a beneficial role in achieving superior surface finishes. Such mechanisms may be particularly advantageous for applications requiring smooth and uniform surfaces, including precision components in tribological systems.

Higher lapping pressure resulted in increased material removal rates (MRRs) without compromising surface finish quality. Specifically, the MRRs obtained at 25 kPa were approximately two to three times greater than those achieved at 20 kPa. However, a progressive decline in MRRs was observed throughout the lapping process, primarily due to the degradation of abrasive cutting efficiency. This reduction is attributed to edge wear of the abrasive grits and contamination of the slurry with alumina debris from the workpieces and aluminum particles from the lapping plate.

Compared to polycrystalline diamond abrasives, single-crystalline diamonds exhibit lower friability, making them less prone to microfracture and less capable of regenerating new cutting edges. This characteristic contributes to the gradual decrease in cutting performance over time, despite their initial effectiveness in producing high-quality surfaces.

The use of distilled water as both the lapping medium and cutting fluid demonstrated that minimal quantities—specifically 25 mL every 30 min—are sufficient to sustain the lapping process, underscoring its potential for environmentally conscious and resource-efficient lapping. This approach eliminates the need for solvent-based cleaning procedures typically required when oil-based fluids are used, as the workpieces exit the process clean. Furthermore, the absence of unpleasant odors and hydrocarbon emissions contributes to a healthier working environment and reduces overall manufacturing costs. These advantages position distilled water as a viable and sustainable alternative for precision abrasive machining of advanced ceramics.

The Fiocchi CNC lap-grinding machine tool, reconfigured specifically for lapping operations, demonstrated the capability to produce high-precision ceramic components despite its relatively low structural rigidity. This performance was enabled by a set of engineering design solutions that ensured consistent parallelism among the surfaces of the workpieces and the lapping plate. These included a self-aligning fixture system, vacuum clamping, and compliant mechanisms that collectively minimized deformation and enhanced process stability.

The findings confirm the process's ability to deliver deterministic, efficient, sustainable, and accurate results using entirely Brazilian-developed technology. This highlights the readiness of national innovation to meet strategic demands in advanced manufacturing and contributes to the strengthening of Brazil's technological and innovative ecosystem.

Future investigations are recommended to explore the effects of several variables that may further enhance the performance of the lapping process and surface integrity of ceramic components. These include:

(a) the use of alternative malleable materials for the lapping plate and different surface texturing strategies to improve the mechanical anchoring of the abrasives,
(b) the evaluation of a broader range of lapping pressures,
(c) the application of various abrasive types and grit sizes,
(d) the use of ecological and biodegradable lapping media and additives, and
(e) the study of different rotation ratios between the workpiece holder and the lapping plate, which influences the cycloidal trajectory patterns of the abrasives.

Such trajectory variations may promote more effective micro-cutting mechanisms and contribute to the homogenization of lapping plate wear, thereby minimizing the introduction of critical surface and subsurface defects. These improvements have the potential to enhance the functional performance of the final product under both static and dynamic operating conditions.

Acknowledgements Special thanks to Prof. Dr. Paulo Noronha Lisboa Filho (FC-UNESP) and Dr. Henry Hwang (FEMEC-UFU) who significantly contributed to the completion of this work; São Paulo State University "Júlio de Mesquita Filho" (UNESP), Post Graduate Program in Mechanical Engineering (PPGEM-UNESP), Department of Mechanical Engineering (FEB-UNESP); Laboratory of Machining Technology (LATUS-FEB-UNESP); Faculty of Science (FC-UNESP), Department of Physics and Meteorology (DFM-UNESP); Laboratory of Advanced Materials and Nanotechnology (FC-UNESP); University of São Paulo (USP), Department of Mechanical Engineering (EESC-USP), Laboratory of Tribology and Composites (LTC USP); Federal University of Uberlândia (UFU), Faculty of Mechanical Engineering (FEMEC-UFU), Post Graduate Program in Mechanical Engineering (PPGEM-UFU), Laboratory for Teaching and Research in Machining (LEPU-UFU), Metrology Laboratory (LEMTRO-UFU), FEMEC MAKER-UFU Advanced Manufacturing Laboratory; São Paulo State Research Foundation (FAPESP) (grants 2008/50233-3 and 2011/22486-7); Minas Gerais State Research Support Foundation (FAPEMIG) (grants APQ-01759-17 and APQ-00403-24); Coordination for the Improvement of Higher Education Personnel (CAPES); National Council for Scientific and Technological Development (CNPq); and EMBRAPII-FEMEC-UFU.

Acknowledgment of AI-Assisted Translation Review The authors acknowledge the use of generative artificial intelligence tools—ChatGPT 5.0 (OpenAI), Gemini 2.5 Pro (Google DeepMind), and Copilot (Microsoft)—to assist in the linguistic and stylistic revision of the English translation of this manuscript. These tools were employed to enhance clarity, technical accuracy, and consistency in terminology. All content generated or suggested by these tools was critically reviewed, validated, and edited by the authors to ensure scientific integrity and compliance with academic standards. The authors remain fully responsible for the accuracy and originality of the information presented.

References

Alao A-R (2024) Review of ductile machining and ductile-brittle transition characterization mechanisms in precision/ultraprecision turning, milling and grinding of brittle materials. Precision Engineering

Bharathi V et al (2021) A review on the challenges in machining of ceramics. Mater Today: Proc 46:1451–1458

Bifano TG, Dow T, Scattergood RO (1991) Ductile-regime grinding: a new technology for machining brittle materials

Brinksmeier E et al (2010) Ultra-precision grinding. CIRP Ann Manuf Technol 59(2):652–671

Bukvic G (2015) Green machining strategy for the design of ceramic parts. Thesis (PhD in Mechanical Design)—São Carlos School of Engineering, University of São Paulo, São Carlos

Chang YP, Hashimura M, Dornfeld D (2000) An investigation of material removal mechanisms in lapping with grain size transition. J Manuf Sci Eng 122(3):413–419

Fiocchi AA (2010) Advances in the development of a finishing process for flat parts subjected to lapping kinematics on a dressing wheel with a coating grade

Fiocchi AA (2014) Science and technology of ultra-precision manufacturing of advanced ceramics: Ud lapping of flat surfaces of yttria-stabilized polycrystalline tetragonal zirconia. 2014. Thesis (PhD in Mechanical Design)—São Carlos School of Engineering, University of São Paulo, São Carlos

Fiocchi AA, Fortulan CA, De Angelo Sanchez LE (2015) Ultra-precision face grinding with constant pressure, lapping kinematics, and SiC grinding wheels dressed with overlap factor. Int J Adv Manuf Technol 79(9–12):1531–1543

Fiocchi AA et al (2016) The ultra-precision Ud-lap grinding of flat advanced ceramics. J Mater Process Technol 231:336–356

ISO/IEC 17025:2017—General requirements for the competence of testing and calibration laboratories. International Organization for Standardization, 2017, Geneva, 3rd edn, p 30

ISO 4288:2008—Geometrical product specifications (GPS)—surface texture: profile method—rules and procedures for the assessment of surface texture. International Organization for Standardization, 2008, 2nd edn

Kim H-M et al (2015) Comparison between sapphire lapping processes using 2-body and 3-body modes as a function of diamond abrasive size. Wear 332:794–799

Pawar P, Ballav R, Kumar A (2015) An overview of machining process of alumina and alumina ceramic composites. Manuf Sci Technol 3(1):10–15

Pyun H-J et al (2018) Fabrication of high-performance copper-resin lapping plate for sapphire: a combined 2-body and 3-body diamond abrasive wear on sapphire. Tribol Int 120:203–209

Rakshit R, Das AK (2019) A review on cutting of industrial ceramic materials. Precis Eng 59:90–109

Touge M, Matsuo T (1996) Removal rate and surface roughness in high-precision lapping of Mn-Zn ferrite. CIRP Ann 45(1):307–310

Zhong ZW (2003) Ductile or partial ductile-mode machining of brittle materials. Int J Adv Manuf Technol 21:579–585

Influence of Dressing Depth on the Roughness and Surface Texture of Aeronautical Alloy During Grinding

Mayara Fernanda Pereira, Bruno Souza Abrão, Amanda Souza Abrão, Anna Beatriz Vaz Martins, Amauri Hassui, and Rosemar Batista da Silva

Abstract In the aerospace industry, the Ti-6Al-4V alloy stands out due to a unique combination of properties such as low density (4 g·cm^{-3}) and mechanical resistance at high temperatures allowing it to be employed in manufacturing of several components, such as engine blades and fuselage parts. Therefore, for the manufacture of these components, it is essential knowledge about the effects of the machining processes involved so that the tolerances specified in the project are met, mainly regarding roughness and dimensional deviations. However, this specific alloy is considered a material that is difficult to grind due to its low thermal conductivity (6.6 W·(m·K)$^{-1}$) and high chemical reactivity with the ceramic materials that make up the grinding wheels. One of the main parameters that influence grinding is dressing, a conditioning operation that directly influences the surface quality of the part being machined. In this sense, this work aimed to evaluate the effects that 2 different dressing depth conditions ($a_d = 20$ μm and $a_d = 40$ μm) can have on the surface quality of the Ti-6Al-4V aeronautical alloy after grinding. The tests were carried out with a silicon carbide grinding wheel (C46K5V5), cutting speed of 38.3 m/s, workspeed 7 m/min and radial depth of cut of 20 μm. Surface quality was evaluated in terms of the roughness parameter Ra and the texture of the samples obtained via Scanning Electron Microscopy (SEM) images. A methodology to evaluate the measurement uncertainty associated with the Ra roughness parameters was developed and the results showed that the change in dressing depth from a severest condition, dressing depth of $a_d = 40$ μm to a soft condition $a_d = 20$ μm resulted up to 25.51% reduction in roughness Ra. From the images of the samples was possible to identify an increase in number of the displacement and detachment of material from the sample, as well as the presence of thermal cracks after machining with under severest dressing condition.

M. F. Pereira (✉) · B. S. Abrão · A. S. Abrão · A. B. V. Martins · R. B. da Silva
Federal University of Uberlândia, Uberlândia, MG, Brazil
e-mail: mayara.pereira@ufu.br

A. Hassui
University of Campinas, Campinas, SP, Brazil

V. E. Beal and A. S. Ribeiro Junior (eds.), *Advances in Manufacturing Engineering Research*, Lecture Notes in Mechanical Engineering,
https://doi.org/10.1007/978-3-032-09858-0_17

Keywords Grinding · Ti-6Al-4V alloy · Dressing · Dressing depth · Finishing · Uncertainty · Measurement

1 Introduction

In the aerospace industry, certain components of aircraft engines, as well as parts of the fuselage, are manufactured from titanium alloys (Leyens and Peters 2003). These materials stand out in such applications due to their low density (4 g·cm^{-3}), high mechanical strength (1200 MPa) at elevated temperatures around 600 °C, and excellent resistance to corrosion and oxidation (Leyens and Peters 2003). Additionally, the Ti-6Al-4V titanium alloy is particularly employed in prostheses and human implants because of its excellent biocompatibility (Anand et al. 2019).

However, titanium alloys have certain characteristics that make them difficult to grind, such as their low thermal conductivity (6.6 W·(m·K)$^{-1}$), which limits the dissipation of heat generated during the cutting process, causing it to remain concentrated on the surface. This can lead to thermally induced damage and render the component unusable (Mello et al. 2017).

Furthermore, due to the high titanium content in their composition, most titanium alloys react chemically with the majority of cutting tool materials at temperatures above 550 °C, especially with the conventional ceramics used in grinding wheels (Motonishi et al. 1987; Silva et al. 2007).

Titanium alloys also exhibit high mechanical strength even at elevated temperatures, which hinders chip formation during machining (Mukhopadhyay and Kundu 2019).

Since the dimensional and geometric tolerances required in aircraft design are extremely tight, for example in turbine blades that require IT03 quality, titanium alloys such as Ti-6Al-4V are usually subjected to grinding in order to ensure low roughness values (Ra below 1.60 μm) combined with strict dimensional tolerances (IT06–IT03) and narrow macrogeometric deviations (Klocke 2009).

These characteristics are ensured in the machined material due to the mechanism of the grinding process, in which the grinding wheel operates at high cutting speeds on the workpiece, which may or may not move simultaneously. From this interaction of movements, chips of small dimensions are generated. The grinding wheel is composed of numerous abrasive grains with irregular geometry but multiple cutting edges, whose sharp edges act on the workpiece to remove material (Malkin and Guo 2008). However, the grinding process, particularly for the Ti-6Al-4V titanium alloy, is challenging for several reasons: both the workpiece and the grinding wheel materials have low thermal conductivity, the chips are small, and it is difficult to ensure cutting fluid jet speeds equal to the grinding wheel's cutting speed so that the fluid reaches the cutting zone and breaks the air barrier that is generally formed around the wheel. As a result, when heat is generated, a large portion of it, about 85% when using conventional grinding wheels, is directed to the workpiece.

When grinding aerospace alloys such as Ti-6Al-4V, greater attention must be given by machining operators, especially regarding the selection of operating parameters, with particular emphasis on dressing, to ensure the workpiece is free of thermal damage and to avoid financial losses, since the value of this alloy is much higher than that of hardened steels and other metals commonly ground (Mukhopadhyay and Kundu 2019).

In this context, the dressing process, considered the key factor in grinding, has a direct influence on the ground surface. This is because the interaction between the dresser (dressing tool) and the rotation of the grinding wheel generates a helical profile in the cutting zone of the grinding wheel, which is directly imprinted on the workpiece surface during grinding (Klocke 2009).

Depending on the dressing parameters selected, the grinding process can be classified as roughing or finishing. In the first case, dressing causes macrofracture of the grinding wheel's abrasive particles and is therefore used for operations involving a high material removal rate. In the second case, the grinding wheel generates microfractures of the abrasive particles and is used to produce surfaces with lower roughness values and narrower dimensional deviations. Among the dressing parameters that directly influence the conditioning of the grinding wheel, the most notable are the dressing overlap ratio and the dressing depth (Klocke 2009).

Several studies have been conducted to evaluate the effects of dressing depth on the surface roughness of titanium alloys after grinding.

In a study conducted by Mukhopadhyay et al. (2018), the authors evaluated the influence of dressing depth conditions of the CGC60K5V8 grinding wheel on the surface quality of ground Ti-6Al-4V. They used five different dressing depth (a_d) values: 5, 10, 15, 20, and 25 μm, and observed that in the mildest condition ($a_d = 5$ μm), the Ra roughness reached values close to 0.7 μm, whereas for higher a_d values, that is, more severe dressing depth conditions, the Ra roughness values were greater, averaging above 1.38 μm. According to the authors, increasing the dressing depth increases the friability of the abrasives, enhancing their cutting ability and, consequently, the grooves generated by the abrasive action on the workpiece tend to be deeper and exhibit greater lateral material displacement, which directly impacts the average surface roughness.

Another study conducted by Pereira et al. (2023) aimed to evaluate the effects of different dressing conditions of a silicon carbide (SiC) grinding wheel on the grindability of the aerospace alloy Ti-6Al-4V. For this purpose, the grinding wheel was dressed under different overlap ratio conditions ($U_d = 3$ and $U_d = 12$) and dressing depth conditions ($a_d = 20$ μm and $a_d = 40$ μm). Based on the analysis of roughness results, the authors found that changing the SiC wheel conditioning through dressing from an aggressive condition ($U_d = 3$ and $a_d = 40$ μm) to a mild condition ($U_d = 12$ and $a_d = 20$ μm) resulted in a 53.74% reduction in the Ra roughness parameter. In terms of the surface texture of the ground workpieces analyzed via Scanning Electron Microscopy (SEM), it was observed that this change in wheel conditioning produced more oriented grooves as well as a reduction in thermally induced damage.

According to Pereira et al. (2023), dressing directly affects the grinding wheel's cutting surface, and a lower overlap ratio generates larger and therefore more aggressive abrasive particles when using a roughing condition, which ensures higher material removal from the workpiece but results in a surface finish composed of more pronounced peaks and valleys, as well as discontinuities, lateral material displacement, and a higher propensity for thermally induced damage. On the other hand, when the SiC grinding wheel was subjected to a milder dressing, the cutting surface presented smaller and less aggressive abrasive particles, consequently producing a better surface finish on Ti-6Al-4V workpieces after grinding.

In this context, the present study aimed to analyze the effects of two different dressing depth conditions ($a_d = 20$ µm and $a_d = 40$ µm) applied to the dressing of a silicon carbide grinding wheel on the Ra roughness and surface finish of ground Ti-6Al-4V alloy. To evaluate Ra roughness, a measurement uncertainty analysis was carried out to quantify the sources of error involved in the measurement process with reliability and metrological traceability. In addition, Scanning Electron Microscopy images of the machined surfaces were obtained to assist in the analysis of workpiece surface texture.

2 Methodology

The grinding tests were carried out on a semi-automatic peripheral surface grinding machine with a resolution of 5 µm in the vertical axis that moves the grinding wheel spindle. The machine operates at a spindle rotation of 2400 rpm with a maximum power of 2.24 kW (3 Hp), model P36, manufactured by MELLO S.A., and is installed in the Conventional Machining Laboratory (LUC) at the School of Mechanical Engineering (FEMEC), Federal University of Uberlândia (UFU).

The material selected for the tests was the aerospace alloy Ti-6Al-4V. Two samples were prepared in a prismatic shape with a rectangular cross-section, with dimensions of 37 mm (length) × 15 mm (width) × 23 mm (height).

The tests were performed using a silicon carbide (SiC) grinding wheel manufactured by Alcar Abrasivos LTDA, with dimensions of 304.8 mm (external diameter) × 25.4 mm (width) × 76.2 mm (internal diameter), with the specification C46K5V5.

Prior to the tests, the grinding wheel was subjected to two different dressing conditions: a mild dressing depth, $a_d = 20$ µm, and a severe dressing depth, $a_d =$ 40 µm. The overlap ratio applied in this operation was $U_d = 12$, and the dressing was carried out using a single-point diamond dresser. All tests were conducted in the presence of a semi-synthetic cutting fluid applied by the conventional method, with an abundant flow rate of 9 L/min.

During the tests, the workpiece speed ($V_w = 7.5$ m/min), radial depth of cut ($a_e =$ 20 µm), and grinding wheel cutting speed ($V_s = 38.3$ m/s) were maintained constant.

As output variables, the Ra roughness parameter and the surface texture of the ground Ti-6Al-4V samples were evaluated using Scanning Electron Microscopy (SEM).

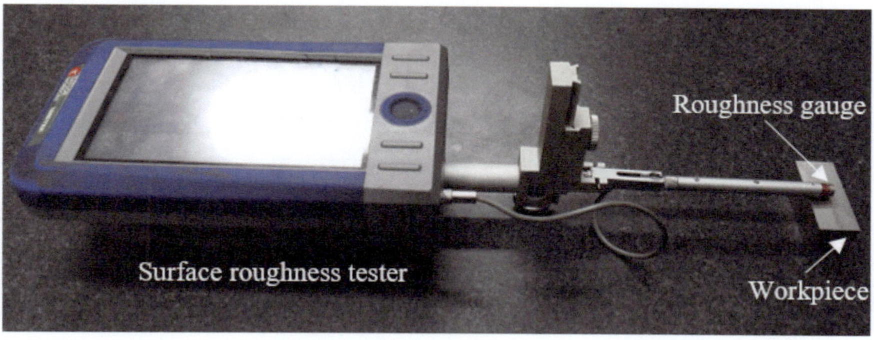

Fig. 1 Roughness measurement on the surface of a Ti-6Al-4V alloy workpiece after grinding with a SiC grinding wheel

After the grinding tests, each sample was taken to the Teaching Metrology Laboratory (LEMETRO at FEMEC—UFU), where, at a controlled temperature of (20 ± 1) °C, roughness measurements in terms of the Ra parameter were performed using a portable surface roughness tester Surtronic S-100 Series (n°. 010720) from Taylor Hobson, with a stylus tip radius of 2 μm and a resolution of 0.01 μm. For the measurements, a sampling length (cut-off) of 0.80 mm and an evaluation length of 4.00 mm were selected. Measurements were taken at five equally spaced regions on the ground surfaces so that the profilometer stylus moved perpendicular to the grinding direction, as shown in Fig. 1.

The Ra roughness parameter was selected for evaluation in this study because it is the most commonly assessed and compared parameter in the literature on the grinding of titanium alloys. This parameter, also known as the arithmetic mean deviation, is defined as the arithmetic mean of the absolute values of the ordinates $Z(x)$ over the sampling length (cut-off), as expressed in Eq. 1 (NBR ISO 4287, 2002).

$$\mathrm{Ra} = \frac{1}{\mathrm{l}} \int |Z(x)| \mathrm{d}x \tag{1}$$

where l = sampling length.

Next, an uncertainty analysis of the Ra roughness measurements was carried out using the profilometer, based on the methodology established in the Guide to the Expression of Uncertainty in Measurement (GUM). According to this methodology, several steps are fundamental for the calculation of measurement uncertainty (INMETRO 2012):

1. Definition of the measurand or output variable;
2. Identification of the variables that may influence the measurement result of the measurand, or input variables;
3. Development of a mathematical model of the measurand as a function of all input variables;
4. Evaluation of the standard uncertainty of the input variables;

5. Evaluation of the combined standard uncertainty related to the measurand;
6. Calculation of the sensitivity coefficients;
7. Evaluation of the expanded uncertainty with reference to the measurand;
8. Measurement result.

The mathematical model established for the uncertainty analysis of the Ra roughness parameter measured with the profilometer was based on five variables (considering that the instrument does not have a calibration certificate), specifically: the variability of the readings indicated by the roughness meter ($\overline{x}$), the resolution of the roughness meter (R), the uncertainty associated with the roughness meter calibration (C), the uncertainty associated with the stylus tip radius (Rp), and the uncertainty associated with the deformation of the workpiece material (D) (Abrão et al. 2023). In this case, the measurand, or output variable, is the Ra roughness parameter, and the mathematical model is given according to Eq. 2.

$$\mathrm{Ra} = \overline{x} + \Delta R + \Delta C + \Delta \mathrm{Rp} + \Delta D \tag{2}$$

To calculate the expanded uncertainty (U), the combined standard uncertainty (a_ c) must first be determined, which is given by Eq. 3:

$$\mathrm{uc2(Ra)} = u2(\overline{x}) + u2(\Delta R) + u2(\Delta C) + u2(\Delta \mathrm{Rp}) + u2(\Delta D) \tag{3}$$

After calculating the combined standard uncertainty (u_c), the effective degrees of freedom (ν_{eff}) are obtained, as given by Eq. 4:

$$\nu_{\mathrm{eff}} = \frac{u_c^4(y)}{\sum_{i=0}^{n} \frac{(u(x_i)\cdot c_i)^4}{\nu_i}} \tag{4}$$

Next, the expanded uncertainty (U(Ra)) can be calculated, as shown in Eq. 5, where k is the coverage factor according to the GUM, and the percentage contribution of each variable, Cont (%), is given by Eq. 6:

$$U(\mathrm{Ra}) = k \cdot \mathrm{uc} \tag{5}$$

$$\mathrm{Cont}(\%) = \frac{(u(x_i) \cdot c_i))^2 \cdot 100\%}{(u_c \cdot y)^2} \tag{6}$$

Table 1 presents the input variables, the estimates, the probability distribution (P.D.), the degrees of freedom (D.F.), the types of uncertainty evaluation (T.E.), the sensitivity coefficients (S.C.), and the calculations of standard uncertainty. In addition, the table includes the sample standard deviation (s), the surface roughness tester (Rrug), the roughness parameter (Prug), and the stylus elastic deformation (D), which is 20 nm (Haitjema 1998; Leach 2001).

Table 1 Estimates for the calculation of Ra roughness measurement uncertainty using a profilometer

Quantity	Estimate	D.P	G.L	T.A	C.S	Standard uncertainty
$\overline{x}$	$\overline{x}$	t-student	4	A	1	$u(\overline{x}) = \frac{s}{\sqrt{n}}$
ΔR	Rrug	Rectangular	∞	B	1	$u(\Delta R) = \frac{\text{Rrug}}{(2\cdot\sqrt{3})}$
ΔC	Rrug	Normal	∞	B	1	$u(\Delta C) = \frac{\text{Rrug}}{k}$
ΔRp	0.02·Prug	Rectangular	∞	B	1	$u(\Delta \text{Rp}) = \frac{(0.02\cdot\text{Prug})}{\sqrt{3}}$
ΔD	D	Rectangular	∞	B	1	$u(\Delta D) = \frac{\text{D}}{\sqrt{3}}$

After the analysis of the Ra roughness parameter, each Ti-6Al-4V workpiece was examined, and images were obtained using a Scanning Electron Microscope (SEM), HITACHI model TM 3000, at a magnification of 1500 times, so that the surface textures after grinding could be observed and analyzed.

3 Results and Discussions

In this section, the results are presented in terms of the Ra roughness parameter, as well as images of the surface textures of Ti-6Al-4V after grinding with a silicon carbide grinding wheel, which was conditioned under two different dressing depth conditions, $a_d = 20$ μm and $a_d = 40$ μm.

3.1 Surface Roughness

After measuring the Ra roughness parameter on the ground Ti-6Al-4V surfaces, an evaluation was carried out in terms of measurement uncertainty to ensure that the results were presented with reliability and metrological traceability. Figure 2 shows the Ra roughness measurement results as a function of the dressing depth conditions.

It is important to note that the measurement results presented in Fig. 2 are expressed in terms of the mean and expanded uncertainty U (error bars). According to Fig. 2, under the milder dressing depth condition ($a_d = 20$ μm), the Ra roughness measurement result was [0.73 ± 0.08] μm. However, when a more severe dressing depth was applied ($a_d = 40$ μm), the measurement result was [0.93 ± 0.15] μm. In both cases, the coverage factor k was 1.96, corresponding to a coverage probability of 95%.

Based on these results, it can be stated that changing the dressing of the SiC grinding wheel from a severe condition ($a_d = 40$ μm) to a mild condition ($a_d = 20$ μm) led to an average reduction of 25.51% in Ra roughness. Although the error

Fig. 2 Ra roughness parameter values as a function of dressing depth on ground Ti-6Al-4V alloy surfaces

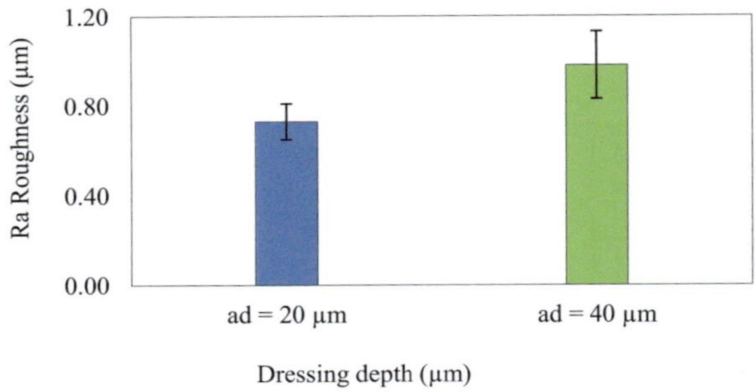

bars overlap at some points, the measurement uncertainty analysis shows that under the milder dressing depth condition, the values exhibit higher repeatability, indicating that this condition, in combination with the other operational parameters selected in this study, is the most suitable for dressing the SiC grinding wheel for Ti-6Al-4V alloy grinding.

Tables 2 and 3 summarize the data obtained in the uncertainty analysis of the Ra roughness parameter measured with a profilometer under the two dressing depth conditions employed, $a_d = 20$ μm and $a_d = 40$ μm.

In both dressing conditions, the variability of the readings was the factor that contributed the most to the measurement uncertainty. The contributions of this factor under the mild and severe dressing conditions were 85.51% and 94.75%, respectively (Tables 2 and 3). These high percentages may be associated with the macro- and micro-effects generated by the dressing after grinding, regarding the randomness of the peaks and valleys produced on the Ti-6Al-4V alloy surface by the action of the silicon carbide abrasives.

Table 2 Uncertainty evaluation data for Ra roughness measurement of Ti-6Al-4V alloy under the condition $a_d = 20$ μm

Quantity	Estimate (μm)	D.P	T.A	G.L	Standard uncertainty (μm)	Contribution (%)
$\overline{x}$	0.73250	t-student	*A*	3	0.03750	85.51452
ΔR	0.01000	Rectangular	*B*	∞	0.00289	0.50675
ΔC	0.01000	Normal	*B*	∞	0.00500	1.52026
ΔRp	0.01465	Rectangular	*B*	∞	0.00846	4.35042
ΔD	0.02000	Rectangular	*B*	∞	0.01155	8.10804
Combined standard uncertainty (u_c)						0.04055
Effective degrees of freedom (υ_{eff})						82,025.00157
Coverage factor (k)						1.95999
Expanded uncertainty (U)						0.07948

Table 3 Uncertainty evaluation data for Ra roughness measurement with a profilometer under the condition $a_d = 40$ μm

Quantity	Estimate (μm)	D.P	T.A	G.L	Standard uncertainty (μm)	Contribution (%)
$\bar{x}$	4.52500	t-student	*A*	3	0.22867	94.75116
ΔR	0.01000	Retangular	*B*	∞	0.00289	0.01510
ΔC	0.01000	Normal	*B*	∞	0.00500	0.04530
ΔRp	0.09050	Retangular	*B*	∞	0.05225	4.94684
ΔD	0.02000	Retangular	*B*	∞	0.01155	0.24160
Combined standard uncertainty (u_c)						0.23492
Effective degrees of freedom (U_{eff})						343.65209
Coverage factor (k)						1.96690
Expanded uncertainty (U)						0.46207

3.2 *Surface Texture After Grinding*

After the experimental grinding tests with a SiC grinding wheel conditioned under two different dressing depth conditions, images of the surface textures of Ti-6Al-4V were obtained, as shown in Figs. 3a and b, using Scanning Electron Microscopy (SEM).

In Fig. 3a, it can be observed that under the milder dressing depth condition of the grinding wheel ($a_d = 20$ μm), the action of the abrasives produced some deeper grooves with lateral material displacement, as well as some regions with discontinuities. However, when a more severe dressing depth was applied ($a_d = 40$ μm), Fig. 3b, the abrasive action resulted in greater material's lateral displacement along the grooves, which exhibited irregular behavior with evident discontinuities.

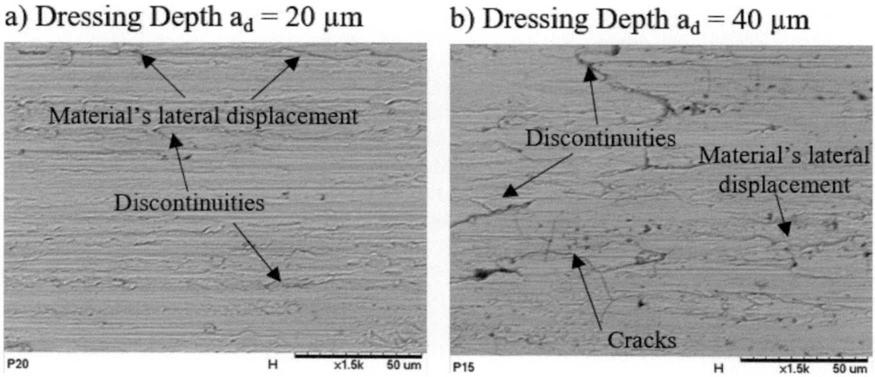

Fig. 3 Surface texture of Ti-6Al-4V alloy ground with a SiC grinding wheel dressed at a dressing depth $a_d = 20$ μm (**a**) and $a_d = 40$ μm (**b**)

In addition, thermally induced cracks perpendicular to the groove direction were observed. This behavior can be explained by the macro-effects generated in the SiC abrasives during dressing under the more severe condition ($a_d = 40$ µm). The SiC abrasives acquire a more friable and irregular geometry, and this cutting surface of the grinding wheel is imprinted onto the workpiece during the grinding process.

Additionally, the Ti-6Al-4V alloy is a difficult-to-machine material due to its low thermal conductivity (6.6 W·(m·K)$^{-1}$), high mechanical strength (1600 MPa) at elevated temperatures, and the high chemical reactivity of titanium with ceramic materials, such as silicon carbide (SiC), which constitutes the grinding wheel abrasives.

4 Conclusions

The following conclusions can be drawn from this study on the influence of SiC grinding wheel dressing depth on the grinding of Ti-6Al-4V alloy:

- Changing from a severe dressing depth condition ($a_d = 40$ µm) to a mild condition ($a_d = 20$ µm) resulted in an average reduction of 25.51% in the Ra roughness parameter.
- In both dressing depth conditions, the measurement uncertainty analysis showed that the values measured under the milder dressing condition exhibited higher repeatability. Therefore, it can be concluded that the condition $a_d = 20$ µm is more suitable for dressing the SiC grinding wheel and, consequently, for grinding the Ti-6Al-4V alloy under the cutting conditions tested in this study.
- Regarding surface texture, SEM images showed that under the milder dressing condition ($a_d = 20$ µm), the grooves were deeper in some areas, with several discontinuities in the marks left by the abrasives. However, under the more severe dressing condition ($a_d = 40$ µm), this behavior was exacerbated, and thermally induced cracks were detected on the surfaces of the Ti-6Al-4V alloy after grinding.

Competing Interests The authors acknowledge CAPES (Finance Code 001) for financial support, FAPEMIG (Call 013/2023, Process PCE-00578-24), as well as the Federal University of Uberlândia (UFU) and the Teaching and Research in Machining Laboratory (LEPU) and Conventional Machining Laboratory (LUC), both Faculty of Mechanical Engineering (FEMEC), for providing the resources necessary for the development of this work. Two of the authors also acknowledge the Graduate Program in Mechanical Engineering at UFU for their scholarship support. Rosemar Batista da Silva acknowledges CNPq for the Research Productivity Grant, process n°.: 312274/2022-0, and FAPEMIG—EDITAL 002/2018—Program Pesquisador Mineiro, PPM XII—Process n°.: PPM-00492-18.

Responsibility Notice The authors are the only responsible for the material included in this work.

References

Abrão AS, Abrão BS, Da Silva RB, Do Nascimento CG, Arencibia RV (2023) Cálculo da incerteza de medição associada à Ra obtida no aço inoxidável 316 uf com superfície interrompida após a retificação. XXV Colóquio de Usinagem. Brusque, Brasil

Anand N, Kumar S, Paul S (2019) Effect of cutting fluids applied in MQCL mode on machinability of Ti-6Al-4V. J Manuf Process 43:154–163

Haitjema H (1998) Uncertainty analysis of roughness standard calibration using stylus instruments. Precis Eng 22:110–119

INMETRO (2012) Avaliação de dados de medição—Guia para a expressão de incerteza de medição—GUM. Brasil, Rio de Janeiro

Klocke F (2009) Manufacturing process 2: grinding, honing, lapping, vol 1, p 1–452. Springer, Aachen

Leach RK (2001) Good Practice n°37—the measurement of surface texture using stylus instruments. National Physical Laboratory, Middlesex, 2ª edição

Leyens C, Peters M (2003) Titanium and titanium alloys: fundamentals and applications, vol 1, pp 1–532. Wiley—VHC

Malkin S, Guo C (2008) Grinding technology: theory and applications of machining with abrasives, vol 2, p 1–372. Industrial Press, New York

Mello AV, Silva RB, Machado AR, Gelamo RV, Diniz AE, Oliveira RFM (2017) Surface grinding of Ti-6Al-4V alloy with SiC abrasive wheel at various cutting conditions. Procedia Manuf 10:590–600

Mukhopadhyay M, Kundu PK, Chatterjee S, Das S (2018) Impact of dressing infeed on SiC wheel for grinding Ti-6Al-4V. Mater Manuf Processes 34:54–60

Motonishi S, Hara Y, Isoda S, Itoh H, Tsumori Y, Terada Y (1987) Study on machining of titanium and its alloys. In: Kobelco technology review, vol 2, p 28–31

Mukhopadhyay M, Kundu PK (2019) Evaluating application potentiality of unconventional fluids for grinding Ti-6Al-4V using alumina wheel. Mater Manuf Processes 34:1151–1159

NBR ISO 4287 (2002) Especificações geométricas do produto (GPS)—Rugosidade: Método do perfil – Termos, definições e parâmetros da rugosidade. Associação Brasileira de Normas Técnicas, Rio de Janeiro, 2ª edição

Pereira MF, Abrão BS, Abrão AS, Machado AR, Jackson MJ, Hassui A, Da Silva RB (2023) Influence of grinding wheel conditioning on the grindability of Ti-6Al-4V alloy. Int J Adv Manuf Technol 125:1531–1542

Rai BR, Mukhopadhyay M, Kundu PK (2019) Evaluating the grinding ratio and surface quality of Ti-6Al-4V under varying grinding pass count and depth of cut. J Phys: Conf Ser 1240:1–8

Silva RB, Machado AR, Ezugwu EO, Bonney J, Costa ES (2007) Evaluation of nano-grain size ceramics when machining Ti-6Al-4V alloy under conventional coolant flow. In: 19th International congress of mechanical engineering, Brasília, Brasil

Determination of Flow Curves for AISI 430 and AISI 430 DDQ Steel

André César Martins de Miranda, Pedro Henrique Rodrigues Pereira, Ariel Rodriguez Arias, and Paulo Roberto Cetlin

Abstract Ferritic stainless steel AISI 430 is one of the most widely produced stainless steels worldwide. It has good corrosion resistance in moderate corrosive environments, good drawability and an aesthetically attractive surface. For applications that require better drawability or weldability, the addition of stabilizing elements promotes the formation of carbonitrides that prevent the formation of martensite and the phenomenon called sensitization. To perform simulations of forming of these steels, either by analytical or numerical methods, the flow curves characterization of these materials is required. In this work, two different methods were used to obtain the flow curves of AISI 430 (standard) and AISI 430 DDQ (niobium stabilized) steels. The first method consisted of tensile testing of cold-rolled samples, each sample being at a different deformation level. The pilot rolling mill at the Aperam South America research facility was used to prepare the samples. The second test consisted of plane-strain compression tests performed in the mechanical testing laboratory of the metallurgy department of the Federal University of Minas Gerais, using devices adapted to a universal tensile machine. The raw materials used were samples of hot-rolled, annealed and pickled coil sheets. In both tests, replicates were performed to provide greater reliability of the results. Through transformations made using the Levy–Mises equations, the flow curves of each steel were plotted and compared. The results showed similarity between the two methods when considering the yield strength of the samples tested by tension at 0.2% strain, with a small variation in the AISI 430 DDQ steel for true strains above 0.5. From the data obtained in the tests, the coefficients of the Hollomon equation were then established for both steels.

Keywords Flow curve · AISI 430 · Plane-strain compression test

A. C. M. de Miranda (✉)
Aperam South America, Timóteo, MG, Brazil
e-mail: andre.martinsdemiranda@aperam.com

P. H. R. Pereira · A. R. Arias · P. R. Cetlin
Universidade Federal de Minas Gerais, Belo Horizonte, MG, Brazil

V. E. Beal and A. S. Ribeiro Junior (eds.), *Advances in Manufacturing Engineering Research*, Lecture Notes in Mechanical Engineering,
https://doi.org/10.1007/978-3-032-09858-0_18

1 Introduction

Stainless steels can be classified into five groups, depending on their microstructure and chemical composition, namely: Austenitic with the addition of manganese (200 series); Austenitic (300 series); Ferritic (400 series); Martensitics (400 series) and Duplex (Davis 1999). 400 series steels can have martensitic microstructure when the chromium has a lower content and the carbon is higher (compared to ferritic steels) (Aperam 2020a).

The most popular ferritic stainless steel is AISI 430, being one of the most produced worldwide (AB 2013). It has chromium content between 16 and 18%, and carbon content less than 0.12%. It features good corrosion resistance in moderate corrosive environments, good cold stamping, aesthetically attractive surface delivered by the manufacturer, and good resistance to high-temperature oxidation. The characteristics of AISI 430 steel make it attractive for applications in the household, cutlery, chimney ducts, daily equipment, decorative components and catering equipment sectors. Its major limitations are in weldability, due to the precipitation of chromium carbonitrides and high grain growth, and in deep drawing (Aperam 2020a).

AISI 430 DDQ steels ("Deep Drawing Quality") usually have the addition of titanium (Ti) and/or niobium (Nb). These elements are called stabilizers, because they have a greater affinity for carbon and nitrogen, promoting the formation of titanium and niobium carbonitrides, and limiting the formation of chromium carbonitrides. The presence of niobium prevents the formation of martensite and limits grain growth, making this steel stabilized and suitable for welding (Aperam 2020a). In addition, the presence of these elements favors the stabilization of the ferritic microstructure during hot rolling, contributing to the recrystallization and increase of the γ fiber, thus increasing normal anisotropy and improving drawability $\overline{r}$ (Penedo 2023; Rodrigues 2017).

2 Materials and Methods

The typical chemical composition of the steels used in the experiment is presented in Table 1 (Aperam 2020b).

The compression test in the plane-strain state, also known as the Ford test (Ford 1948), consists of a test in which the flat specimen of width w is compressed by punches of length *b*, and with a width greater than that of the specimen as shown in

Table 1 Typical chemical composition values (Aperam 2020b)

AISI	IN	US	Class	C	Yes	Mn	Cr	Nb
430	1.4016	S43000	Standard	0.04	0.35	0.3	16.5	–
430	1.4016	S43000	DDQ	0.02	0.35	0.3	16.5	0.3

Fig. 1 Illustration of device used for the plane-strain compression test (Ford test). RD: Rolling direction

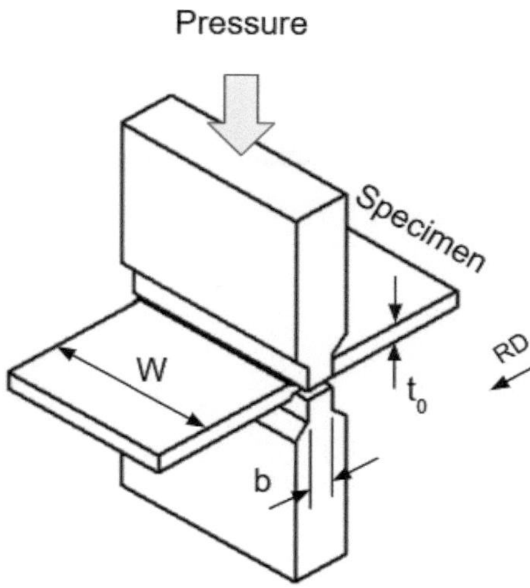

Fig. 1. To ensure the condition of plane-strain state (negligible lateral deformation), the width of the specimen should be between 6 and 10 times larger than the length of the punch, and in order for the strain to be homogeneous, the ratio between the length of the punch (*b*) to the thickness of the specimen (*t*) should be between 2 and 4 (Dieter et al. 2003).

To perform the Ford test, an Instron universal traction testing machine model 5582 available at the mechanical testing laboratory of the metallurgy department of UFMG was used, as well as the apparatuses shown in Fig. 2. For each steel (AISI 430 and AISI 430 DDQ) 6 samples of sheets in the dimensions of $t_0 = \sim 3$ mm and $w_0 = \sim 20$ mm were used, from hot-rolled, annealed and pickled coils. Figure 2e shows a photo of a sample after the test.

To maintain the proper *b*/*t* ratio (length of the punch *b* over the thickness of the sample *t*) and to promote good lubrication at the punch/sample interface, the test was done in 16 steps with the application of progressive loads as shown in the Table 2, using two punch configurations: 5 mm and 2 mm length. After each compression step, the punches and plate were cleaned, the width and thickness of the deformed region of the sample were measured, and relubrication were performed to perform the next step. Lubrication was performed with molybdenum disulfide-based paste. Carrying out the test in this way prevented the measurements from having any influence from the elastic deformation of the equipment and apparatus.

Since the ratio between the width of the plate *w* and the length of the punch *b* was less than 6 until step eight ($w/b = 4$), the condition of plane-strain state was not perfectly satisfied, and a small spreading of the material in the width direction was expected. Therefore, after each compression step shown in the Table 2, in addition to the measurements of the deformation $d\varepsilon_3$ (thickness reduction), measurements of the width increase $d\varepsilon_2$ in the deformed region were also made. From the results of the test, to obtain the flow curve, the Levy–Mises equations were used (Bhaduri 2018).

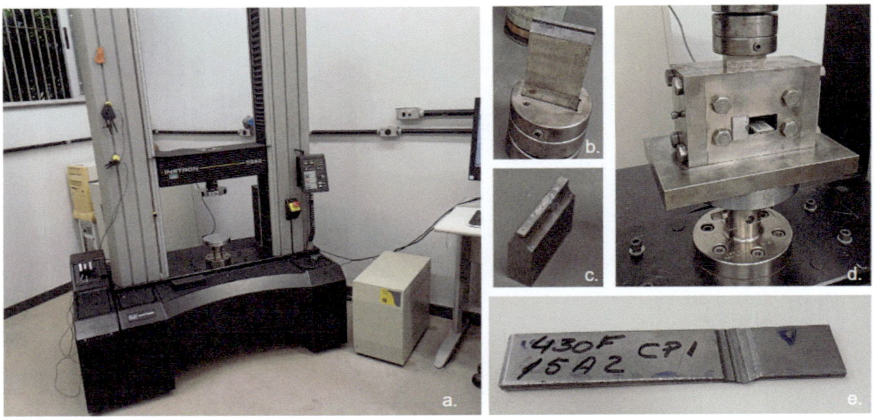

Fig. 2 Instron tensile testing machine model 5582 (**a**), upper punch (**b**), lower punch (**c**), die (**d**) and sample of AISI 430 steel after the test has been carried out (**e**)

Table 2 Steps for performing the compression test in the plane-strain state

Step	Punch length, *b* (mm)	Force (kN)	Ratio *w*/*b*	Ratio *b*/*t*
1	5	54	4	1.8
2	5	59	4	1.9
3	5	64	4	2.0
4	5	69	4	2.2
5	5	74	4	2.5
6	5	79	4	3.1
7	5	84	4	3.6
8	5	89	4	4.0
9	2	37	10	2.4
10	2	38	10	3.0
11	2	39	10	3.3
12	2	40	10	4.0
13	2	41	10	5.1
14	2	43	10	5.9
15	2	45	10	7.2
16	2	47	10	8.2

$$d\varepsilon_1 = \frac{d\overline{\varepsilon}}{\overline{\sigma}}\left[\sigma_1 - \frac{1}{2}(\sigma_2 + \sigma_3)\right] \tag{1}$$

$$d\varepsilon_2 = \frac{d\overline{\varepsilon}}{\overline{\sigma}}\left[\sigma_2 - \frac{1}{2}(\sigma_1 + \sigma_3)\right] \tag{2}$$

$$d\varepsilon_3 = \frac{d\overline{\varepsilon}}{\overline{\sigma}}\left[\sigma_3 - \frac{1}{2}(\sigma_1 + \sigma_2)\right] \tag{3}$$

where $\sigma_1, \sigma_2, \sigma_3$ and $\varepsilon_1, \varepsilon_2, \varepsilon_3$ correspond to the principal plastic stresses and strains, and $\overline{\sigma} e \overline{\varepsilon}$ the effective stress and strain, respectively. When considering the boundary condition $\sigma_1 = 0$, the Eqs. (1)–(3) were developed to obtain the effective stress equation by:

$$\overline{\sigma} = \frac{1,5\sigma_3 d\varepsilon_{ef}/d\varepsilon_2}{1 + 2d\varepsilon_3/d\varepsilon_2} \tag{4}$$

where the deformation ε_2, ε_3 and $\overline{\varepsilon}$ are given by the equations:

$$d\varepsilon_3 = \ln\left(\frac{t}{t_0}\right) \tag{5}$$

$$d\varepsilon_2 = \ln\left(\frac{w}{w_0}\right) \tag{6}$$

$$d\overline{\varepsilon} = \frac{\sqrt{2}}{3}\left[(d\varepsilon_1 - d\varepsilon_2)^2 + (d\varepsilon_2 - d\varepsilon_3)^2 + (d\varepsilon_3 - d\varepsilon_1)^2\right]^{1/2} \tag{7}$$

To perform the tensile test, for each steel, 10 samples of ~ 3 × 50 × 300 mm were cold rolled to different progressive reductions as shown in the Table 3. As the initial w/t ratio in the rolling was ~ 16, the condition of plane-strain state was obtained in which $d\varepsilon_2 = 0$. The samples used for this test were collected from the same batch of material used in the Ford tests.

Table 3 Pass schedule of cold-rolled samples submitted to tension test

Sample	430		430 DDQ	
	h (mm)	ε_3	*h* (mm)	ε_3
1	2.663	0.11	2.691	0.10
2	2.391	0.21	2.425	0.20
3	2.130	0.33	2.130	0.33
4	1.912	0.44	2.004	0.39
5	1.734	0.54	1.709	0.55
6	1.352	0.78	1.366	0.77
7	1.103	0.99	1.063	1.02
8	0.884	1.21	0.921	1.17
9	0.595	1.60	0.582	1.63
10	0.424	1.94	0.402	2.00

Fig. 3 4-High rolling mill from the Aperam research center (**a**), cold-rolled samples (**b**) and Instron tensile testing machine (**c**)

The cold rolling was carried out in the 4-Hi mill, model Fenn 4-085, available at the Aperam South America research department, as shown in Fig. 3a. The lubrication of the samples was performed before each rolling pass, applying a semisynthetic lubricant on both sides of the sample. For each rolling pass, two samples were collected and prepared for the tensile test according to the ASTM 6892-1:201 standard (ISO 2019). The tests were carried out on the Instron tensile testing machine model 5985 of the metallurgy laboratory of Aperam South America shown in the Fig. 3c.

As the rolled specimens were deformed in the plane-strain state, the previous effective strain of these specimens before the tensile test was calculated by the equation (Bhaduri 2018):

$$d\bar{\varepsilon} = \frac{2}{\sqrt{3}} \ln\left(\frac{t}{t_0}\right) \tag{8}$$

As the result of the tensile test represents the engineering stress–strain curve, the conversion to the true stress–strain curve was performed by the equations:

$$\varepsilon = \ln(1 + e) \tag{9}$$

$$\sigma_{\mathrm{pl}} = S(1 + e) \tag{10}$$

where S and e are the engineering stress and strain, respectively, and E the modulus of elasticity of the specimen (210 GPa). From the results of the Ford test, adjusted by the Eqs. (4) and (7), and the results of the rolled followed by tensile testing, adjusted by the Eqs. (9) and (10), the flow curve of the material was obtained by nonlinear regression according to Hollomon's model, represented by the equation (Hollomon 1945):

$$\sigma = K\varepsilon^n \tag{11}$$

where K and n are a constant and the hardening coefficient of the material, respectively.

3 Results and Discussion

Figures 4 and 5 show the result of the compression test in the plane-strain state for the samples of AISI 430 and AISI 430 DDQ steel, respectively. The dashed lines shown in the graphs indicate the flow curves obtained by nonlinear regression, with the respective equations and coefficient of determination R2.

The results of the tensile tests of the cold-rolled specimens are presented in the Figs. 6 and 7, in which the true stress–strain curves of each cold-rolled sample were plotted up to the yield strength at $\varepsilon = 0.2\%$. From these results, the envelopes of the yield stress values were generated by nonlinear regression (dashed line), with the respective equation and coefficient of determination R2.

It is noticed in Fig. 6 and 7, that for high strains ($\varepsilon > 1.5$ for AISI 430 steel, and $\varepsilon \approx 2$ for AISI 430 DDQ steel), the curves obtained by the tensile test presented atypical yield stress values, much higher than the envelope obtained by nonlinear regression.

Figure 8 shows in a comparative way the results obtained for each steel and type of test. It can be seen in Fig. 8a that for AISI 430 steel, the results obtained were equivalent between the two methods: Ford test and tensile test of rolled specimens. For AISI 430 DDQ steel, however, it was observed that the flow curves obtained by the two methods are initially coincident, but distance themselves with the increase

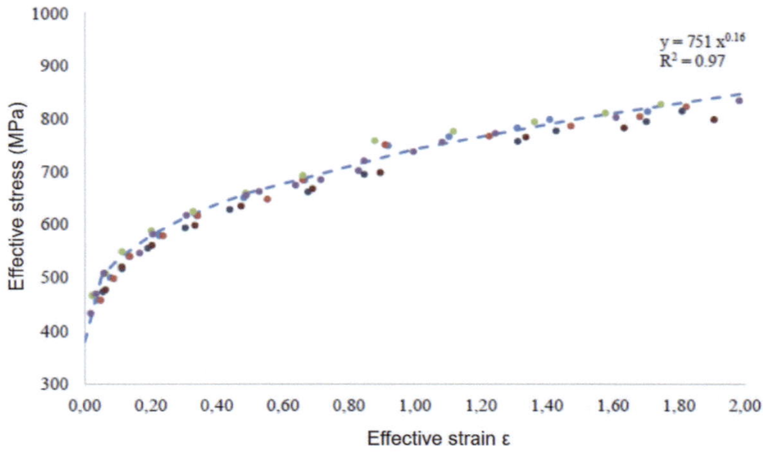

Fig. 4 Ford test result. AISI 430 steel

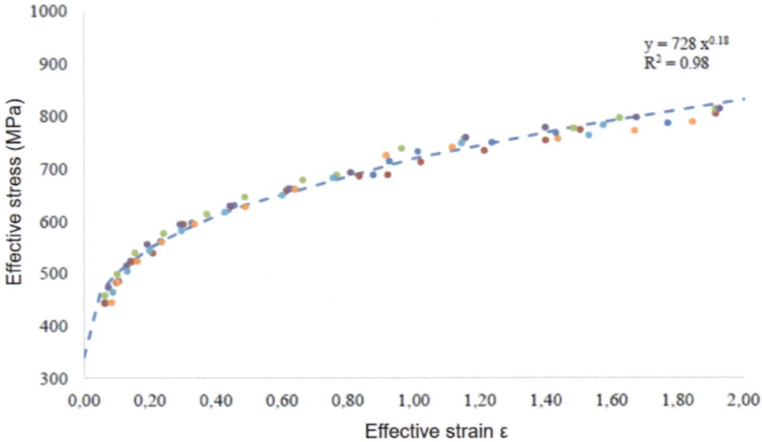

Fig. 5 Ford test result. AISI 430 DDQ steel

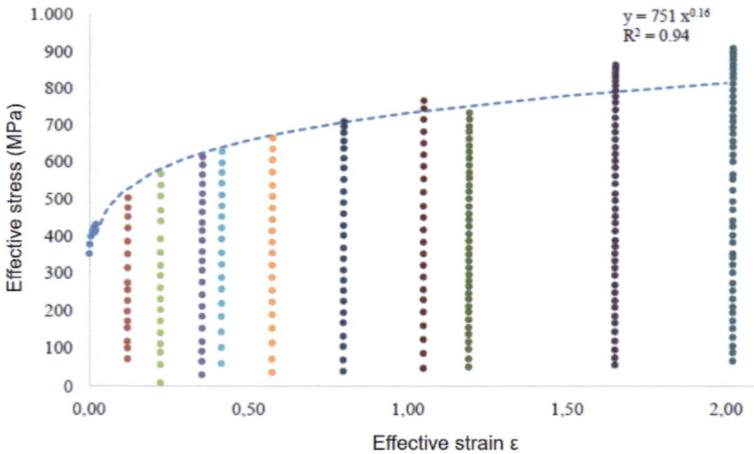

Fig. 6 Result of the tensile test of cold-rolled samples. AISI 430 steel

in deformation as shown in Fig. 8b, reaching a relative difference between them of ~ 4%.

It was expected that the curves of the cold-rolled samples submitted to tensile testing would be slightly below than the samples tested by the Ford test, since in this method there is no influence of friction between the punch and the plate as in the Ford test. A factor that contributed to the curves of the two tests being close is the fact that, for the samples tested by tensile, the values of yield stresses were based on a plastic deformation of $\varepsilon = 0.2\%$. Thus, the yield stress results used have a slight increase in plastic deformation of 0.2%.

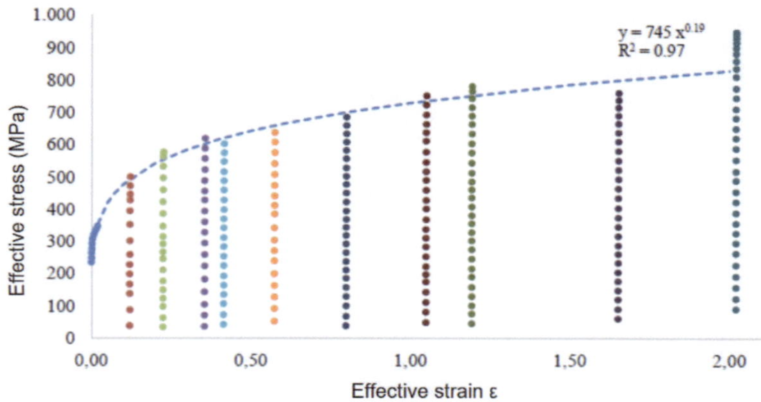

Fig. 7 Result of the tensile test of cold-rolled samples. AISI 430 DDQ steel

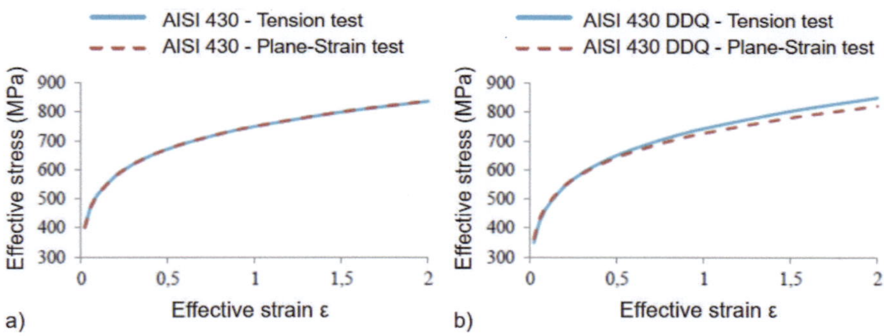

Fig. 8 Results of the tensile and Ford tests for: **a**. AISI 430 steel and **b**. AISI 430 DDQ

In addition, it was observed that the results obtained by nonlinear regression in the tensile samples were very sensitive to the atypical stress values that occurred at high strains. If the sample deformed by $\varepsilon \approx 2$ shown in Fig. 7 is neglected, the new flow curve obtained by nonlinear regression of the tensile test coincides with that obtained by the Ford test as shown in Fig. 9.

To understand this behavior, complementary tests are necessary, such as analysis of the crystallographic texture of the samples by the EBDS technique ("Electron Back Scatter Diffraction"). Figure 10 shows the comparative result of the flow curves of AISI 430 and AISI 430 DDQ steels obtained by the Ford test. It is noticed that AISI 430 steel has a strength initially higher than AISI 430 DDQ, about ~ 12%. However, due to the hardening factor of AISI 430 DDQ steel being higher than that of AISI 430 steel ($n = 0.18$ versus $n = 0.16$), this difference is reduced with increasing deformation, reaching almost the same stress strength for $\varepsilon = 2$, where the relative difference in strength between these steels is ~ 1%.

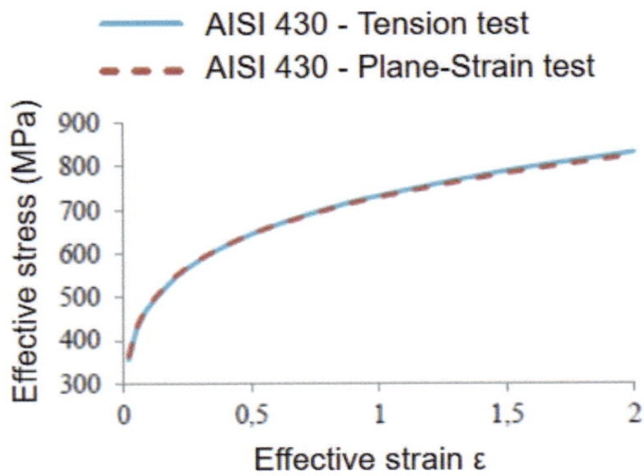

Fig. 9 Result of tensile tests (adjusted) and Ford test—AISI 430 DDQ steel

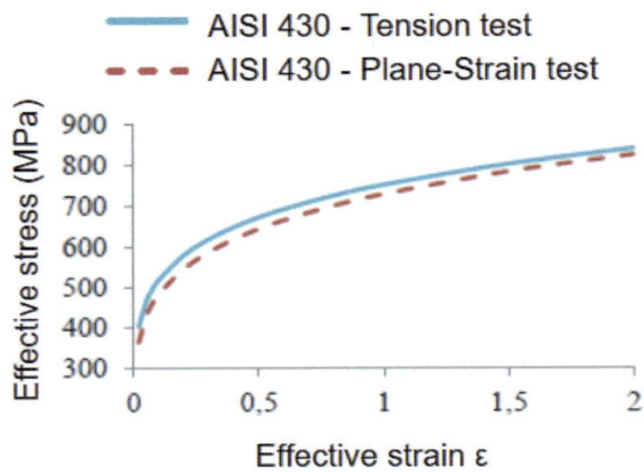

Fig. 10 Comparison of the flow curves of AISI 430 and AISI 430 DDQ steels obtained by the Ford test

4 Conclusions

The objective of this study was to obtain the flow curves of AISI 430 and AISI 430 DDQ steels by two different methods: a. tensile test of cold-rolled samples and b. compression test in the near plane-strain state (Ford test). From the results, it can be stated:

- The two methods have shown equivalent results for AISI 430 steel. The similarity of the results obtained between the two methods was attributed to the fact that the result of the tensile test contemplated an increase in plastic deformation of 0.2%, like the contribution caused by the friction effect in the Ford test.
- For AISI 430 DDQ steel, the flow curve obtained by the tensile test was slightly higher than that obtained by the Ford test. This behavior may have been influenced by the sensitivity of the nonlinear regression to the atypical stress obtained in the tensile test of the cold-rolled specimen to $\varepsilon \approx 2$.
- AISI 430 DDQ steel was initially softer than AISI 430 steel, and its strength increases at a slightly higher rate than AISI 430, approaching the strength of AISI 430 when it is subjected to large deformations ($\varepsilon \approx 2$).

Acknowledgements The authors would like to thank the Robotics, Welding and Simulation Laboratory (LRSS) of the mechanical engineering department of UFMG, the mechanical testing laboratory of the metallurgy department of UFMG, and the research and metallurgy laboratories of Aperam South America for their support in the development of this work. This work was carried out with the support of the Coordination for the Improvement of Higher Education Personnel—Brazil (CAPES)—Financing Code 001.

Responsibility for Information The authors are solely responsible for the information included in this work.

References

AB OS (2013) Handbook of stainless steel. Sandvikens Tryckeri AB, Avesta

Aperam (2020a) Aço Inox—Especificações Técnicas

Aperam (2020b) KARA ferritic stainless steel—K30-K30ED

Bhaduri A (2018) Mechanical properties and working of metals and alloys. Springer, Singapore

Davis JR (ed) (1999) ASM specialty handbook: stainless steels. ASM Int

Dieter GE, Kuhn HA, Semiatin SL (2003) Handbook of workability and process design. ASM Int

Ford H (1948) Researches into the deformation of metals by cold rolling. Proc Inst Mech Eng 159(1):115–143

Hollomon JH (1945) Tensile deformation. Trans. AIME 162:268

ISO 6892-1:201 (2019) Metallic materials—tensile testing, part 1: method of test at room temperature. International Organization for Standardization

Penedo R (2023) Microestrutura, textura e propriedades de estampagem de aços inoxidáveis ferríticos estabilizados ao titânio e ao nióbio após processos de laminação a quente, a frio e recozimento. Universidade Federal de Minas Gerais

Rodrigues DG (2017) Influência da redução a frio e da microestrutura e da textura iniciais na microestrutura, textura e propriedades de um aço inoxidável ferrítico estabilizado ao nióbio. Universidade Federal de Minas Gerais

Application of Active Learning Methodologies and Skills Development in the Metrology Laboratory of FEM UNICAMP

Robert Eduardo Cooper Ordoñez and Olívio Novaski

Abstract Given the need for higher education to adapt to contemporary demands and the requirements of the labor market, it is essential to explore alternative teaching methods that can effectively prepare students for the challenges of the twenty-first century. Furthermore, with the onset of the fourth industrial revolution, the need to ensure manufacturing conditions for mass production, so that variability in manufacturing processes can be controlled to guarantee part interchangeability, has become more evident. Therefore, the objective of this work is to present a proposal for the reformulation of the metrology discipline at UNICAMP's School of Mechanical Engineering, to harmonize a programmatic content with sufficient depth to connect to the main theoretical engineering disciplines, encouraging complex thinking and preparing students for the labor market. To achieve this objective, a 100% experimental discipline was developed, where students enter the laboratory to put into practice the concepts they learned in previous disciplines such as technical drawing and mechanical technology, so that they can verify theory and expand the application of these concepts to other disciplines such as machine element design and manufacturing engineering, where conditions for the success of a product design are analyzed in terms of interchangeability, high quality, and good cost–benefit ratio. As a result, three active methodologies were considered in the design of the metrology discipline: Problem-Based Learning (PBL), Flipped Classroom, and competency-based teaching applied to the Learning Factory. A metrology laboratory infrastructure was also developed that can accommodate up to 33 students, in eleven different experiments, performed by groups of three students at the same time, in a rotation system. As support material, an experiment script, a handout, and a laboratory website were created and updated. The contribution of this work lies in the fact that by including the active learning model in the metrology laboratory class, the student, in addition to reading and listening, must actively participate, solve problems, discuss, and have a critical look at the experiments performed, which contrasts with traditional teaching where the student is a passive element.

R. E. C. Ordoñez (✉) · O. Novaski
University of Campinas—UNICAMP, Campinas, SP, Brazil
e-mail: recooper@unicamp.br

V. E. Beal and A. S. Ribeiro Junior (eds.), *Advances in Manufacturing Engineering Research*, Lecture Notes in Mechanical Engineering,
https://doi.org/10.1007/978-3-032-09858-0_19

Keywords Active learning · Metrology · Teaching methodology · Measurement systems analysis · Manufacturing engineering

1 Introduction

Industry has evolved with innovative techniques and technologies, especially during industrial revolutions, which transformed technological standards. Today, the biggest challenge is to understand and shape the new technological revolution, capable of transforming society. We are at the beginning of an era that changes how we live, work, and relate.

Likewise, the industry increasingly demands qualification to deal with complex problems, highlighting the need for advanced skills and the importance of fast and efficient educational methods for human capital development. Thus, it is essential to adopt alternative teaching methods that prepare students for market demands. The various teaching standards must evolve in parallel with the rapid advancement of technologies, tools, and techniques related to production (International Monetary Fund 2013).

For this, it is necessary to adapt training content and its delivery mechanisms to the new requirements of knowledge-based manufacturing. Education for the manufacturing environment must concern itself with providing integrated engineering competencies and multidisciplinary training (Chryssolouris et al. 2016).

One of the sciences that has accompanied the development of industry is metrology. The relationship between metrology and industrial revolutions is essential and profound, as metrology played a crucial role in technological advancement, increased productivity, and the standardization necessary for each stage of industrial transformation.

Thus, the present work aims to present a proposal for the application of active learning methodologies and competency development in the metrology laboratory of FEM (School of Mechanical Engineering) – UNICAMP, with the aim of contributing to the training of professionals, through alternative teaching methods necessary to adequately prepare students for the current reality and market demands.

2 Theoretical Concepts

2.1 Learning Factory

Industry 4.0, enabled by technologies such as Artificial Intelligence, Internet of Things, Big Data, Cloud Computing, and Advanced Robotics, has brought advances, but also new risks for companies, requiring the development of innovative approaches to risk management (National Confederation of Industry 2016; Tupa et al. 2017).

Regarding professional qualification, it is clear that the introduction of new advanced manufacturing concepts will demand the following aspects: training of professionals already working in the industry, emergence of new professional profiles, and consequently, the need for changes in courses to improve professional training.

In this context, Learning Factories, or Teaching Factories, emerge as an alternative to integrate learning and work environments, from which more realistic experiences arise (Chryssolouris et al. 2016).

According to Wagner et al. (2012), the term Learning Factory combines teaching and learning with the simulation of a real production environment. Originated in 1994 at the National Science Foundation (NSF) in the USA, according to Abele et al. (2015), its objective is to promote technological innovation and competitiveness through the development of skills to deal with complex situations in research, training, and education contexts.

The concept of Learning Factory can range from the production of physical products, with a real value chain and in-field learning, to services, with a virtual value chain and remote teaching, depending on the approach adopted (Enke et al. 2016).

2.2 *Metrology*

Metrology plays a fundamental role in various sectors, ensuring precision and reliability in industrial, commercial, and healthcare processes. In different contexts, its crucial role in guaranteeing quality, safety, and efficiency in various areas of activity is evident, and it is fundamental for scientific, technological, and industrial development, ensuring that measurements are consistent and comparable in different contexts.

According to the International Vocabulary of Metrology (2012), metrology is the science of measurement and its applications. That is, metrology is the science that studies measurements, covering all theoretical and practical aspects that guarantee their accuracy and reliability.

The importance of metrology can be observed in different contexts, such as:

- **Dimensional control applications** in industry, universities, and specialized laboratories, more specifically concerning instrument calibration, essential to guarantee accurate measurements and protect the consumer (Neto 2012).
- In the context of **industrial metrology**, as a function of Quality Management and in the implementation of precise measurement systems to optimize manufacturing processes and ensure product conformity with established standards (Alves 2003).
- Focusing on **measuring instruments and measurement systems**, highlighting the importance of measuring instruments and their calibration in industry.

It is also important to highlight how metrology ensures the necessary precision in production processes, contributing to the quality and efficiency of final products (Souza and Sobrinho 2005).

According to Albertazzi and Souza (2008), metrology is divided into three main categories:

- **Scientific Metrology**: Focuses on the development of new measurement methods and the maintenance of fundamental reference standards.
- **Industrial Metrology**: Deals with measurement applications in production processes and product quality assurance.
- **Legal Metrology**: Relates to measurements that directly affect transparency in commercial relations, public health and safety, ensuring compliance with regulations and standards.

In Brazil, the National Institute of Metrology, Quality and Technology (INMETRO) is the entity responsible for coordinating and executing activities related to metrology, ensuring the conformity of products and services with established standards.

2.3 Teaching Methodologies

Active learning has several interpretations, but, in general, it involves the active participation of students in activities such as problem-solving, discussions, and critical analyses, in contrast to traditional teaching, where the student is passive. According to authors such as Prince (2004) and Bonwell and Eison (1991), and corroborated by Konopka et al. (2015), active methodologies help engage students and break with the "knowledge sponge" vision. These methodologies should be constructivist, collaborative, interdisciplinary, critical, reflective, investigative, and motivating.

One of these characteristics, collaborative practice, involves the interaction between students and teachers in small groups, with a focus on solving problems or creating projects applying course concepts. According to Smith and MacGregor (1992), this approach teaches how to deal with differences, work in teams, resolve conflicts, and value group voices, essential skills for community living, which are developed both in the classroom and in campus activities.

Considering the active teaching methodologies, those considered most suitable for the metrology laboratory context are: Problem-Based Learning (PBL); Flipped Classroom; and competency-based teaching applied to the Learning Factory. Below is a brief explanation of each of them.

Barrows (1985) introduced PBL as a problem-solving-based teaching methodology, where students take responsibility for their own learning, and the teacher acts as a facilitator. The process occurs in cycles: Students analyze complex problems, identify necessary skills, work individually and in groups, and develop solutions. This approach promotes greater knowledge retention in the medium and long term (Newstetter 2006).

The flipped classroom is a methodology in which traditional activities are swapped: Students study content at home, through direct instructions, and perform interactive group activities in the classroom, focusing on interaction. This approach prepares students in advance, allowing the teacher to conduct collaborative activities and students to deepen their learning, create questions, review materials, and consolidate knowledge (Bishop and Vergeler 2013; Abeysekera and Dawson 2015).

Regarding competency-based teaching applied to Learning Factories, it can be said that it has been gaining prominence, defined as cognitive dispositions acquired in specific contexts, which allow the individual to deal with challenges in specific areas, such as learning a native language or creating mathematical models (Klieme and Leutner 2006).

Tisch et al. (2013) developed a guide whose purpose is to assist in the systematic approach to designing the action-oriented and competency-based Learning Factory, called the LFC-Guide or Learning Factory Curriculum Guide. This guide provides a factory design focused on the development of the competencies necessary for a target audience that will use the Learning Factory. Therefore, the intended competencies are a central key for the construction of the Learning Factory and are the result of a systematic analysis of the organization and personnel conditions (purpose, type of production, and target audience).

3 Proposed Teaching Methodology

As previously stated, three active methodologies were considered in the design of the metrology discipline: Problem-Based Learning (PBL), Flipped Classroom, and competency-based teaching applied to the Learning Factory. The PBL methodology is used to present students with a problem to be solved. In it, students must work on the problem independently and then cooperatively to build knowledge and present a solution.

The flipped classroom methodology is used so that, at first, students prepare conceptually and in advance, using the support material to understand the experiment before entering the laboratory; in a second moment, in the laboratory, interactive group activities can be carried out, focusing on the student.

The Learning Factory concept is used because each of the experiments is independent of each other and simulates a workstation, with its own equipment. In this way, the set of experiments transforms into a teaching and learning platform to provide an environment that simulates a real production environment, allowing participants to understand complex and unfamiliar situations.

The personal competencies (soft skills) aimed at with the application of active methodologies are of four types: **Argumentation and critical thinking** (promoting a debate among students about a proposed solution to a problem); **Problem-solving** (based on theoretical concepts, students are challenged to compare measurement results with theory and propose explanations for inconsistencies in results); **Systemic thinking and decision-making** (linking the results obtained to industrial applications, as well as their relationship with use in other disciplines of the mechanical engineering course, such as machine elements, mechanisms, strength of materials, and mechanical vibrations); and **Teamwork** (students perform different roles in the experiment, such as coordinator and inspector; similarly, they participate in the experiment planning and report elaboration).

The interaction between these three methodologies and the laboratory environment is represented through a five-step process (Fig. 1) similar to that presented by Tominaga and Cooper (2023) in a proposed teaching methodology in a Learning Factory within the scope of Industry 4.0:

1. **Preparation of the Subject**: Aims at preparing the subject to be taught, based on competency-based teaching.
2. **Student Preparation**: Aims at getting to know the student, their individual experiences and their theoretical and practical background in relation to the content to be applied.
3. **Contextualization**: Students and teacher, together, discuss the questions to be solved and what kind of knowledge needs to be mastered.
4. **Practice**: This is the stage where the discussed content is put into practice and, in this case, the use of the Learning Factory (equipment and measuring instruments) as a basis for active learning.
5. **Evaluation**: Through the submission of the report (script), and its correction, this is the moment when the student demonstrates that they have acquired the knowledge and competencies, therefore, understood and absorbed the content.

The discipline (Industrial Metrology) is 100% practical. The experiments were designed to understand concepts about measurement, geometric tolerances, and dimensional tolerances, linked to mechanical manufacturing processes. Classes are divided into groups of three students who will simultaneously perform the 11 experiments according to the schedule delivered at the beginning of the semester. The schedule is done in a "rotation" format, so that all groups work simultaneously, changing the type of experiment each week. The printed report is filled out as student's progress with the experiment and is delivered at the end of the class, being stored in a specific locker. The report does not leave the laboratory, to avoid data sharing, and students receive feedback on the results at the beginning of the next class.

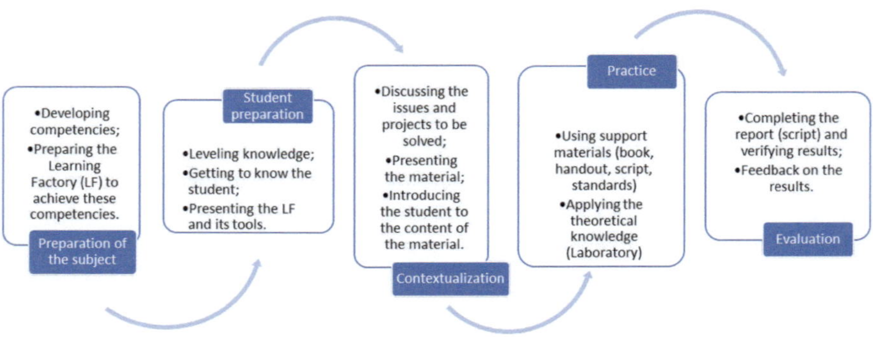

Fig. 1 Stages of application of the teaching methodology for the metrology laboratory

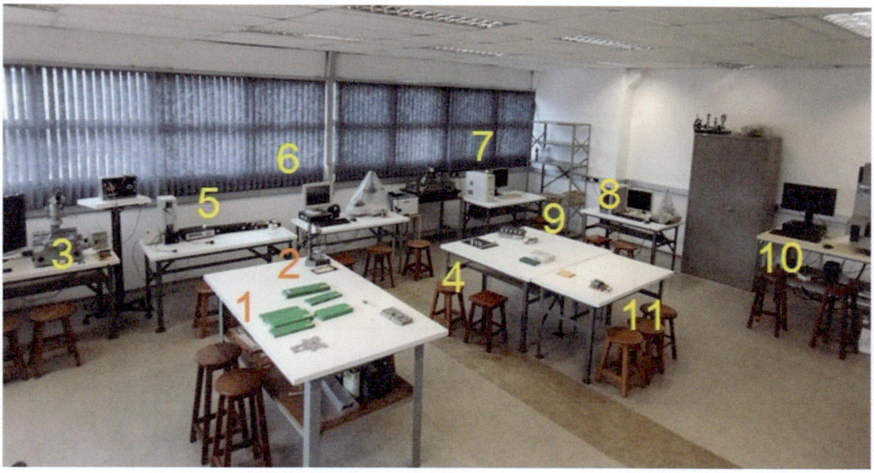

Fig. 2 View of the metrology laboratory area

3.1 Metrology Laboratory Infrastructure

The metrology laboratory is installed in an area of 60 m^2, located in block “M” of the undergraduate laboratories building of FEM (School of Mechanical Engineering – UNICAMP). It has equipment, self-manufactured devices, and measuring instruments adapted for each of the 11 designed experiments, which were assembled with equipment from the former Technology Center (CT) of UNICAMP.

Figure 2 shows a view of the laboratory, and Fig. 3 shows the layout of the equipment. The sequence of experiments follows the order specified below. In this work, each experiment includes a link to access the complete script, by simply clicking on the desired experiment.

Experiment 1: Dimension transfer and angular measurement
Experiment 2: Measurement uncertainty
Experiment 3: Calipers
Experiment 4: Position of the actual dimension in the tolerance field
Experiment 5: Gauge blocks
Experiment 6: Geometric tolerances: Form deviations
Experiment 7: Measurement systems analysis
Experiment 8: Roughness
Experiment 9: Position and tolerance field
Experiment 10: Coordinate Measuring Machine (CMM)
Experiment 11: Dimensional and geometric control

This infrastructure supports the offering of undergraduate and postgraduate disciplines. The Industrial Metrology discipline (EM-035) is currently an elective discipline for the undergraduate course in mechanical engineering, and in this context,

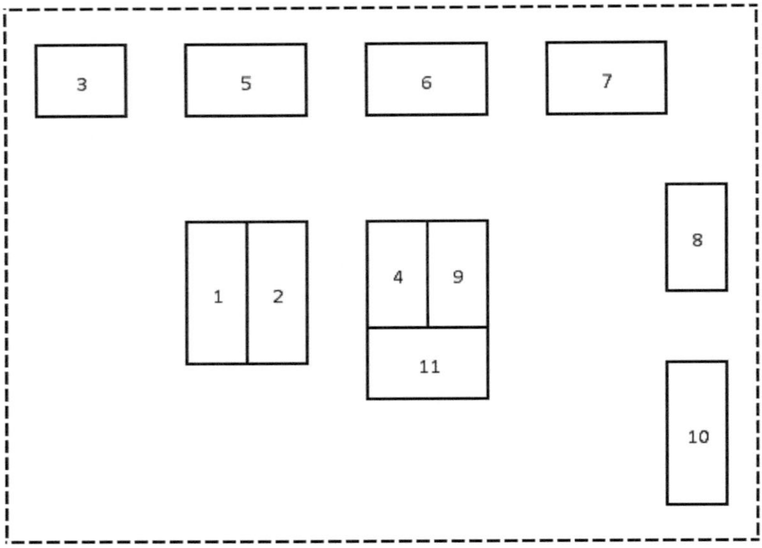

Fig. 3 Layout of equipment in the metrology laboratory area

activities of instrument calibration and verification, conformity tests, and uncertainty analysis are carried out to ensure the precision of industrial measurements.

In the context of postgraduate studies, the metrology discipline is offered under the acronym IM-118 (Metrology Topics: laboratory experiences), and in addition to the activities carried out in undergraduate studies, students are encouraged to explore new themes and develop experiments so that at the end of the semester they can elaborate a scientific article.

Table 1 describes the identification of the experiments, the objective, and the support materials that are used in each experiment, which must be consulted by the student before class, as indicated in stages 3 and 4, of contextualization and practice. In addition to the experiment scripts, the support materials are a book (Novaski 2013a, b) and a handout (Novaski 2013a), which is in its seventh version.

4 Final Considerations

The following items are organized in the form of questions to facilitate the identification of the contribution intended to be achieved with the proposed solution. The objective is to highlight how these alternative teaching methods can adequately prepare students for the current reality and market demands.

What Innovation Was Implemented in the Metrology Laboratory? A 100% experimental discipline was developed, considering active learning concepts and techniques, where students enter the laboratory to put into practice the concepts

Table 1 List of experiments, objectives, and support material

No.	Experiment	Objective	Support material
1	Dimension transfer and angular measurement	Determine and verify tolerances and clearances by means of dimension transfer. Measure angles directly and indirectly	Book "Introdução à Engenharia de Fabricação Mecânica," Chap. 6; Handout, Chaps. 1, 2, 3 and 13
2	Measurement uncertainty	Determine the measurement uncertainty of a dial gauge	Handout, Chaps. 1, 3, 4, 5 and 7
3	Calipers	Find tolerances, measure the dimensions of standardized calipers, and compare the measurements with the measuring system	Book "Introdução à Engenharia de Fabricação Mecânica," Chap. 5; NBR Standard – 6406
4	Position of the actual dimension in the tolerance field	Define the position of rings and pins in their tolerance field	Book "Introdução à Engenharia de Fabricação Mecânica," Chaps. 1, 2 and 3
5	Gauge blocks	Classify and calibrate gauge blocks by the comparison method	DIN Standard – 861. Handout, Chaps. 4, 6 and 7
6	Geometric tolerances: form deviations	Measure the circularity and radial runout of the cylindrical/conical shaft	Book "Introdução à Engenharia de Fabricação Mecânica," Chap. 7; Handout, Chap. 8; NBR – Standard 6409
7	Measurement systems analysis	Evaluate and compare two measurement systems by means of R&R studies	Book "Introdução à Engenharia de Fabricação Mecânica," Chap. 9; Handout, Chap. 9
8	Roughness	Measure surface roughness of parts with different manufacturing processes	Book "Introdução 'a Engenharia de Fabricação Mecânica," Chap. 8; Handout, Chaps. 10 and 12
9	Position and tolerance field	Determine the tolerance field of pins and rings applying the concept of reverse engineering	Book "Introdução à Engenharia de Fabricação Mecânica," Chaps. 2, 3 and 4
10	Coordinate Measuring Machine (CMMs)	Verify if the part made according to the drawing meets the specified tolerances	Book "Introdução à Engenharia de Fabricação Mecânica," Chap. 7; Handout, Chap. 11
11	Dimensional and geometric control	Verify if the measurement systems used are suitable for such measurement Verify if the measured elements of the part meet the specifications	Book "Introdução à Engenharia de Fabricação Mecânica," Chap. 7; Handout, Chap. 2

they learned in previous disciplines such as technical drawing and mechanical technology, so that they can verify theory and expand the application of these concepts to other disciplines such as machine element design and manufacturing engineering, where conditions for the success of a product design are analyzed in terms of interchangeability, high quality, and good cost–benefit ratio.

Why Can the Teaching Methodology be Considered Innovative? The innovation lies in the fact that by including the active learning model in the laboratory class, the student, in addition to reading and listening, must actively participate, solve problems, discuss, and have a critical look at the experiments performed, which contrasts with traditional teaching where the student is a passive element. Thus, in addition to discussions of technical concepts, students are guided to develop collaborative practices and for the development of technical and personal competencies (soft skills).

In the design of the metrology discipline, three active methodologies are used: Problem-Based Learning (PBL), Flipped Classroom, and competency-based teaching applied to the Learning Factory concept. The PBL methodology is used to present students with a problem to be solved. In it, students must work on the problem independently and then cooperatively to build knowledge and present a solution. The flipped classroom methodology is used so that, at first, students prepare conceptually with the help of support material to develop the experiment before entering the laboratory; in a second moment, in the laboratory, interactive group activities can be carried out, focusing on the student. The Learning Factory concept is used because each of the experiments is independent of each other and simulates a workstation, with its own equipment and instruments. In this way, the set of experiments transforms into a teaching and learning platform to provide an environment that simulates a real production environment, allowing participants to understand complex and unfamiliar situations.

The personal competencies that are aimed at with the application of active methodologies are of four types: argumentation and critical thinking (promoting a debate among students about a proposed solution to a problem); problem-solving (based on theoretical concepts, students are challenged to compare measurement results with theory and propose explanations for inconsistencies in results); systemic thinking and decision-making (linking the results obtained to industrial applications, as well as their relationship with the use in other disciplines of the mechanical engineering course, such as machine elements, mechanisms, strength of materials, and mechanical vibrations); and teamwork (students perform different roles in the experiment, such as coordinator and inspector; similarly, they participate in the experiment planning and report elaboration).

What Motivated the Search for this Innovation? Given the need for higher education to adapt to contemporary demands and the requirements of the labor market, it is essential to explore alternative teaching methods that can effectively prepare students for the challenges of the twenty-first century. Furthermore, with the onset of the fourth industrial revolution, the need to ensure manufacturing conditions for mass production, so that variability in manufacturing processes can be controlled to guarantee part interchangeability, has become more evident. In this way, it was necessary to propose a reformulation of the metrology discipline, to harmonize a programmatic content with sufficient depth to connect to the main theoretical engineering disciplines, encouraging complex thinking and preparing students for the job market.

What Impacts Were Generated with the Reactivation of the Laboratory? Due to internal organizational changes, the metrology laboratory ceased to operate in 2015 and was going to be dismantled. However, in 2018, the proposal for its reactivation and relocation emerged, which was previously in the CT (UNICAMP Technology Center) and was transferred to the FEM undergraduate laboratories building. From then on, a great effort was made to install the equipment, retrofit some machines, and update equipment and software. In this sense, as a result, the approval and execution of two projects with financial contributions can be presented:

- Project for the Revitalization of laboratories and the Implementation of new Undergraduate Teaching strategies-2019 (CGU Call 01/2019, UNICAMP).
- Improvement of teaching methodologies in the metrology laboratory of FEM (FAEPEX-Support for Teaching Call 2022, UNICAMP).

Additionally, a master's dissertation, two final course projects, and two scientific initiation projects were developed in the laboratory. Two scientific articles were published in scientific journals and two papers were presented at scientific congresses, in addition to establishing a collaboration with CTI Renato Archer for research development in additive manufacturing.

Some of the laboratory experiments were included in the mechanical technology disciplines, offered to the mechanical engineering and control and automation courses at FEM, with excellent receptivity from the students.

What Are the Future Developments? The methodology presented in this work is currently implemented, and since the reactivation of the laboratory in 2019, the EM-035 discipline has been taught three times in undergraduate studies and the IM-118 discipline has been taught once in postgraduate studies. In addition, the laboratory infrastructure provides technical support for the development of standard procedures and participation in research to advance measurement techniques, contributing to the quality and conformity of manufactured products.

Considering future actions, it can be said that starting from the second semester of 2025, an evaluation of the development of students' technical and personal competencies is planned, and a questionnaire will be created to monitor the level of satisfaction with the discipline and to collect suggestions from students.

Regarding the link with other areas of the course, it is intended to create a project that combines the study of metrology with the application of the concept of complexity in engineering, seeking to establish closer connections with other disciplines of the curriculum, to position metrology as a comprehensive and interdisciplinary discipline, transitioning between different areas of knowledge, such as physics and statistics, solid mechanics, and manufacturing processes, providing an integrated view of engineering problems.

Responsibilty Notice

The authors are the only responsible for the printed material included in this paper.

The translation of the original article written in Portuguese into English was carried out using Google's Notebook LM tool.

Acknowledgements The authors express their gratitude to CGU (General Coordination of the University—UNICAMP) and to the Support Fund for Teaching, Research and Extension—FAEPEX/UNICAMP, for the financial support received.

At the time of the reactivation of the metrology laboratory, the authors express their gratitude to the technicians of the undergraduate laboratories of FEM, Thiago Gonzaga and Marcos Saramago. In particular, to the FEM maintenance technicians, Edmur Antunes and Jair, and to Sergio Luis Zarpellon, for all the help and efforts made.

The author Robert Cooper expresses his recognition and gratitude to Prof. Dr. Olívio Novaski, for planning the implementation of the laboratory and sparing no effort for its implementation at the former UNICAMP Technology Center, for creating the support materials and inspiring generations of young people in the application of metrology concepts.

References

Abele E, Metternich J, Tisch M, Chryssolouris G, Sihn W, Elmaraghy H, Hummel V, Ranz F (2015) Learning factories for research, education, and training. The 5th conference on learning factories. Proc CIRP 32:1–6

Abeysekera L, Dawson P (2015) Motivation and cognitive load in the flipped classroom: definition, rationale and a call for research. Higher Educ Res Dev 1:1–14

Albertazzi GA, Souza AR (2008) Fundamentos de metrologia científica e industrial. Editora Manoele

Alves MF (2003) ABC da Metrologia Industrial. Instituto Superior de Engenharia do Porto, 2ª Edição

Barrows HS (1998) The essentials of problem-based learning. J Dent Educ 62(9):630–633

Bishop JL, Verleger MA (2013) The flipped classroom: a survey of the research. In: ASEE national conference proceedings, Atlanta, GA, vol 30

Bonwell CC, Eison JA (1991) Active learning: creating excitement in the classroom, ASHEERIC Higher Education Report n.1, George Washington University, Washington, DC

Chryssolouris G, Mavrikios D, Rentzos L (2016) The teaching factory: a manufacturing education paradigm. 49th CIRP conference on manufacturing systems (CIRP-CMS 2016). Proc CIRP 57:44–48

Confederação Nacional Da Indústria. Desafios Para Indústria 4.0 No Brasil. 2016. Brasília, CNI. 30 Maio 2019. http://www.Portaldaindustria.Com.Br/Publicacoes/2016/8/Desafios-Para-Industria-40-No-Brasil

Enke J, Tisch M, Metternich J (2016) Learning factory requirements analysis—requirements of learning factory stakeholders on learning factories. 5th CIRP global web conference research and innovation for future production learning. Proc CIRP 55:224–229

IMF (International Monetary Fund) (2013) World economic outlook—hopes, realities, risks, world economic and financial survey. International Monetary Fund. 30 maio 2019 https://Www.Imf.Org/External/Pubs/Ft/Weo/2013/01/

Klieme E, Leutner D (2006) Competence models for assessment of individual learning outcomes and the evaluation of educational processes. Zeitscrift fur Padagogik 52:876–903

Konopka CL, Adaime MB, Mosele PH (2015) Active teaching and learning methodologies: some considerations. Creat Educ 6(14):1536–1545

Neto SJC (2012) Metrologia e controle dimensional. Elsevier, Rio de Janeiro

Newstetter WC (2006) Fostering integrative problem solving in biomedical engineering: the PBL approach. Ann Biomed Eng 34(2):217–225

Novaski O (2013) Introdução à Engenharia de Fabricação Mecânica. 2ª edição, Blucher, São Paulo

Novaski O (2013a) Metrologia Industrial. Apostila. 7ª edição; Faculdade de Engenharia Mecânica, UNICAMP

Prince M (2004) Does active learning work? A review of the research. J Eng Educ 93(3):223–231. https://doi.org/10.1002/j.2168-9830.2004.tb00809.x

Smith BL, MacGregor JT (1992) What is collaborative learning? National Center on Postsecondary Teaching, Learning, and Assessment at Pennsylvania State University

Souza PK, Sobrinho MD (2005) Instrumentos de Medida e Sistemas de Instrumentação. Apostila. Faculdade de Engenharia de Ilha Solteira, UNESP

Tisch M, Hertle C, Cachay J, Abele E, Metternich J, Tenberg R (2013) A systematic approach on developing action-oriented, competency based learning factories. Proc CIRP 7:580–585

Tominaga LKG, Cooper RE (2023) Proposta de metodologia de ensino em learning factories no âmbito da indústria 4.0. In: XLIII Encontro Nacional de Engenharia de Produção. Fortaleza, Brasil

Tupa J, Simota J, Steiner F (2017) Aspects of risk management implementation for industry 4.0. 27th international conference on flexible automation and intelligent manufacturing, Modena, Italia. Proc Manuf 11:1223–1230

Vocabulário Internacional de Metrologia: Conceitos fundamentais e gerais e termos associados (VIM 2012) (2012) INMETRO, Duque de Caxias, RJ, 94 p

Wagner U, Algeddawy T, Elmaraghy H, Müller E (2012) The state-of-art and prospects of learning factories. 45th CIRP conference on manufacturing systems. Proc CIRP 3(1):109–114

Study of Residual Stresses in Laser Cutting of High Strength Welded Tubes for Use in Vehicle Impact

Luis Fernando Folle, Tiago Nunes Lima, Matheus Passos Sarmento Santos, Renato Batista da Cruz, Marco Antônio Colosio, and Rodrigo Santiago Coelho

Abstract Vehicle safety devices play a crucial role in protecting occupants and reducing the severity of traffic accidents. Airbags, protection bars, seat belts, the Anti-lock Braking System (ABS), and others perform specific functions to mitigate damage in the event of collisions. Side protection bars, also known as door beams, are important components in vehicles, designed to enhance occupant safety in the event of a side collision. These bars are generally made of high-strength tubular or stamped steel and are installed inside the vehicle doors. Among the design configurations, the focus was on those made with martensitic steel tubes welded to stamped parts known as brackets. These high-strength tubes must be laser cut, a technology widely used in the automotive industry, especially for cutting high-hardness sheets, such as those with high martensite content. However, there are drawbacks associated with this type of cutting, such as splashes of molten material expelled by gas pressure, irregular cutting roughness, and laser residue when the cut is made on part of a tube. This requires careful planning of the cut to avoid compromising the performance of the produced part. Another drawback associated with laser cutting and subsequent welding is the residual stresses present in the material after processing due to the associated temperature gradient. Therefore, in this study, an analysis of this residual was conducted to quantify its value and region, where it was found that they can reach values on the order of 1000 MPa. This is not exactly a problem, as corresponding values have been found in the literature, and the material, being predominantly martensitic (with a maximum stress of 1600 MPa), can accommodate this level of stress.

Keywords Steel tube · 22MnB5 · Numerical simulation · Laser cutting

L. F. Folle (✉)
Federal University of Santa Maria—UFSM, Santa Maria, RS, Brazil
e-mail: folle.luis@ufsm.br

T. N. Lima · M. P. S. Santos · R. B. da Cruz · R. S. Coelho
SENAI CIMATEC University, Salvador, BA, Brazil

M. A. Colosio
General Motors Do Brasil, São Caetano Do Sul, SP, Brazil

V. E. Beal and A. S. Ribeiro Junior (eds.), *Advances in Manufacturing Engineering Research*, Lecture Notes in Mechanical Engineering,
https://doi.org/10.1007/978-3-032-09858-0_20

1 Introduction

Laser cutting is a well-established tool in the industry. However, there are drawbacks, such as residual stresses due to the high temperature gradient in the cutting region. This problem is difficult to measure where the most commonly used method is X-ray diffraction, which is expensive and complex. Another way to assess this type of condition is through numerical simulations, but these are susceptible to errors due to a lack of information about the material, model adjustment, laser data, and other factors.

Most technical articles correlating lasers with residual stresses focus on welding, used for joining materials, or as an additive manufacturing technology. However, some studies have investigated specific cutting situations, such as the work of Arif et al. (2009), which studied the effect of cutting on a 15 mm sheet through simulation and X-ray diffraction (XRD) measurements. A similar study was conducted by Yilbas and Akhtar (2011), utilizing a different material and a 2 mm wall thickness. In the study by Maeda et al. (2015), a method of light edge pressing was proposed to reduce the effect of residual stresses in steel sheets that would be subsequently stamped. These stresses cause distortion in the part after stamping, resulting in an undesirable shape. In the work by Yilbas et al. (2010), the effect of laser cutting on parts with sharp edges was analyzed, where it was found that the temperature is elevated at the sharp edge when the laser beam is aimed at this region. This, in turn, decreases the cooling rate and reduces the localized von Mises stress. In another study, Mori et al. (2021) compared the residual stresses generated by punching versus laser cutting in 1.5 GPa ultra-high-strength sheets. They observed that the residual stress at the press-cut edges of the sheet is highly tensile, while the residual stress found at the laser-cut edges is compressive. All the studies mentioned used both numerical simulation and XRD measurements to evaluate the results. However, studies on tubes are scarce, motivating this work.

In the paper presented at the 77th Annual ABM International Congress, (Folle et al. 2024) simulations were performed based on data extracted from cut and welded tubes already in use in vehicles. Preliminary tests were also performed on tubes from a new supplier that would serve as a parameter for the manufacture of side protection bars. At that time, it was impossible to obtain data directly from the cut made in the weld region, as the welding process could mask the cutting results. Therefore, more accurate data were obtained from the cut region before welding to understand its effect on the welding process.

Within this context, this article presents the results of tests performed on a tube to define the zone and dimension of the residual cut, which is the effect of the heating generated by the laser on the opposite side of the tube's cut region. This study also presents computer simulations to validate the results, which had already been presented but contained missing information. Therefore, this study highlights some adjustments that should be made to the simulation to achieve better results, as well as an analysis of whether laser cutting or the residual cut effect can in any way influence the welding results of the final part.

2 Methodology

2.1 Determining the Shape of the Laser Beam for Cutting

A previous study (Folle et al. 2024) had already simulated laser cutting, analyzing the residual effect of this cut, that is, the region impacted by the laser on the other side of the tube. It was found that the surface material melted without reaching the tube's thickness. However, because the component analyzed was a complete piece, no data were available on the cut region before welding. This resulted in a tube that was simply cut for analysis in terms of microstructure, hardness, and the residual cut region. Figure 1 shows the analyzed region of the tube, where the cut residue was observed. In section AA, only one side has this residue, indicating that the cut was made slightly higher than it should have been according to the CAD drawing. This can be seen in Fig. 2, where the cut residue is higher than the opposite side. Therefore, when the laser passes over the opposite side, there is no residue, as it does not reach any material due to its higher region.

From images of the actual part, it was possible to determine the shape of the laser. Literature shows that the laser is a double cone, as shown in Fig. 3, where the focal point is the part where the cone has the smallest diameter. Based on the cut thickness, residual, and tube dimensions, the approximate dimensions of this cone were determined, as shown in Fig. 4c

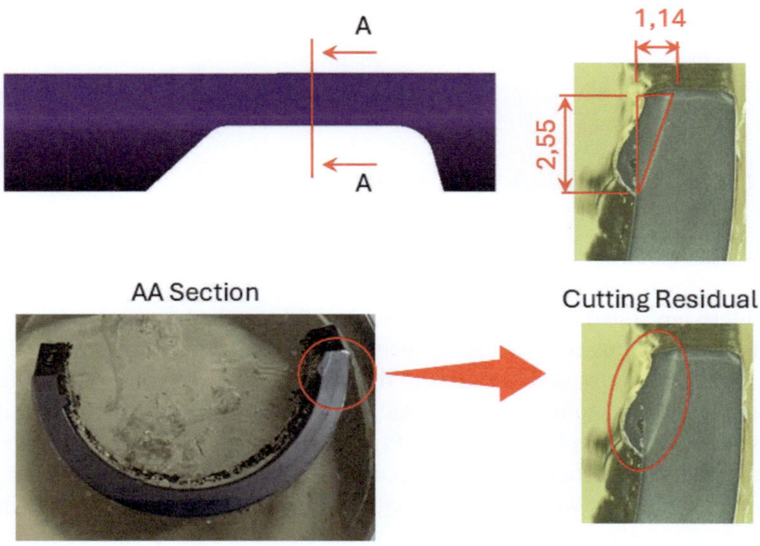

Fig. 1 Laser-cut region of the tube before welding (dimensions in millimeters)

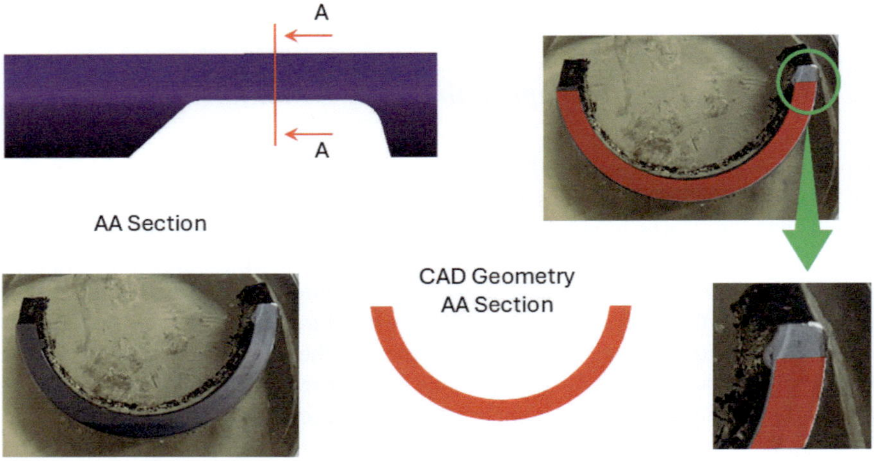

Fig. 2 Comparison between the laser cut made on the real part and the CAD file

Fig. 3 Diagram of laser focal position. *Source* (Moradi et al. 2021)

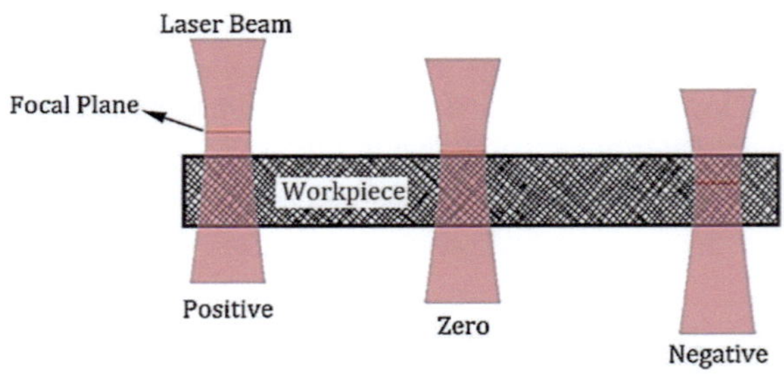

2.2 *Cutting Simulation in Simufact Welding*

The cone geometry obtained from the experimental data was loaded into Simufact Welding, correcting the previously adopted shape. However, this cone geometry did not result in a residual shape in the simulation that matched the real one. It merely demonstrated that this cone can reach the other side of the pipe, as shown in Fig. 5a and c, where the bottom corresponds to the cut residual. Therefore, the simulation was modified to use another option available in the software: the surface heat source. This is shown in the yellow image in Fig. 6. This way, the simulation can more faithfully match the actual shape of the cut residual, as shown in Fig. 7. The mesh used in the simulation was tetrahedral with a maximum element size of 2 mm and a twofold refinement in the regions close to the cut. The cutting speed adopted was 8 mm/s, and the laser power was 3000 W with 30% efficiency.

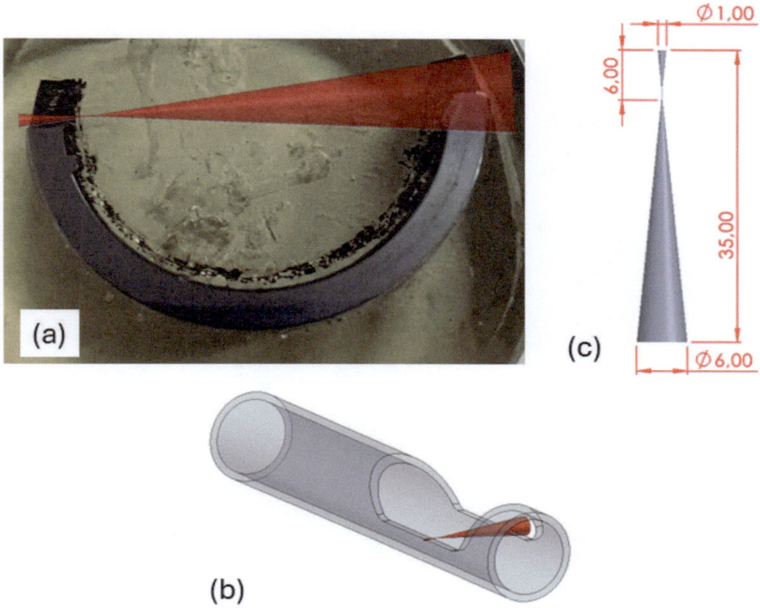

Fig. 4 **a** Actual part cut with laser cone geometry, **b** laser geometry on CAD-cut tube, **c** laser cone dimensions (dimensions in millimeters)

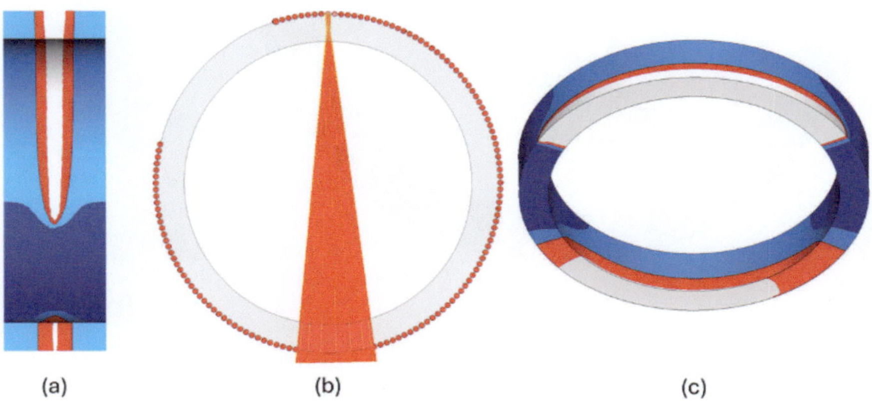

Fig. 5 Simulation performed in Simufact with the experimentally determined cone shape. **a** Side view of the tube, **b** front view of the laser cone and **c** inclined view of the molten region. The white regions correspond to the part that will be melted, the red region corresponds to the part that will be pasty in the material and the blue ones are the solids at high temperature

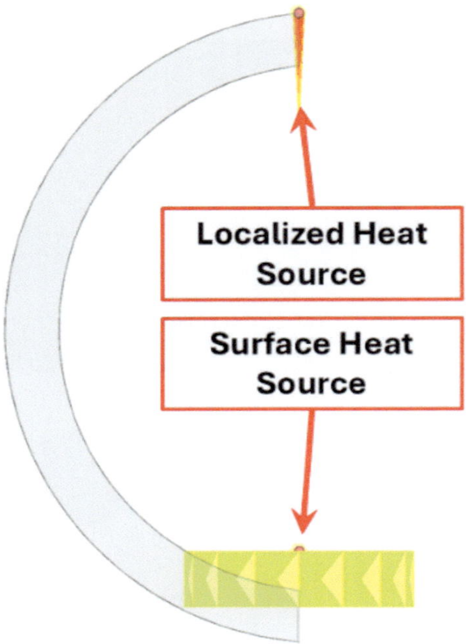

Fig. 6 Approximation made by Simufact to simulate the cutting residual

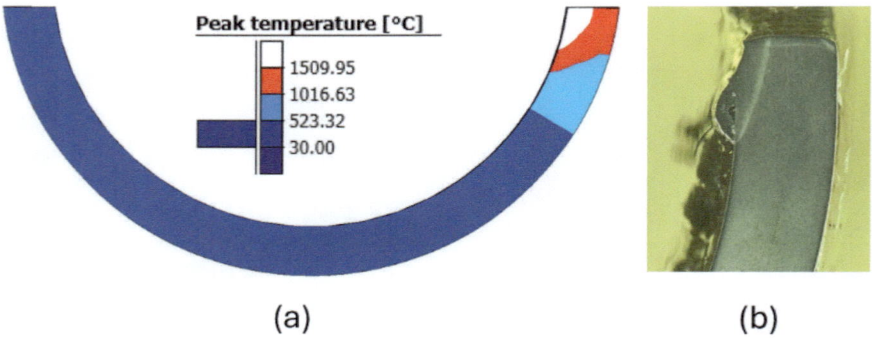

Fig. 7 Simulation with heat source correction made in the software (**a**), real cutting residual (**b**)

2.2.1 Material Selection for Simulation

After correcting the laser geometry, some simulations were performed to determine which material would be best to adopt in the results. Only the cutting part on one side of the tube was simulated. The Simufact Welding library has some types of steels, and the ones that most closely matched the properties of the tube steel were S690QL and 22MnB5. Thus, a correlation was made between the hardness results for each material with that measured in the cut part. Figure 8 shows the hardness data measured in the cut part, and Fig. 9 shows the simulation performed with the two materials from the software library. The tube before the simulation, in the software, was configured as 100% martensitic, as observed in the measured metallographies. It is possible to observe from the figures that the material that most closely matched the hardness results was S690QL. However, the rest of the part had a lower than normal hardness. This probably occurred because the software could not attribute a transformation to the regions that were not austenitized, considering the hardness of the softer phase, but which, for the necessary analyses, does not bring considerable impact, since the regions of interest are those affected by the cut. This could be easily corrected with a simulation for a previous heat treatment in the part and obtaining the correct hardness results in the rest of the part, but this was not done, as it would generate more computational time for the simulations, in addition to possible convergence problems to be solved. The difference observed between the hardness results for the two materials comes from the properties loaded into the software, which are the CCT curves of each material.

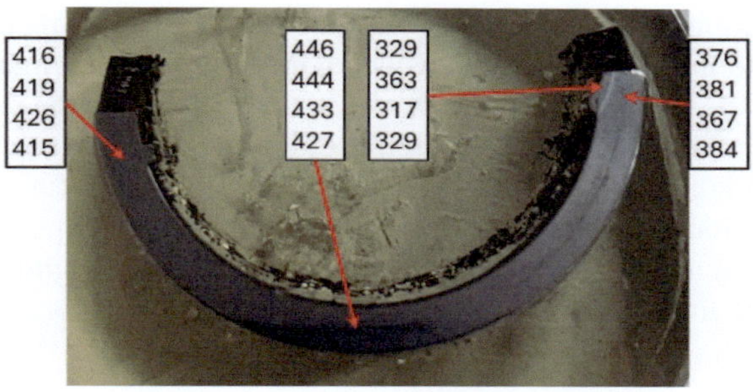

Fig. 8 Vickers hardness data obtained in each region of the cut part

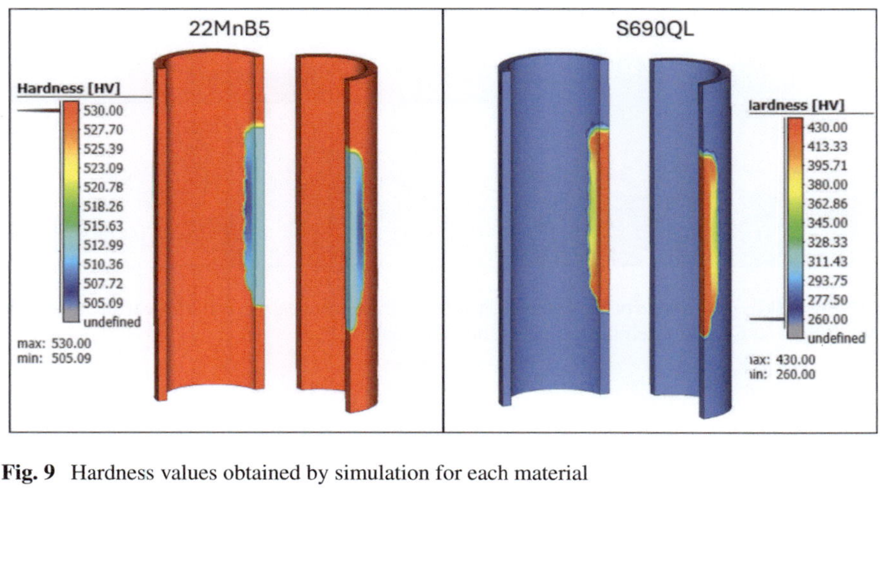

Fig. 9 Hardness values obtained by simulation for each material

2.3 Welding Simulation in Simufact Welding

During the complete evaluation of the component generated through the weld between the tube and the bracket, the hypothesis arose that the laser cutting of the tube, which precedes welding, with the residual, could influence the results of the joining weld. Thus, 4 variations were made in the simulation, 2 with the tube from the cutting simulation and two with an unsimulated geometry, that is, without the laser cutting results, as if the part were cut by some method that did not generate variations in the material, such as electro-erosion, for example. The other variation is the order of weld application, that is, starting from the left or from the right. The meshes used were the same as already mentioned, and the weld and bracket material was selected as S235, which was the material with properties closest to those used in the study. The simulation was divided into 3 stages, the first 2 being the application of the weld bead, each 48 mm long, and lasting 4.8 s without waiting time, finishing at 9.6 s. This application time was based on the welding speed of 10 mm/s. The voltage used was 20 V and the current was 150 A according to measurements. The weld efficiency was 90%. In the third stage, a cooling time of up to 5 min was applied to the entire assembly, which generally reached a final weld temperature of around 45 °C. The bead geometries were obtained from the prototyped parts and are shown in Fig. 10a. The Goldak ellipsoid was defined based on the bead geometries shown. Figure 10b shows the result of the molten regions (in white in the upper figure) in the simulation and in the real welded part. There is a good correspondence between the simulation and the real, although the shape is slightly different, but this is justified in finite element comparisons. Thus, the simulation can be considered validated.

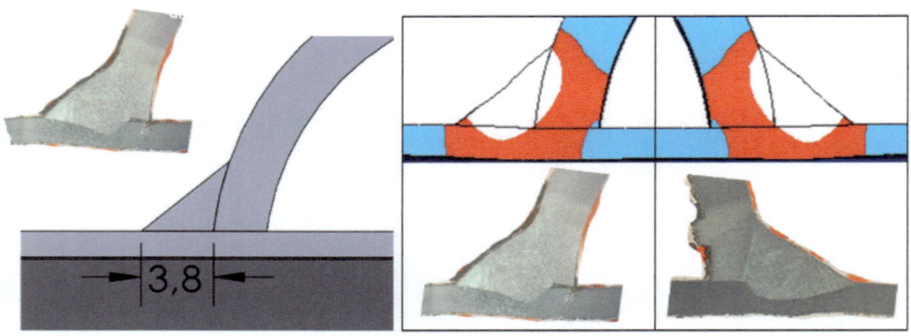

Fig. 10 Weld region **a** weld bead geometry (height has the same measure as the base with dimension in millimeters) **b** weld penetration in simulation and real

3 Results

3.1 Simulations for Evaluating Residual Stress in the Final Tube Geometry After Laser Cutting

The simulations performed during the development of the work methodology were only to calibrate the software results for the best conditions. After that, the tests were performed more realistically, that is, with the real cutting profile and with the cut and the residual passing through the entire cut zone of the part. The main objective of the cutting simulation is to observe whether there would be residual stresses that could affect or even compromise subsequent welding. Thus, Fig. 11 shows that in the cutting region, the stresses are relatively low, but there is a small region where the stresses can reach very high values, exceeding 1000 MPa. However, the hypothesis is that this alone would not be a problem, since this entire region will be welded and the material, in this region, will be reheated again. On the other side of the cut, the results are practically the same as those shown in Fig. 11.

Another observation is that there is a certain amount of other phases in the cutting residual, as shown in Fig. 12, where the main one is the bainitic phase, which according to the simulation, can reach 83%. This is in accordance with the measurements, since the observed metallography shows probable bainite islands present in the cutting residual. Figure 12c shows the hardness result in the part with the total geometry, where it is observed that the results are practically the same as in Fig. 9.

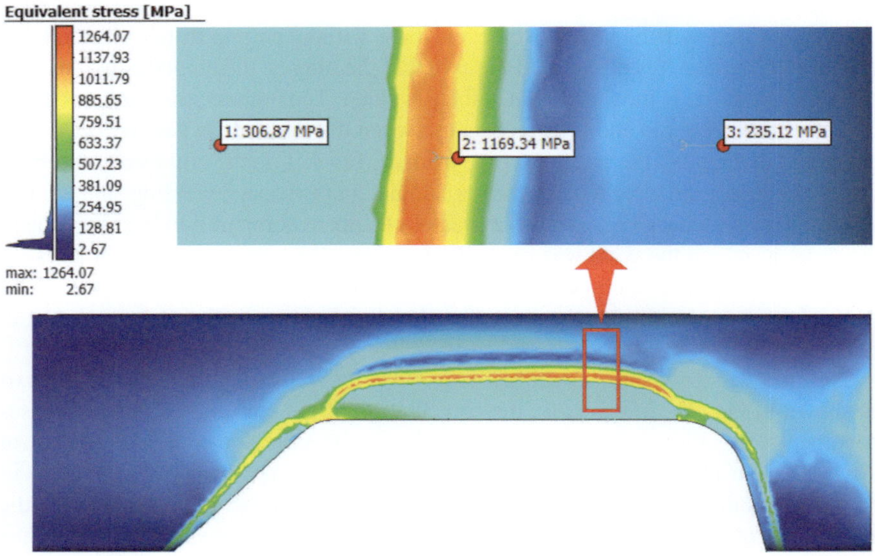

Fig. 11 Residual stress obtained by simulation after cutting

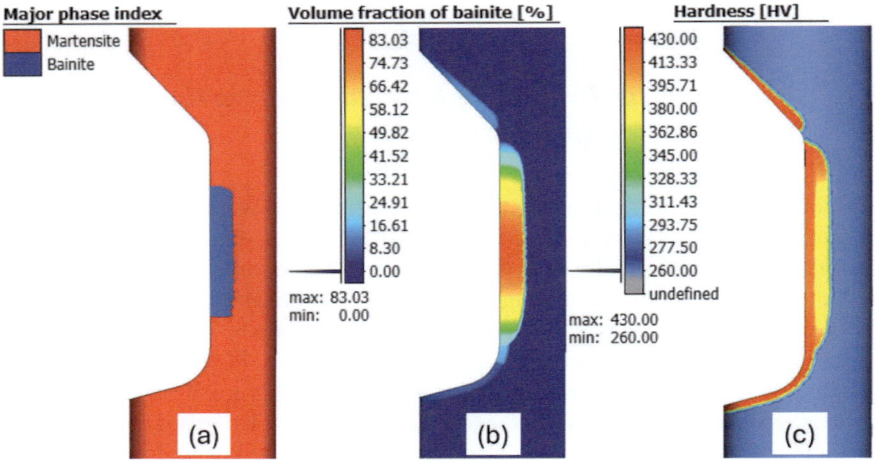

Fig. 12 **a** and **b** Phases present in the cutting residual according to the simulation and **c** hardness result in the complete part

3.2 Welding Simulations on the Tube After Laser Cutting

The main analysis was regarding residual stresses due to the difficulty in measuring them throughout the part. Figure 13 shows the stress results for each applied condition, making it possible to identify that there is a small difference on each side of the sheet, but that there are practically no changes between the stress values observed in the parts welded with the laser-cut tube and those without cutting from a visualization of the weld region. It is possible to observe that the biggest difference is in the order of welding and not in the fact that the tube is welded. The highest stress observed was in the weld bead at the end. This is because when the first side is welded, a stress is generated that tends to separate the tube from the bracket on the other side, requiring greater stress for the weld to keep the parts joined in the second welding. This can be seen in Fig. 14, where Fig. 14a has a greater clearance compared to Fig. 14b due to the order in which the weld was made, first (a) then (b), generating greater stress in (a). This same phenomenon can be observed on the underside of the bracket where the highest stresses are exactly those found in the regions welded last, as shown in Fig. 15.

Another way to visualize residual stresses is by analyzing a section of the part as shown in Fig. 16. In this case, the contribution of the cut to the residual stresses becomes more evident, increasing the maximum stress region in the upper part of the tube, right after the weld bead. This, however, does not mean that the cut will be detrimental, since highly concentrated stresses in a small region can favor the appearance of cracks, as there is less mass to dissipate these stresses. The difference between stresses with and without cutting is about 17%, but there is a larger area of action (almost double), that is, more material to accommodate these stresses.

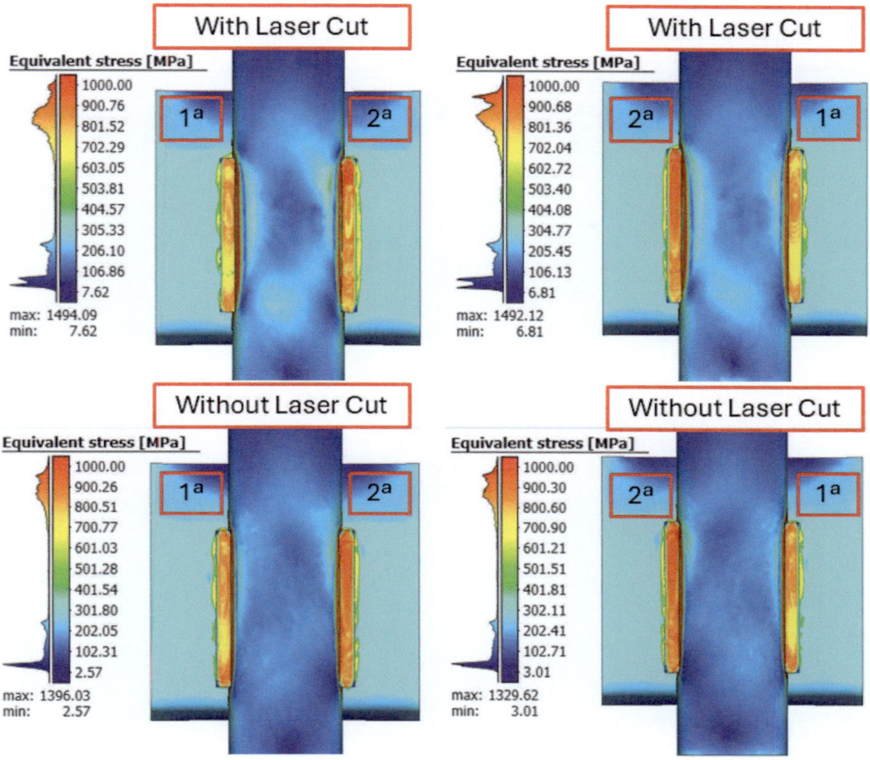

Fig. 13 Residual stresses with reduced scale

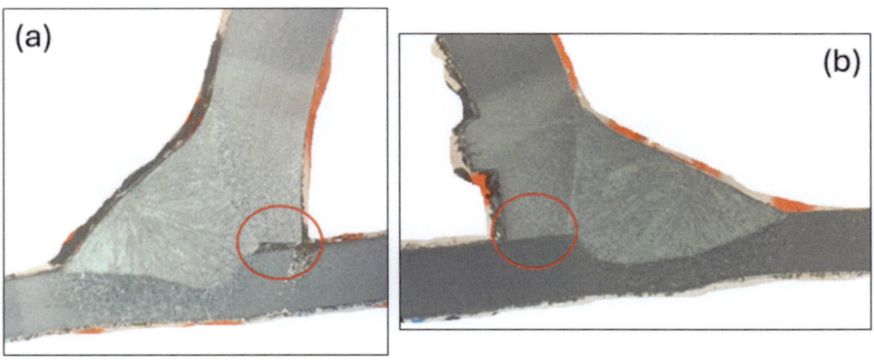

Fig. 14 Macrograph of the beads on each side of the bracket **a** left side and **b** right side

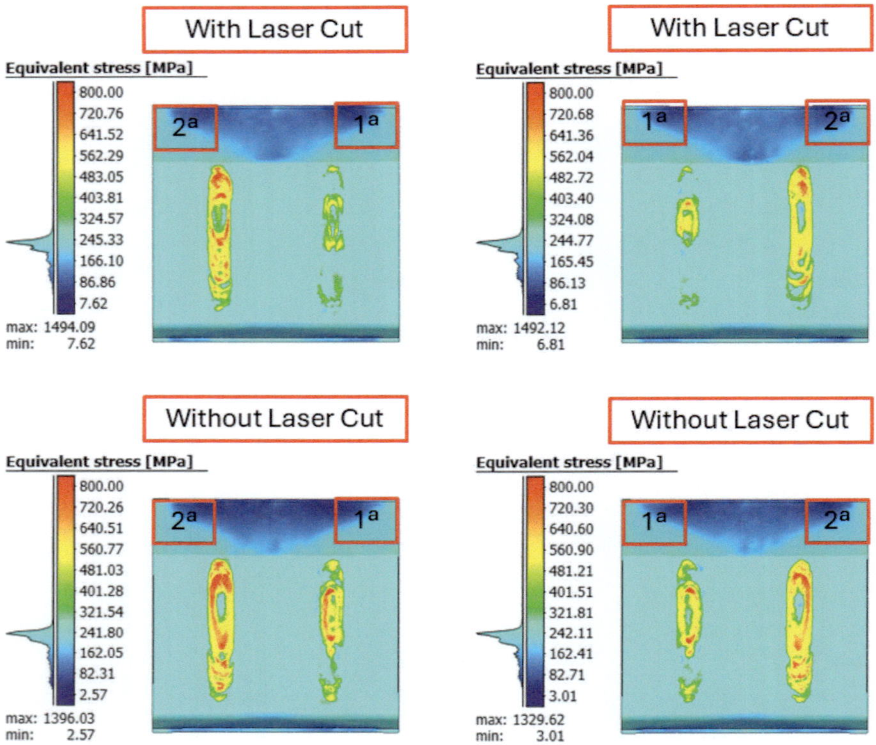

Fig. 15 Residual stresses on the underside of the bracket

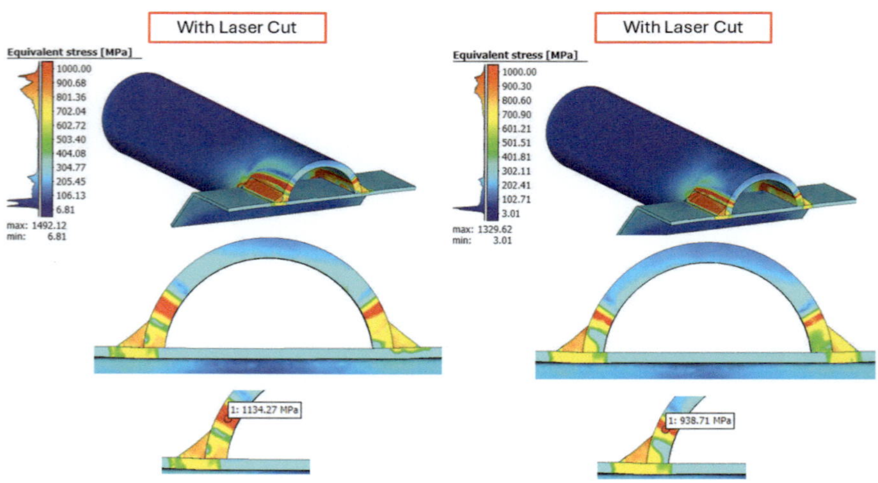

Fig. 16 Comparison between the residual stresses of parts with and without laser cutting in a section

Another comparison made was regarding hardness, both in the weld and in the tube and bracket. These results serve more for comparison or validation than for understanding the associated phenomena, as there is little correspondence with residual stresses, since more appropriate tests exist for this. However, hardness provides good information regarding the phases present in the steel. Figures 17 and 18 show the comparison between simulation results and those measured in the real part. As seen, there is good correspondence both for the weld bead and the bracket, differing in the tube. This difference in the tube, previously mentioned, probably comes from the way the software loads this information from the initial tube that did not undergo a heat treatment simulation. However, this does not prevent other analyses, such as those performed in Fig. 19, which shows the most predominant phases in each region as well as the dosage of each phase present. According to the figure, the weld region including bracket and tube has a greater presence of bainite with approximately 55%, followed by pearlite 10% and ferrite with 35%. The bracket and tube that were not affected by heat remained with their original structures, ferrite and martensite, respectively.

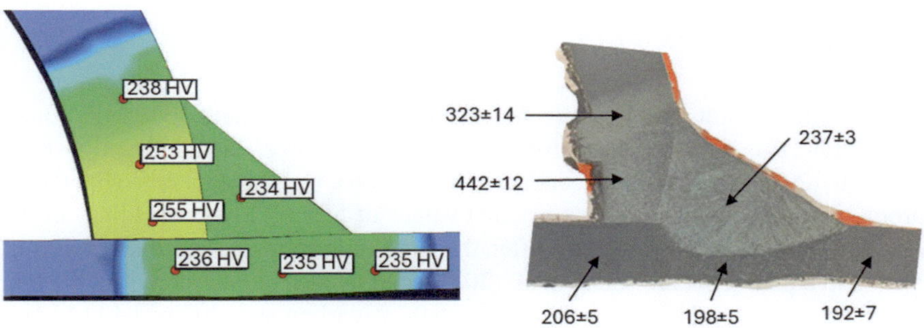

Fig. 17 Comparison between measured and simulated hardness (Vickers) of the weld on the right side

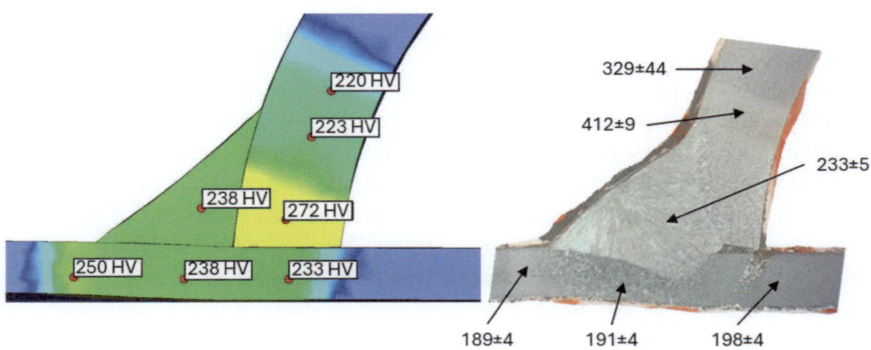

Fig. 18 Comparison between measured and simulated hardness (Vickers) on the left side

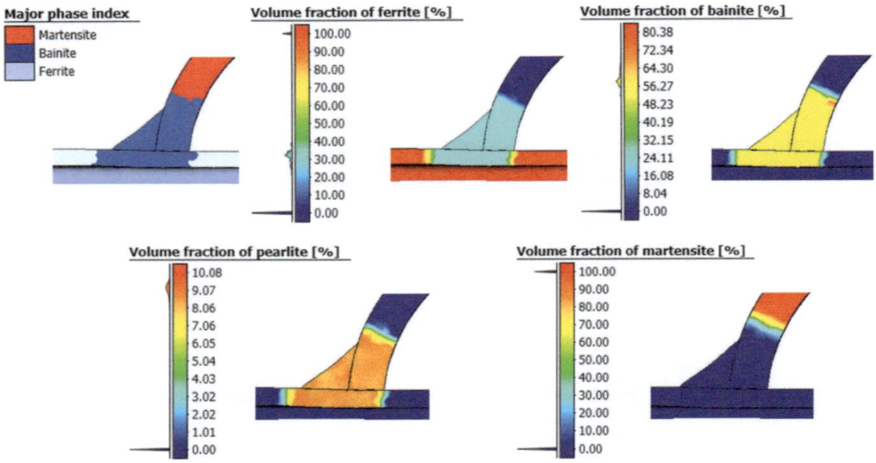

Fig. 19 Phases present in welding according to the simulation

3.3 Correspondence with Literature

Residual stress results can be much higher than expected (around 1000 MPa) in the weld region. This is characteristic for most steels regardless of chemical composition. Wen et al. (2013), for example, conducted a study on a welded steel tube with 0.04% carbon, where the residual stress reached values of 530 MPa after welding. In this case, the stresses were smoothed after the tube underwent an internal expansion process to equalize them. Zhang et al. (2022), conducted a study regarding the type of path the weld would follow in a steel with 0.2% carbon and obtained 400 MPa in some regions. Wang et al. (2017), conducted a study on a high-strength ferritic steel with 0.09% carbon to evaluate the type of material model to predict residual stresses. They reached values of more than 1000 MPa in virtual simulations and these were validated by measurements. Knoedel et al. (2017), also conducted a study using steel with 0.13% carbon and found residual stresses of the order of 600 MPa, close to measurements made on the part.

From the works presented above, it is clear that residual stresses do not necessarily depend on how resistant the material is, but rather on the aspect of welding and the restriction that is made. In the case of the tube welded to the brackets, restrictions were placed on both the tube and the bracket near the weld, as the total geometry of the side protection bar would take too long to simulate due to the size of the part in relation to the mesh, making the analyses unfeasible. It is believed that the stresses presented very high values due to the restrictions that were placed near the welds and that are not necessarily real. In any case, high residual stress values are not impossible to obtain after the welding process, as other works have already demonstrated this possibility; however, it is more likely that these values will be lower in the real part that will be produced.

Another important observation is that the tube made of 22MnB5 has a martensitic structure and remains martensitic after welding in the zone of maximum residual stresses, and in this case, these stresses are within the maximum stresses in the tensile test (which can reach 1600 MPa, depending on how the material is treated) and can transform into deformation to accommodate this level of stress and not cause localized cracks.

4 Conclusions

The interest in evaluating residual stresses from the welding process, through finite element (FE) simulation, has increased in recent decades. The prediction of residual stresses and deformations is important for the industry, as it can ensure a more efficient mechanical design and manufacturing process, also reducing the time for post-welding treatments. More specifically, in structural engineering, welding residual stresses can have a detrimental effect on fatigue behavior or stability.

Welding simulation is a complex subject, and the application of simplified models to calculate residual stresses is quite common in practice. For this reason, the results often deviate significantly from experimentally measured residual stresses in real welds. In this work, several analyses were performed through measurements and simulations to understand the phenomenon of laser cutting and to verify the possibility of reducing process problems. It was found that there will be a laser cutting residual on the opposite side of the tube's cutting region, but that this is practically impossible to eliminate. Also, it was verified that there will be residual stresses amplified by laser cutting on the tube side, but that this may not necessarily be a problem. It is also important to mention that there is a need to validate these results through further experimental measurements, as the simulations were calibrated only through indirect tests such as hardness, and thus validate the proposed method.

Responsibility Notice The authors are the only responsible for the printed material included in this paper.

References

Arif AFM, Yilbas BS, Aleem BJA (2009) Laser cutting of thick sheet metals: residual stress analysis. Opt Laser Technol 41(3). https://doi.org/10.1016/j.optlastec.2008.07.006

Folle LF, Lima TN, Santos MPS, Sobrinho CMDO, Colosio MA (2024) Simulação númerica do corte a laser de tubos de 22mnb5 tratados termicamente. In: ABM proceedings, pp 964–975. https://doi.org/10.5151/2594-5327-40625

Knoedel P, Gkatzogiannis S, Ummenhofer T (2017) Practical aspects of welding residual stress simulation. J Constr Steel Res 132. https://doi.org/10.1016/j.jcsr.2017.01.010

Maeda A, Jin Y, Kuboki T (2015) Light press of sheet metal edge for reducing residual stress generated by laser cutting considering mechanical properties and intensity of residual stress. J Mater Process Technol 225. https://doi.org/10.1016/j.jmatprotec.2015.05.019

Moradi M, Moghadam MK, Shamsborhan M, Beiranvand ZM, Rasouli A, Vahdati M, Bakhtiari A, Bodaghi M (2021) Simulation, statistical modeling, and optimization of CO_2 laser cutting process of polycarbonate sheets. Optik 225. https://doi.org/10.1016/j.ijleo.2020.164932

Mori KI, Nakamura N, Abe Y, Uehara Y (2021) Generation mechanism of residual stress at press-blanked and laser-blanking edges of 1.5 GPa ultra-high strength steel sheet. J Manuf Process 68. https://doi.org/10.1016/j.jmapro.2021.05.047

Wang Q, Liu XS, Wang P, Xiong X, Fang HY (2017) Numerical simulation of residual stress in 10Ni5CrMoV steel weldments. J Mater Process Technol 240. https://doi.org/10.1016/j.jmatprotec.2016.09.011

Wen S, Dong H, Zhang SY, Bannister A, Connelly M (2013) Neutron diffraction measurement of weld residual stresses in an UOE linepipe subjected to mechanical expansion. In: Proceedings of the international conference on offshore mechanics and arctic engineering—OMAE, 3. https://doi.org/10.1115/OMAE2013-11257

Yilbas BS, Akhtar SS (2011) Laser cutting of alloy steel: three-dimensional modeling of temperature and stress fields. Mater Manuf Process 26(1). https://doi.org/10.1080/10426914.2010.501092

Yilbas BS, Arif AFM, Abdul Aleem BJ (2010) Laser cutting of sharp edge: thermal stress analysis. Opt Lasers Eng 48(1). https://doi.org/10.1016/j.optlaseng.2009.03.006

Zhang H, Ouyang Z, Li L, Ma W, Liu Y, Chen F, Xiao X (2022) Numerical study on welding residual stress distribution of corrugated steel webs. Metals 12(11). https://doi.org/10.3390/met12111831

Mathematical Modeling, Mechanical Properties, Fatigue Resistance of ASTM A572 Steel and Welded Joints Obtained by GMAW and FCAW

Fernando Vidal dos Santos, José Rubens Gonçalves Carneiro, and João Pedro Carneiro

Abstract This work characterized the microstructure and evaluated the mechanical behavior of ASTM A572 steel before and after GMAW and FCAW welding processes for base metal, fusion zones, and heat-affected zones. The microstructures were observed using optical and scanning electron microscopy. Characterization techniques included optical microscopy, scanning electron microscopy, and backscattered electron diffraction. The mechanical behavior was evaluated by tensile tests, hardness tests, fracture toughness tests (linear elastic fracture mechanics), and fatigue crack propagation tests ($da/dN \times \Delta K$). The fatigue tests were performed with ratios ($R = 0.1$) applying empirical models from Paris (1963), Bergner (2000), and the more recent one proposed by Jones et al. (2012) to predict the material's behavior. Simulation models for the tensile test were implemented using the finite element method with Abaqus software, and simulation models for the fatigue test were implemented with Ansys software. The simulation results were compared with the tensile tests to evaluate the validity of these models. For the ASTM A572 steel, the crack propagation rate in the fusion zone of both welded joints was lower than in the base metal. The base metal and the welded joints showed an R-ratio dependence on fatigue. The empirical models were satisfactory for fitting the experimental data. Comparing the empirical and numerical results with the experimental ones, a similar behavior is observed between the curves for the base metal, where the maximum error obtained was 17%, but in both welded specimens, the error was above 20%. Therefore, a mathematical adjustment was made to the constants for the Jones model, with this change in the constant 0.036 varying with the crack propagation threshold value ($\Delta K_{\text{threshold}}$) fitting the curves and providing an error of less than 2%. The methodology used for life prediction in this work can help in deciding and determining maintenance schedules for equipment and in project development for more reliable life predictions.

F. Vidal dos Santos (✉) · J. R. G. Carneiro · J. P. Carneiro
Pontifícia Universidade Católica de Minas Gerais, Belo Horizonte, MG, Brazil
e-mail: fv.vidal@hotmail.com

V. E. Beal and A. S. Ribeiro Junior (eds.), *Advances in Manufacturing Engineering Research*, Lecture Notes in Mechanical Engineering,
https://doi.org/10.1007/978-3-032-09858-0_21

Keywords ASTM A572 · GMAW and FCAW welding · Corrosion resistance · Finite elements

1 Introduction

The ASTM A572 Grade 50 structural steel is a micro-alloyed steel produced in the form of hot-rolled plates and profiles. Its fundamental requirements are higher mechanical strength and greater toughness compared to carbon steels. These properties have enabled the use of thinner sections, leading to weight reduction and the maintenance of the structure's service life, and several studies have been conducted for this application (Silva 2018; Lee and Choi 2021).

The safety and reliability of joints require the evaluation of fatigue life due to their sensitivity to loading and the presence of these discontinuities, with the help of fracture mechanics theory (Jie et al. 2018). A welded joint consists of three regions from a microstructural point of view: the base metal, the heat-affected zone, and the fusion zone. These regions depend on the parameters used in the welding process and have complex microstructures that make it difficult to evaluate mechanical properties such as resistance to fatigue crack propagation (Wang and Wang 2022).

Analytical simulation uses experimental data from the evolution of crack length with the number of cycles, which involves the threshold and Paris regimes. In the threshold regime, the $\Delta K_{\text{threshold}}$ is a crucial parameter for obtaining maximum loads under non-crack propagation conditions, being influenced by the load ratio, residual stresses, environment, and microstructure.

Some authors (Cui and Zhang 2024) have studied crack growth behavior in the base metal, heat-affected zone, and weld metal regions. The results indicated that the lowest crack propagation rate (da/dN) and the highest ($\Delta K_{\text{threshold}}$) were obtained for the weld metal compared to the base metal and heat-affected zone. In particular, the grain size, the percentage of the second phase, and the type and distribution of inclusions present in the microstructure, influence fatigue crack growth in the threshold region. With the need to simulate the mechanical behavior of materials, commercial finite element programs such as Ansys emerged. If the welded joint is continuous, the finite element technique can be used to model fatigue crack propagation and compare it to existing analytical models in the literature (Alshoai and Fageehi 2024).

Analytical models available in the literature are studied and compared with the crack propagation behavior of ASTM A572 steel with a notch in welded joints obtained by GMAW and FCAW in the fusion zones, when subjected to cyclic loading with a stress ratio of $R = 0.1$ at room temperature. Simultaneously, the microstructures of the ASTM A572 steel and the welded joints are characterized, tests are performed on the crack propagation stages, and fatigue test simulation models are implemented using Ansys.

The analysis of components undergoes development through studies of fatigue, structural reliability, and numerical methods of structural analysis in industries (Defiltro and Gomes 2024). The durability of a piece of equipment can be defined as the structure's ability to maintain its mechanical performance throughout its life; therefore, there is a close relationship between durability and safety (Makhutov and Gadenin 2020). With the adjustment of the analytical model and finite element modeling, it was possible to find the average error in the crack length versus number of cycles curve, which serves as a reference for promoting interventions in structures and equipment that use ASTM A572 steel and welded joints.

2 Materials and Experimental Procedures

2.1 Materials

ASTM A572 steel and welded joints, with a thickness of 6.3 mm, were studied. The joints had a "V" groove type, a 60° angle, and a weld opening of 3 mm, in accordance with AWS D1.1 (2002). The wires (AWS A5.28-96) used were "A-GMAW" for the ER 70S-6 wire with the GMAW welding process and "B-FCAW" for the 81 T1-A1 wire with the FCAW welding process, both with diameters of 1.2 mm. The chemical composition of the welded joints was determined by optical emission spectrometry (OES) and is presented in Table 1.

The base metal is characterized by low-carbon content with the presence of microalloying elements Nb and Ti. The microalloying elements can form second-phase particles, reduce grain size, and retard recrystallization. The filler metals have low-carbon and alloying elements that can promote phase transformation during solidification in the weld pool. Comparing Table 1 with that of ASTM (ASTM International 2024), it can be stated that the ASTM A572 steel is a low-alloy steel.

Table 1 Chemical composition of steel rails (% by weight)

Material	C	Mn	Si	P	S	Cr	Ni
ASTMA572	0.1800	1.51	0.35	0.0016	0.0080	0.020	0.010
A-GMAW	0.0830	1.49	0.88	0.0219	0.0120	0.029	0.011
B-FCAW	0.0037	0.88	0.34	0.0210	0.0160	-	0.200
	N	Mo	V	Al	Cu	Nb	Ti
ASTMA572	0.006	–	0.0020	0.002	0.010	0.014	0.014
A-GMAW	–	0.0026	0.0027	0.002	0.022	–	–
B-FCAW	–	0.5100	–	–	–	–	–

2.2 Microstructural Analysis

The samples used for microstructural examination were cut, ground, polished, and etched with a 2% Nital solution. The techniques of optical microscopy, scanning electron microscopy, and EBSD were used to characterize the base metal, heat-affected zone, and fusion zone. The samples were analyzed in a confocal microscope, obtaining stitching images, and an optical microscope for detailed analysis of the base metal (BM), heat-affected zone (HAZ), and fusion zone (FZ) regions. The samples were also prepared for analysis of contours and texture of the HAZ and FZ regions in a scanning electron microscope (SEM).

2.3 Mechanical Properties

The specimens used in the mechanical tests were obtained by cutting the plates in the direction perpendicular to the rolling direction, along the width of the plate, with extra material for subsequent machining. The micro tensile specimens were taken from the welds of the welded joints themselves, as shown in Fig. 1.

2.4 Vickers Micro Hardness and Nano Hardness Tests

The Vickers micro hardness and nano hardness profiles were taken on the embedded samples of the BM, "A-GMAW" and "B-FCAW" welded joints of ASTM A572 steel. In the Vickers micro hardness tests, the applied load was 98.07 mN ($HV_{0.01\ kgf}$). Ten random impressions were made on a 10 mm × 10 mm sample of the embedded and chemically etched BM. On the "A-GMAW" and "B-FCAW" welded joints, 3 measurements were made on the weld bead, spaced 0.5 mm apart and 0.3 mm from each other, as shown in Fig. 2a. In the nano hardness tests, the indentation load applied was 20 mN. Fifteen indentations were made along the depth of the cross-section of the welded joints as shown in Fig. 2b.

2.5 Tensile Tests

The tensile and micro tensile specimens follow the reference standard ASTM E8M (2013). In Fig. 1b–c, the dimensions of the specimens manufactured for the BM, "A-GMAW" and "B-FCAW" welded joints of ASTM A572 steel are shown. The strain rate used in the tensile tests was $6.7 \times 10^{-4}\ s^{-1}$ up to the yield load, and subsequently, the strain rate was increased to $1.8 \times 10^{-2}\ s^{-1}$ until the specimens fractured. The gauge length used for elongation evaluation was 50 mm. The strain rate used in the

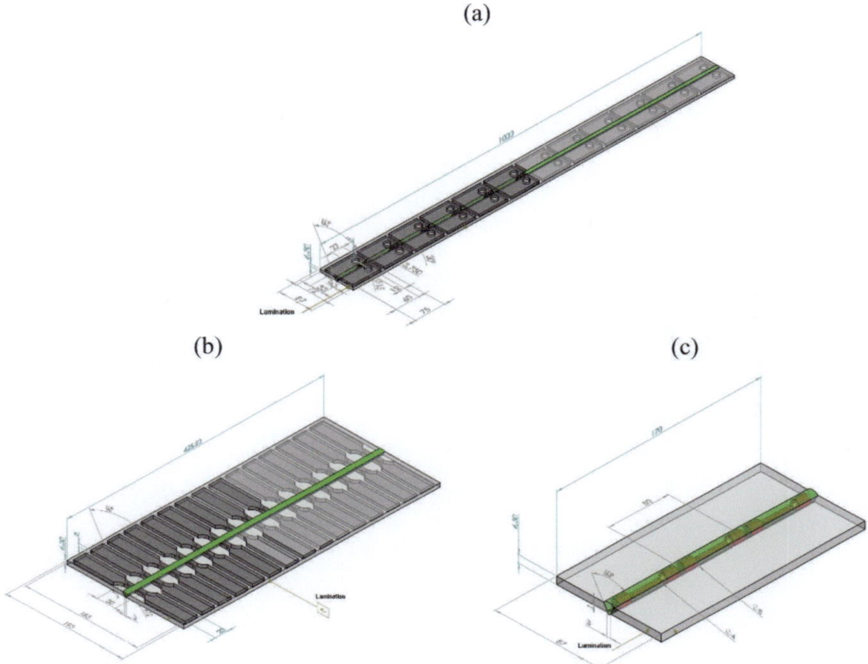

Fig. 1 Specimens used in the tests: **a** fatigue specimens, **b** tensile specimens, and **c** micro tensile specimens

micro tensile test was 9.1×10^{-4} s^{-1} up to the yield load, and subsequently, the strain rate was increased to 2.6×10^{-2} s^{-1} until the specimens fractured.

2.6 Simulations in Abaqus Software of Tensile Tests

The modeling and computational simulation in Abaqus of the tensile test of the base metal and welded joint specimens followed the flowchart shown in Fig. 3. For this modeling, Abaqus/CAE 2019 version was used.

2.7 Fracture Resistance Tests

The fracture resistance tests were performed using pre-cracked C(T) specimens, with a constant load amplitude of approximately 4.5 kN and sinusoidal variation, a loading frequency of 25 Hz, and a load ratio of $R = 0.1$ in accordance with the ASTM E647

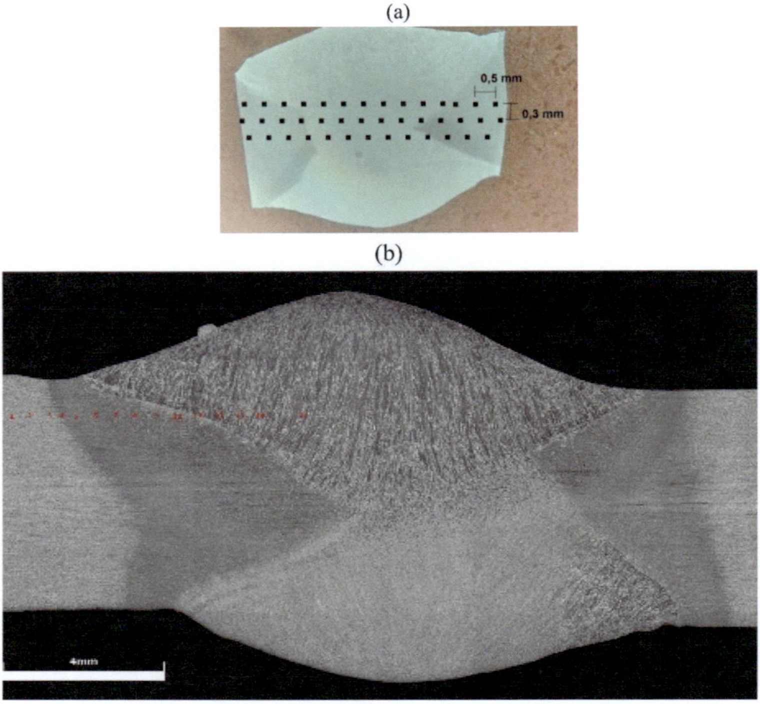

Fig. 2 Measurement scheme tests of welded joints of ASTM A572 steel: **a** micro hardness and **b** nano hardness

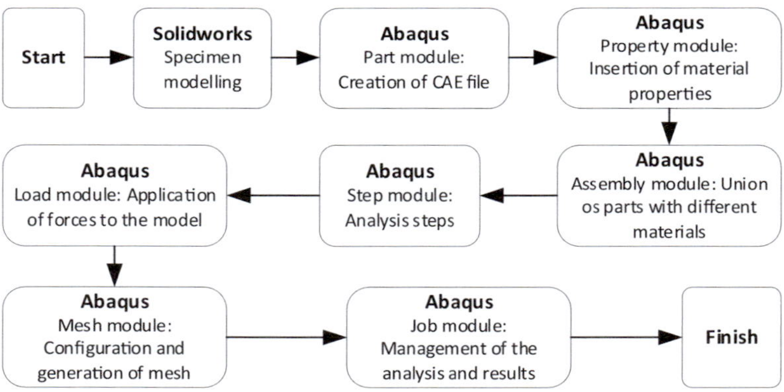

Fig. 3 Flowchart of the modeling and computational simulation in Abaqus of the tensile tests

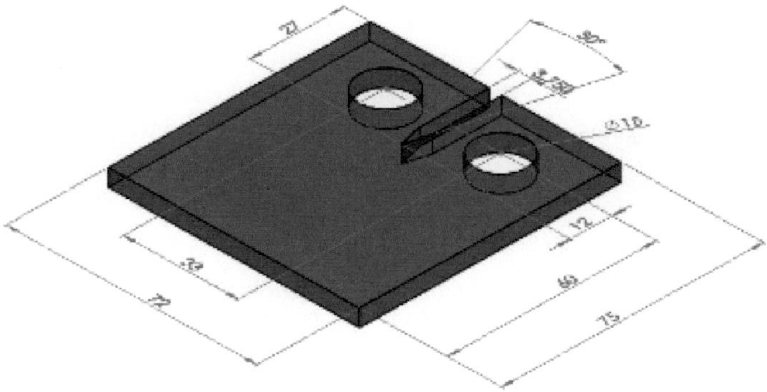

Fig. 4 Fatigue specimens of ASTM A572 steel (drilling and notching by wire EDM)

(2024) standard. The crack size measurement was performed using an Instron clip-gauge, model 2670-116, with a fixed opening of 10 mm and an amplitude of ± 4 mm. The standard establishes that the specimen must have a pre-crack with a minimum length of (0.1 B), 0.06 W, or 1 mm. Tests were performed on the BM, "A-GMAW" and "B-FCAW" welded joints of the ASTM A572 steel. The adopted dimensions are shown in Fig. 4.

The procedure adopted to obtain the $\Delta K_{threshold}$ value followed the ASTM E647 (2013) standard. The values of the geometric parameters of specimen W, B, the initial crack length of 3 mm, the notch length of 6 mm, the yield stress, and a load amplitude of 0.45 P(a) were used to create "versus" plots. The initial value of ΔK for the decreasing load tests was 20 MPa $m^{1/2}$ for the base metal and 35 MPa $m^{1/2}$ for the fusion zone and was used to calculate the crack growth rate using the expression. The procedure of K-reduction was adopted, which limits the normalized K gradient, with Eq. (1), at −0.08 mm. The crack length and the number of cycles were recorded by an acquisition system connected to the computer.

$$C = \left(\frac{1}{K}\right) \times \left(\frac{\mathrm{d}K}{\mathrm{d}a}\right). \tag{1}$$

2.8 Analytical Models Used

For comparison with the results of the fatigue crack propagation test, models were proposed to describe its behavior (from the literature), involving material constants, loading rate, and the level of acting stress, as well as the finite element method model using a commercial software such as Ansys®.

Paris and Erdogan demonstrated that fracture mechanics could be successfully used to characterize fatigue crack growth. Thus, they proposed a relationship to describe fatigue crack growth and used a power-law equation, which is an empirical relationship, determined according to Eq. (2), where C is the power-law coefficient; a material constant, m_f is the Paris exponent, a material constant, a is the crack size, N is the number of cycles, and ΔK is the range of the stress intensity factor.

$$\frac{da}{dN} = C(\Delta K)^{m_f}. \tag{2}$$

In the model proposed by Bergner and Zouhar, a new perspective on the correlation between C and m_f, proposed by the Paris and Erdogan Model, is presented. Their result shows a power-law coefficient (C_o) that is statistically independent of the Paris exponent (m_f).

Thus, through this knowledge, they present a new equation by dividing ΔK by a normalization factor (or scaling factor), (ΔK_N), obtaining Eq. (3).

$$\frac{da}{dN} = C_n\left(\frac{\Delta K}{\Delta K_N}\right)^{m_f}. \tag{3}$$

The Jones concept derives from Hartman–Schijve, which establishes that i is dependent on the amount by which it exceeds $\Delta K_{threshold}$ under the test conditions. The constant α is equal to 2, as suggested by Liu (1984) on Eq. (4). He also performs a quantitative analysis of structure-sensitive fatigue crack growth in steels. He revealed that da/dN should always be proportional as presented in Eq. (5), where E is the modulus of elasticity and σ_y is the yield stress.

$$\frac{da}{dN} = (\Delta K - \Delta K_{\text{limiar}})^2, \tag{4}$$

$$\frac{da}{dN} = \left(\frac{0.036}{E\sigma_y}\right)(\Delta K - \Delta K_{\text{threshold}})^2. \tag{5}$$

2.9 Simulations in Ansys Software of the Fatigue Tests

A computational model in Ansys 2020 R1 software was developed based on the geometry of the C(T) specimen for numerical crack propagation, based on the Paris Equation. The load amplitude and load ratio adopted in the experimental tests were considered in the computational analysis. The model is composed of second-order tetrahedral elements (SOLID 187), with mesh refinement (element size equal to 0.4 mm) performed at the crack front to increase the accuracy of the results.

The Ansys SMART "Separating, Morphing Adaptive and Remeshing Technology" tool was adopted for automatic crack propagation, following the crack growth

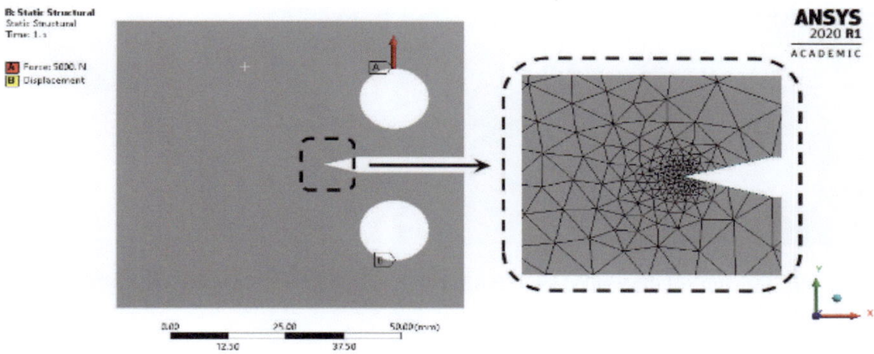

Fig. 5 Model of the C(T) specimen in Ansys software

steps defined prior to the start of the simulation. This tool performs an adaptive adjustment as the crack propagates along the specimen. Figure 5 illustrates the computational model used, indicating the applied load and the zero-displacement boundary condition, as well as the mesh refinement at the crack front.

Since crack propagation occurs perpendicularly to the applied load (mode I) and is confined to the weld metal in the welded specimens, it was decided to use the same computational model for both the base metal and the weld metal, varying only the constants of the Paris Equation for the entire specimen depending on the propagation medium under study.

Regarding the type of crack adopted in the Ansys software, a pre-meshed crack was adopted because it is suitable for the geometry of the C(T) specimen, which has an indentation to represent a pre-crack. In the software, it is possible to define regions associated with the crack front and the upper and lower faces of the crack, with nodes of the solid tetrahedral model mesh being assigned to each of the regions.

3 Results and Discussions

3.1 Microstructural Analysis

In the microstructural analyses performed by optical microscopy, it is observed that the BM region, as expected due to the chemical composition of the steel, is composed of a banded structure of ferrite and pearlite.

In the HAZ, there was a change in the banded structure with recrystallization and phase transformations. Finally, the FZ regions present in the "A-GMAW" joint are composed of columnar grains, typical of an as-cast structure in one of the beads and recrystallized grains in the lower part, while in the "B-FCAW" welded joint, the two beads present FZs with the appearance of an as-cast structure.

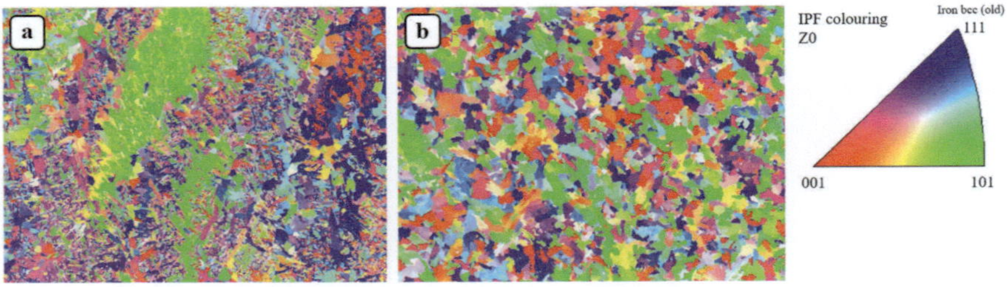

Fig. 6 EBSD texture of the welded joint (HAZ) subjected to the: **a** "A-GMAW" and **b** "B-FCAW" welding processes with ASTM A572 steel

The results of the EBSD analysis indicate that in the HAZ region of the "A-GMAW" sample (Fig. 6), there is a slightly more intense texture than in the "B-FCAW" sample, and the texture intensity is quite low for both samples. Furthermore, the GMAW sample shows signs of recrystallization due to the presence of the cubic ⟨111⟩ parallel x direction, which can be explained by the fact that this sample was welded with two passes.

3.2 *Vickers Micro Hardness and Nano Hardness*

In Fig. 7, a gradual increase in hardness is observed from the base metal to the weld metal, and within the weld metal, from the root to the face of the bead. The observed range of variation is from 130 to 200 HV. The maximum hardness value is observed near the top of the bead and does not indicate the presence of brittle constituents. This hardness profile was expected, as the filler metal has superior mechanical properties to the base metal. This fact is justified by the higher cooling rate in this region, which was not observed in the samples analyzed. It can be suggested that this difference in behavior is because the welding procedure was of the multipass type.

An increase in the hardness range is observed in the HAZ, such that the points with the highest hardness values reached 200 HV. This value is still relatively low and is associated with the finishing pass, which possibly had lower dilution, and consequently, higher Mn and Si contents. To also verify these points, it was decided to perform the nano hardness test. As with the micro hardness, it is possible to observe in the nano hardness results (Tables 2 and 3) that in both welding processes, there is a gradual increase in hardness from the base metal to the weld metal. The observed range of variation is from 272 to 507 HV. The maximum hardness value is observed in the HAZ / FZ, since the welding process leads to the presence of a typical martensitic microstructure in these regions. The modulus of elasticity varies with the microstructure.

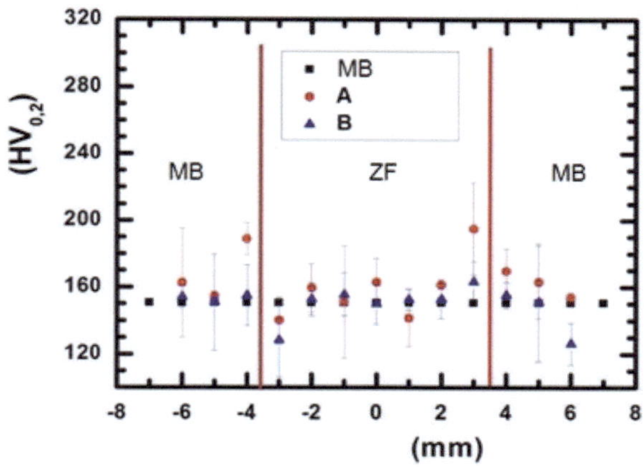

Fig. 7 Micro hardness profile of the base metal "A-GMAW" and "B-FCAW"-type joints with ASTM A572 steel

Table 2 Mechanical prop. obtained in the micro hardness tests of the "A-GMAW"-type joint with ASTM A572 steel

Designation	Hardness HV (MPa)	*E* (GPa)
ASTM A572 "A-GMAW" (BM) Average values of points 1 to 3	272.1 ± 21.1	205.0 ± 21.3
Joint "A-GMAW" (BM/HAZ) Average values of points 4 to 6	342.8 ± 30.5	190.3 ± 14.6
Joint "A-GMAW" (HAZ) Average values of points 7 to 9	331.6 ± 20.7	154.1 ± 4.0
Joint "A-GMAW" (HAZ/FZ) Average values of points 10 to 11	507.8 ± 48.9	177.4 ± 5.8
Joint "A-GMAW" (FZ) Average values of points 12 to 15	319.1 ± 79.8	171.9 ± 32.8

Table 3 Mechanical prop. obtained in the nano hardness tests of the "A-GMAW"-type joint with ASTM A572 steel

Designation	Hardness HV (MPa)	*E* (GPa)
ASTM A572 "B-FCAW" (BM) Average values of points 1 to 3	285.6 ± 38.4	231.1 ± 26.8
Joint "B-FCAW" (BM / HAZ) Average values of points 4 to 6	350.1 ± 35.6	233.5 ± 8.2
Joint "B-FCAW" (HAZ) Average values of points 7 to 9	454.1 ± 146.8	264.8 ± 48.1
Joint "B-FCAW" (HAZ / FZ) Average values of points 10 to 11	486.9 ± 104.7	272.2 ± 24.7
Joint "B-FCAW" (FZ) Average values of points 12 to 15	257.9 ± 18.5	187.6 ± 13.6

Table 4 Mechanical properties from the tensile tests of "A-GMAW" and "B-FCAW" joints with ASTM A572 steel

Designation	σ_V (MPa)	σ_r (MPa)	δ (%)
ASTM A572	420	581	32
Joint "A-GMAW"	430	598	30
Joint "A-GMAW" (micro tensile)	455	556	23
Joint "B-FCAW"	433	602	37
Joint "B-FCAW" (micro tensile)	445	558	30

3.3 Tensile Test

For the ASTM A572 steel, a yield plateau typical of low-carbon steel was obtained. The elongation value evidences the good plastic deformation capacity of this ASTM A572 steel and is in accordance with the ASTM standard for the steel (Table 4). The true strain and true stress were calculated according to the Hollomon equation. The value of the work-hardening coefficient confirms the good plastic deformation capacity of the ASTM A572 steel.

The fractures of the GMAW and FCAW welded joint specimens occurred in the base metal, which may indicate the absence of weld discontinuities. The evolution of true stress with true strain for the welded joints presents the same aspect as the ASTM A572 steel, and the work-hardening coefficient value of the ASTM A572 steel was slightly higher than that observed in the "A-GMAW" and "B-FCAW" welded joints. The disappearance of the yield plateau may have contributed to the lower plastic deformation of the welded joints when compared to the ASTM A572 steel.

In the micro tensile test of the Fusion Zone material, it was found that the yield strength, ultimate tensile strength, and elongation were similar for the "A-GMAW" and "B-FCAW" type welded joints. Since the mechanical properties are similar in both welded joints, when comparing the types of shielding used in the two welding processes, GMAW and FCAW, it is verified that the shielding flux in the FCAW process was more efficient in protecting the fusion zone in the welded joint, given its elongation, which was 30% greater. The work-hardening coefficient of the "B-FCAW" welded joint was higher when compared to the "A-GMAW" welded joint, which reveals a greater deformation capacity.

Table 4 contains a summary of the mechanical properties obtained for the ASTM A572 steel and the "A-GMAW" and "B-FCAW" type welded joints.

3.4 Simulations of the Tensile Tests

The results of the Abaqus simulation on the ASTM A572 steel and welded joint specimens for the analysis of tensile tests after deformation are presented in Fig. 8.

The results of the real tested and simulated specimens of ASTM A572 steel after deformation are presented to scale in Fig. 9.

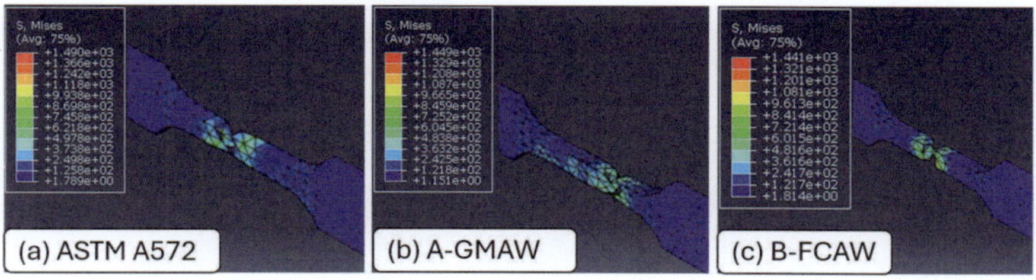

Fig. 8 ASTM A572 steel and welded joints—virtual specimen after deformation (Abaqus simulation)

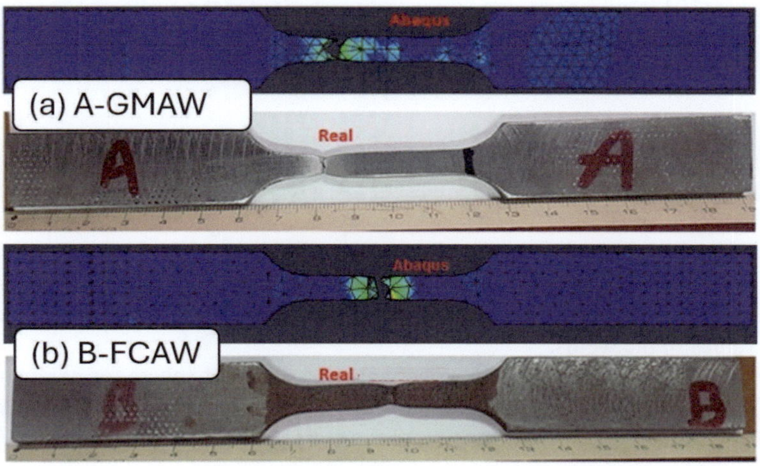

Fig. 9 ASTM A572 steel and welded joints—real tested and simulated specimens after deformation

Necking occurred in similar locations on the specimens, as shown in Fig. 9, which did not demonstrate the presence of a stress concentrator. Figure 9 compares the real tested specimens versus the simulated ones and shows that the real necking is similar to what was found in the simulations performed in Abaqus.

The finite element numerical simulation of the tensile test using Abaqus for the ASTM A572 steel base metal and welded joints proved to be efficient and reliable when compared with the tensile tests performed on the universal testing machine. The curves followed a parallel path in the elastic regime, with slightly lower values that favor the safety limit. Thus, the effects of internal failures become less significant when using this tool (Lee and Choi 2024).

In both cases, the models used were capable of reproducing the phenomenon of increased stiffness and ultimate tensile strength of the material. Ultimate tensile strength values were obtained that were slightly higher than those obtained experimentally. The fracture of the specimens in the simulation occurred at lower strain levels than those observed in the mechanical tests, with the greatest discrepancy being 3.0%.

3.5 *Fracture Resistance Tests*

In Fig. 10, it is noted that the tests showed considerable dispersion. It is verified that the increase in crack length up to 100,000 cycles showed a low crack growth rate, and from 200,000 cycles onward, this rate was higher, especially for the base metal and the "B-FCAW"-type joint.

Warley (2018), also performed three tests and adopted the curve that "best fit" the analytical models, disregarding these deviations.

3.5.1 Paris and Erdogan Model

From the previous fatigue resistance test curves, it was possible to obtain the evolution of da/dN as a function of the stress intensity factor range ΔK for the base metal (Fig. 11), in accordance with the ASTM E647 standard. The Paris model, fitted to the experimental data, was presented in these curves. The exponent "m" for the base metal and the fusion zones had very similar values, while the log intercept (C) showed lower values in the welds.

For ASTM A572 steel, the three tests yielded different values for parameters m and C, and it was not possible to identify the variability in these tests. For the data obtained, it was not possible to determine the end of stage I and the beginning of stage III. The parameters m and C of the ASTM A572 steel are close to those cited in the literature for Ferritic–Pearlitic steels (Dowling 1999).

The average value for parameter C found for ASTM A572 steel by (Warley 2018) was 1.56×10^{-8} mm/cycle. This value is on the order of 36.5 times smaller than that found in the experiments performed (5.68×10^{-9} mm/cycle). The average value

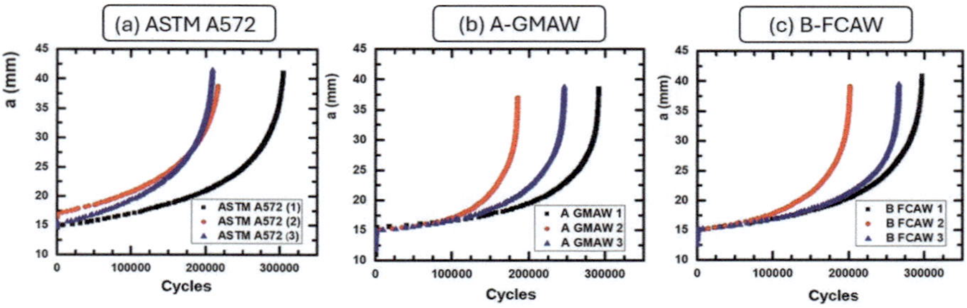

Fig. 10 Evolution of crack length as a function of the number of cycles for ASTM A572 steel and joints

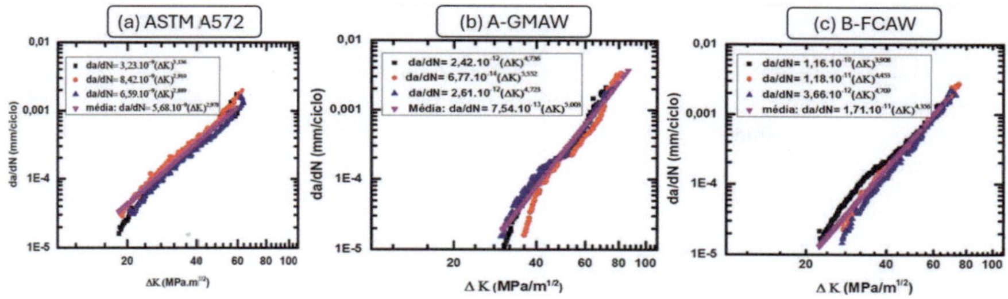

Fig. 11 Evolution of the crack growth rate as a function of the number of cycles for ASTM A572 steel and joints

for parameter m found for ASTM A572 steel by Moreira, W. L., 2018, was 2.43. This value is in the order of 22% smaller than what was found in the experiments performed (2.98).

The fatigue crack propagation behavior is different in the base metal and welded joints. For the welded joints, it can be said that the variation of coefficients *C* and *m* depends on the microstructure, grain size, second-phase particles, and precipitates (Warley 2018). The increase in the value of m favors crack growth, which decreases the fatigue life of the material; however, the increase in the value of C is more sensitive to increasing the life of the material. In this way, we can highlight that the welded joints have a longer average life when compared to the base metal.

3.5.2 Bergner Model

The Bergner and Zouhar model was calculated from the constants obtained by the Paris and Erdogan model for base metal specimens and the fusion zones. It is observed that the Bergner model is able to model stage II of the d*a*/d*N* curve for all three tests. The Bergner model highlights, from coefficients C and m, the difference in crack propagation behavior between the ASTM A572 steel and the "A-GMAW" and "B-FCAW"-type welded joints. It is observed that in stage II, the "A-GMAW" and "B-FCAW"-type joints have similar crack propagation rates, which are distinct from that of the base metal (ASTM A572) (Fig. 12).

Table 5 shows the values obtained for the crack propagation rate d*a*/d*N*, the threshold stress intensity factor $\Delta K_{\text{threshold}}$, and the fracture toughness value K_{IC}.

Just as in the Paris model, in the Bergner model for ASTM A572 steel and welded joints, the three tests yielded different values for parameters m and C. These parameters had the same values as those obtained for the Paris model. The fracture toughness value (273 MPa $\sqrt{m}$) for ASTM A572 steel is about 10% higher than for similar steels cited in the literature (Dowling 1999), which is important for defining the premature failure of a cracked structure.

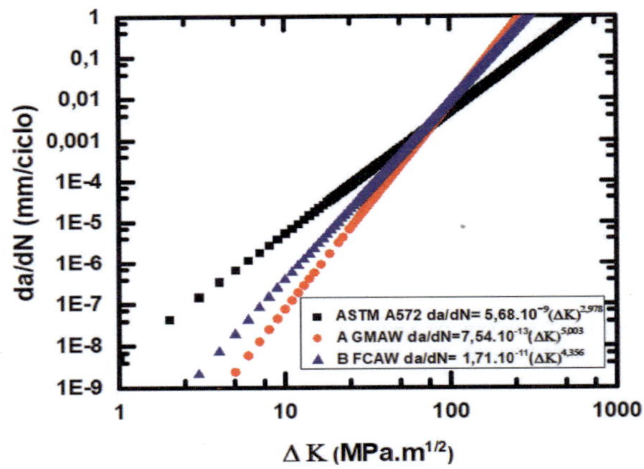

Fig. 12 Three crack growth rate regions as a function of the stress intensity factor range ΔK obtained by extrapolation of the Bergner equation from stage II for ASTM A572 steel and joints

Table 5 Values obtained for K_{IC} and $\Delta K_{\text{threshold}}$ by extrapolation of the equation proposed by the Bergner model in stage II for ASTM A572 steel and joints

Designation	Equation stage II da/dN	%1 $\Delta K_{\text{threshold}}$ MPa $\sqrt{m}$	%1 K_{IC} MPa $\sqrt{m}$
ASTM A572	$5.68 \times 10^{-9} \Delta K^{2.97}$	3 ± 0.5	273 ± 30.6
Joint A-GMAW-Wire ER 70S-6	$7.54 \times 10^{-13} \Delta K^{5.00}$	–	167 ± 13.3
Joint B-FCAW-Wire 81 T1-A1	$1.71 \times 10^{-11} \Delta K^{4.35}$	–	176 ± 15.5

3.6 Computational Model of Crack Propagation

Figure 13 presents the displacements obtained in the computational model of crack propagation on the C(T) specimen for crack lengths of 16 mm (1st propagation step), 30 mm (15th propagation step), and 45 mm (30th propagation step). It is possible to visualize the mesh adaptation at the crack front as the Ansys software performs the propagation using the SMART tool. Thirty propagation steps were adopted, with increments of 1 mm each.

Figure 14 shows the results related to the crack propagation simulations on the C(T) specimen for the base metal "BM" and the "A-GMAW" and "B-FCAW"-type welded joints, respectively. In all figures, a comparison is made of the average experimental curves of crack growth "a" as a function of the number of cycles "N" with the theoretical formulation presented by the ASTM E647 standard and with the numerical results obtained from the simulations in Ansys software.

From the three experimental curves of crack length as a function of the number of cycles, the average curve was obtained. This curve has the average number of cycles (N) and the average crack length (a), with the calculated error in Eqs. (6) and (7).

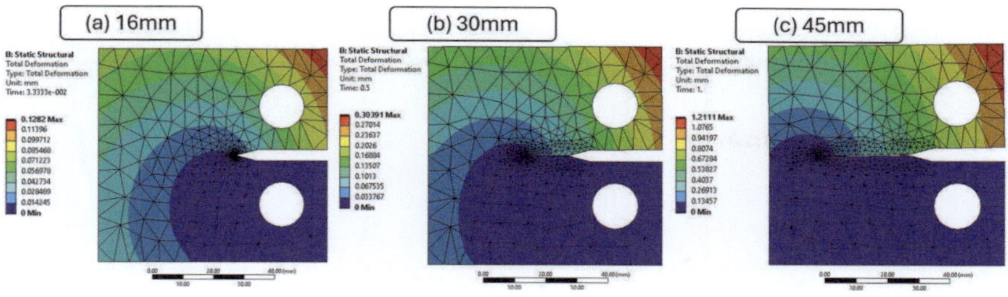

Fig. 13 Displacements in the computational model of crack propagation on the C(T) specimen for lengths **a** 16, **b** 30, and **c** 45 mm

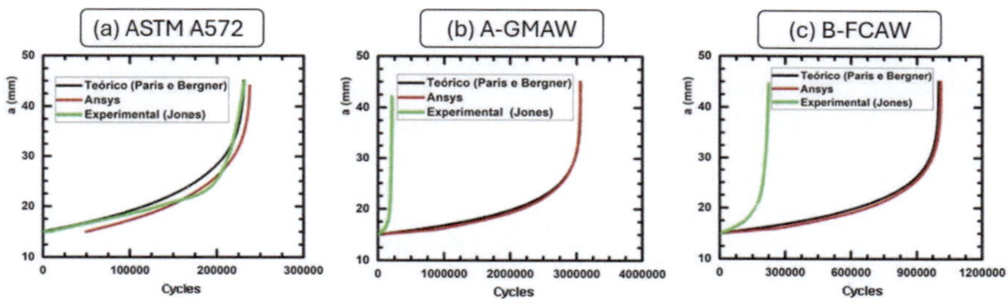

Fig. 14 Comparison between the crack growth curves as a function of the number of cycles on the C(T) specimen for ASTM A572 steel and welded joints

$$\text{Error } (a)(\%) = \left(\frac{a_{\text{experimental}} - a_{\text{modeled}}}{a_{\text{experimental}}}\right) \times 100\% \,, \tag{6}$$

$$\text{Error } (N)(\%) = \left(\frac{N_{\text{experimental}} - N_{\text{modeled}}}{N_{\text{experimental}}}\right) \times 100\% \,. \tag{7}$$

For the base metal, in the comparison of the crack growth curves, the numerical results (Ansys) are consistent with the theoretical results (Paris/Bergner) for all simulated specimens. The maximum error obtained is 3% for the base metal in the final portion of crack propagation. When comparing the theoretical and numerical results with the experimental ones, a similar behavior is observed between the curves for the base metal, with a maximum error of 17% (Jones experimental vs. theoretical).

For the welded joints, an error of over 20% was found when comparing the theoretical and numerical results with the experimental ones for both welded specimens. We believe it's necessary to perform a mathematical adjustment on the constants for the Jones model to account for this. Figure 15 shows the mathematical adjustments made in the comparison of the crack growth curves as a function of the number of cycles on the C(T) specimen for the base metal and the GMAW and FCAW-type welded joints.

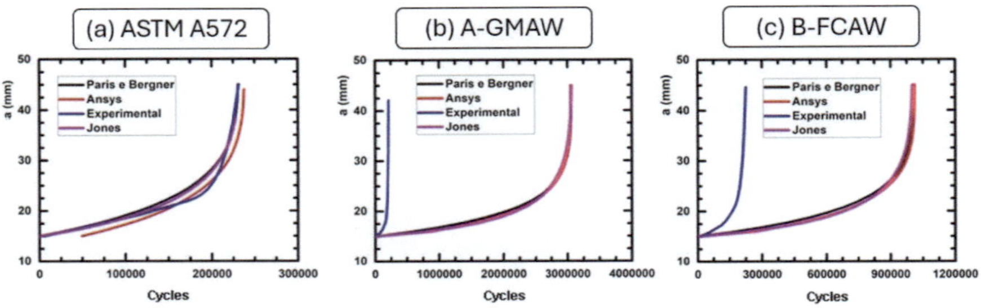

Fig. 15 Comparison between the crack growth curves as a function of the number of cycles on the C(T) specimen, adopting the mathematical adjustment in the Jones model for ASTM A572 steel and welded joints

We didn't obtain the experimental value for $\Delta K_{\text{threshold}}$ in the welded joints. For the "GMAW" welded joint, by adopting an experimental $\Delta K_{\text{threshold}}$ value of 14.95 (Eq. (8)), the constant for adjusting the curve has a value of 0.0093.

$$\frac{da}{dN} = \left(\frac{0.093}{E\sigma_y}\right)(\Delta K - 14.95)^2. \tag{8}$$

Adopting the Jones constant of 0.036, the $\Delta K_{\text{threshold}}$ should be 17.14 to adjust the curve, that is expressed on Eq. (9).

$$\frac{da}{dN} = \left(\frac{0.036}{E\sigma_y}\right)(\Delta K - 17.14)^2. \tag{9}$$

For the "FCAW" welded joint, adopting the experimental value of the $\Delta K_{\text{threshold}}$ of 14,23 the constant for adjusting the curve would have a value of 0.0223, resulting in Eq. (10).

$$\frac{da}{dN} = \left(\frac{0.0223}{E\sigma_y}\right)(\Delta K - 14.23)^2. \tag{10}$$

And, finally, adopting the Jones constant of 0.036, the $\Delta K_{\text{threshold}}$ of 15.52 to adjust the curve, gives the Eq. (11).

$$\frac{da}{dN} = \left(\frac{0.036}{E\sigma_y}\right)(\Delta K - 17.52)^2. \tag{11}$$

The mathematical adjustment adopted for the Jones model, with its constant of 0.036 varying with the $\Delta K_{\text{threshold}}$ value, adjusted the curves, providing an error of less than 2% for both the base metal and the welded joints. We observed that the values of constants C and m are very sensitive, and any small variation in the experimental data can reflect the linear regression, which will directly affect these values.

4 Conclusions

The microstructure of the ASTM A572 base metal consists of ferrite and pearlite grains in a banded structure. The welding process led to a known and expected microstructural alteration due to the thermal input, with the welding energy employed being 17% higher in the FCAW process. The HAZ/FZ interface showed micro hardness and nano indentation values typical of a martensitic structure. In the tensile and micro tensile tests, the mechanical properties of the chosen wires were compatible with the mechanical strength of the ASTM A572 steel, without compromising the mechanical strength of the welded joints. The three-dimensional computational modeling with Abaqus software reproduced the tensile test and proved to be efficient in modeling these tests, with a discrepancy of less than 3%.

The experimental results of the crack propagation rate (da/dN) as a function of the number of cycles were used to validate the analytical models of Paris, Bergner, and Jones, R. It was proven that the Bergner model could explain that microstructural variations impact constants C and m. The crack propagation rate in the fusion zone of both joints was lower than in the base metal.

The use of the finite element method with Ansys software to model crack length as a function of the number of cycles (Paris/Bergner) for ASTM A572 steel approximated the empirical results with an error of 6%.

The modeling of crack length as a function of the number of cycles for the welded joints using the analytical model of Jones, *R*., yielded a high error of over 20%. The alteration of the constant 0.036, varying with the value of $\Delta K_{\text{threshold}}$ in the Jones model, adjusted the curves, providing an error of less than 2%.

Despite the differences found, the methodology used is important for the life prediction performed in this work (especially for ASTM A572 steel). It can also help in the decision and determination of deadlines for maintenance on equipment and in the development of projects for more reliable life predictions.

5 Responsibility for the Information

The authors are solely responsible for the information included in this work. Translated with the Gemini AI version 2.0 Flash.

References

Alshoai AM, Fageehi YA (2024) Advances in finite element modeling of fatigue crack propagation. Appl Sci 14(20):9297. https://doi.org/10.3390/app14209297

ASTM International (2012) ASTM A572/A572M - 12: Standard specification for high-strength low-alloy structural steel. ASTM International, West Conshohocken, PA. Available at: https://www.astm.org/a0572_a0572m-12.html. Accessed 17 Oct 2024

ASTM International (2022) ASTM E647-22: Standard test method for measurement of fatigue crack growth rates. ASTM International, West Conshohocken, PA. Available at: https://webstore.ansi.org/standards/astm/astme64722. Accessed 17 Oct 2024

Cui YH, Zhang JL (2024) Influence of stress corrosion cracking on crack tip creep and crack propagation rate in welded joints. Strength Mater 56:610–616. https://doi.org/10.1007/s11223-024-00675-z

Defiltro VF, Gomes WJS (2016) Structural reliability analysis of problems based on solid mechanics. Electronic J Civil Eng 12(2). Available at: https://revistas.ufg.br/reec/article/download/37390/pdf/0. Accessed 17 Oct 2024

Dowling NE (1999) Metal fatigue in engineering, 2nd edn. Prentice Hall, Englewood Cliffs, NJ

Jie ZY, Li YD, Wei X, Zhuge P (2018) Fatigue life assessment of inclined welded joints in steel bridges subjected to combined normal and shear stresses. Int J Adv Steel Constr 14(4):620–633. https://doi.org/10.18057/IJASC.2018.14.4.6

Lee S-H, Choi B-J (2021a) Mechanical properties of ASTM A572 grades 50 and 60 steels at high temperatures. Appl Sci 11(24):11833. Available at: https://www.mdpi.com/2076-3417/11/24/11833. Accessed 17 Oct 2024

Lee S-H, Choi B-J (2021b) Mechanical properties of ASTM A572 grades 50 and 60 steels at high temperatures. Appl Sci 11(24):11833. https://doi.org/10.3390/app112411833

Makhutov NA, Gadenin MM (2020) Integrated assessment of the durability, resources, survivability, and safety of machinery loaded under complex conditions. J Mach Manuf Reliab 49:292–300. https://doi.org/10.3103/S1052618820040093

Silva JR (2018) Equivalent of ASTM A572 grade 50, Gr 65, 60 steel properties. Brazilian J Eng Mater 12(3), 45–52

Wang X, Wang D (2022) Microstructure and mechanical properties of welded joint of X80 pipeline steel before and after ultrasonic impact treatment. J Mater Eng Perform 31:1465–1477. https://doi.org/10.1007/s11665-020-05321-8

Warley LPM (2018) Study of fatigue crack propagation in ore reclaimer buckets. University of São Paulo, São Paulo

Analysis of Corrosion Resistance in Salt Solution of Welded Joints of AISI 409 Ferritic Stainless Steel

Henrique Rodrigues Braga, Leonel Dimas de Abreu, Marcela Rabelo Menezes Vilela, and Pedro Paiva Brito

Abstract Ferritic stainless steels are used in the "cold end" of automotive exhaust systems, in the form of welded tubes. The use of this class of steel in the mentioned application is due to its competitive price, acceptable mechanical strength at high temperatures and excellent corrosion resistance. In addition to the oxidation problems at high temperatures caused by the flow of gases from the engine in the exhaust system tubes, the external surface of stainless steels is subject to atmospheric corrosion, which is aggravated by the presence of chloride ions, which can be observed in regions located near the coast. In the present work, AISI 409 ferritic stainless steel sheets were subjected to autogenous welding by the GTAW process, with different welding parameters, with the objective of evaluating the influence of the morphology of the weld region on the corrosion resistance of the joints formed. The welds obtained were characterized according to their microstructure and mechanical properties (Vickers microhardness), and electrochemical corrosion tests were performed in an aqueous solution containing 3.5% NaCl (weight). To evaluate corrosion resistance, electrochemical impedance spectroscopy and potentiodynamic polarization techniques were used. After performing the tests, it was possible to observe hardening in the weld region, in addition to the formation of large grains in the fusion zone and in the heat affected zone. It was also possible to observe that GTAW welding affected the resistance and corrosion pattern compared to the base metal, with the direction of the weld bead in relation to the sheet rolling direction, the welding speed, and consequently the weld grain profile were decisive factors in this behavior.

Keywords Welding · Ferritic stainless steel · Corrosion · Grain profile · Seam welded tube

---

H. R. Braga (✉) · L. D. de Abreu · M. R. M. Vilela · P. P. Brito
Pontifícia Universidade Católica de Minas Gerais, Belo Horizonte, MG, Brazil
e-mail: henrique.braga.1129800@sga.pucminas.br

V. E. Beal and A. S. Ribeiro Junior (eds.), *Advances in Manufacturing Engineering Research*, Lecture Notes in Mechanical Engineering,
https://doi.org/10.1007/978-3-032-09858-0_22

1 Introduction

A typical automotive exhaust system is composed of two sections, known as hot and cold ends. Through the hot-end, gas flow temperatures are found to be in the 750–950 °C range and the materials of choice are normally austenitic stainless steel with elevated high temperature oxidation resistance (Inoue and Kikuchi 2003). In the cold end components, gas temperatures usually fall below 400 °C and the most common materials employed are ferritic stainless steels (Loto 2019), that possess relatively lower Cr contents in comparison to austenitic stainless steels while meeting the demands for corrosion resistance, high temperature strength as well as ductility (necessary for forming in the shape of pipes).

Corrosion is a critical factor that affects the lifetime of exhaust gas pipes, since the materials are subject to high temperature gas flows and the environmental conditions (Oliver and Sephton 2003). In coastal or maritime regions, where the presence of relatively high chloride ion concentrations can be found in the atmosphere, exhaust gas components are more susceptible to corrosion. The weld seams found in these elements are the most sensitive regions due to microstructural modifications generated by the welding process, such as composition heterogeneities, residual stresses, and/or chromium carbide precipitation, which may lead to sensitization, among other issues that may negatively affect corrosion resistance (Van Niekerk et al. 2012).

In this context, a series of works have sought to analyze the corrosion resistance of different stainless steel grades used in automotive exhaust gas systems, such as (Menezes et al. 2010), who discussed the electrochemical corrosion mechanisms of ferritic stainless steels in synthetic fuel (ethanol) condensate solutions. A similar study involving gasoline condensate solutions was also carried out (Huttunen-Saarivirta et al. 2014). However, comparatively fewer investigations have been carried out on welded joints of stainless steel grades employed in exhaust gas systems, which is relevant since these components are manufactured as seam pipes.

Based on these premises, the objective of the present work is to verify the influence of welding parameters on the microstructure formation of welded joints produced on an AISI 409 ferritic stainless steel sheet, and the consequent electrochemical corrosion resistance in a simulated marine environment.

2 Methodology

The materials used in the present work were cold-rolled AISI 409 stainless steel sheets with 1.6 mm thickness, which are commonly employed for manufacturing cold end components of automotive exhaust gas systems in the form of seam pipes (Loto 2019). The sheets were submitted to autogenous GTAW (Gas Tungsten Arc Welding) using two welding speeds (100 and 400 mm/min) while adjusting welding current and tension values in order to ensure similar heat input values for each case. Furthermore, the welds were performed along both longitudinal (rolling) and transversal directions, for a total of four distinct experimental conditions which are presented in Table 1 (heat input values were calculated considering an efficiency of 80%). The welds were produced with 100% Ar as shielding gas, with a constant flow of 12 l/min. A copper backing plate was used in order to avoid oxidation of the weld root, as reported in Fig. 1a. An overview of a typical weld is presented in Fig. 1b.

After performing the joining procedures, the welds were characterized in terms of microstructure and microhardness distribution. The analyses were conducted on the top surface of the welds. Microstructure analysis was performed by optical microscopy using a Zeiss Axio A.1 microscope after conventional metallographic preparation procedures (grinding and polishing followed by etching with Villela's reagent). The microhardness measurements were performed using a Shimadzu HMV-2 microhardness tester with an applied load of 980.7 mN, load time of 15 s and 0.5 mm distance between consecutive indentations.

The electrochemical behavior of the base material and welded samples was analyzed by open circuit potential (OCP) monitoring tests, electrochemical impedance spectroscopy (EIS) and potentiodynamic polarization. All tests were conducted in 250 ml of a 3.5wt.% NaCl solution, to simulate a marine environment, at room temperature. A three-electrode cell was employed, with a Pt wire as counter electrode, Ag/AgCl reference electrode and the samples as working electrodes. The tests performed on the welded materials were conducted on the face of the welds, and the samples were cut in the shape of small rectangles containing only the heat affected zone (HAZ) and fusion zone (FZ), as illustrated in Figs. 1b and 2. Initially, the samples were exposed to the electrolyte for a period of 5400 s, in

Table 1 Welding parameters used for each condition

Sample ID	Weld direction	Current (A)	Voltage (V)	Speed (mm/min)	Heat input (J/mm)
LG 400	Longitudinal	116.0	12.2	400	169.8
TR 400	Transversal	116.0	11.7	400	162.9
LG 100	Longitudinal	39.0	9.0	100	168.5
TR 100	Transversal	39.0	10.2	100	190.9

Fig. 1 Overview of the welding procedure: **a** steel plate fixed on the copper plate, **b** appearance of a typical weld bead with an extracted sample

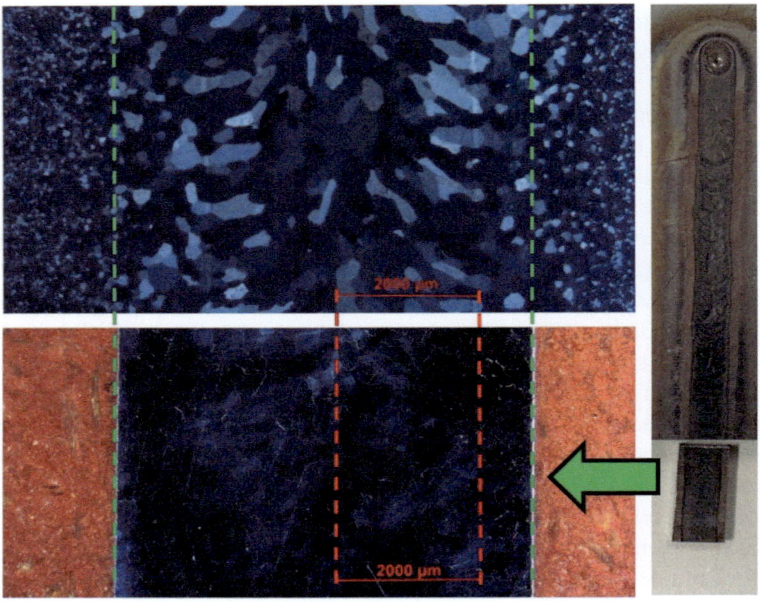

Fig. 2 Exemplification of the isolation of HAZ and FZ contained in the samples

order to stabilize the OCP. Then, EIS measurements were conducted at the stabilized OCP using a 10 mV sinusoidal signal with a frequency range of 100,000 to 0.01 Hz. Finally, the samples were submitted to potentiodynamic polarization with a scan range of − 0.25 V relative to the stabilized OCP value to 0.5 VAg/AgCl with a scan rate of 0.2 mV/s. Following electrochemical analyses, the exposed surfaces of the welds and base material were analyzed by optical microscopy in order to evaluate the morphology of the corroded materials.

3 Results and Discussion

3.1 Microstructure and Mechanical Properties

The microstructure of the upper surface of each weld was analyzed by optical microscopy, as shown in Fig. 3a–d in which the results obtained from the LG400, TR400, LG100 and LG400 samples are shown, respectively. In each case, the base material (BM), heat affected zone (HAZ) and fusion zone (FZ) are identified. The significant grain growth observed in the FZ and HAZ regions is visible in all situations, compared to the initial BM. The microstructure of the FZ is strongly influenced by the heat flow during solidification. In the central line of the FZ of the samples welded at 400 mm/min, medium-sized equiaxed grains are observed, while on the sides, columnar grains orthogonal to the bead direction can be seen. The samples welded at 100 mm/min present elongated grains orthogonal to the weld bead at the interface between the HAZ and FZ that converge to the center of the FZ aligned with the welding direction. In both cases, grain growth during solidification occurs parallel to the direction of heat flow (Villafuerte and Kerr 1990). In all welds, the transition from the FZ to the HAZ is characterized by equiaxed grains that are coarser relative to the BM, which exhibits fine and equiaxed grains.

The mechanical behavior of the welds was evaluated in terms of Vickers microhardness distributions, and the results are presented in Fig. 4. It can be observed that in all analyzed samples the hardness values are slightly higher in the HAZ/FZ transition region (corresponding to approximately 2.5 mm from the center of the weld) and part of the FZ (corresponding to approximately 0.5 mm from the center of the weld), compared to the BM. The exception is the central region of the welds, which exhibited similar hardness relative to the BM. However, the increase in hardness is not sufficient to indicate the formation of martensite in the welds, in agreement with the microstructural analysis presented in Fig. 3. The martensite formed in ferritic stainless steel welds tends to appear in the form of lamellae at the grain boundaries (Anttila et al. 2013), and these characteristics were not found in this investigation.

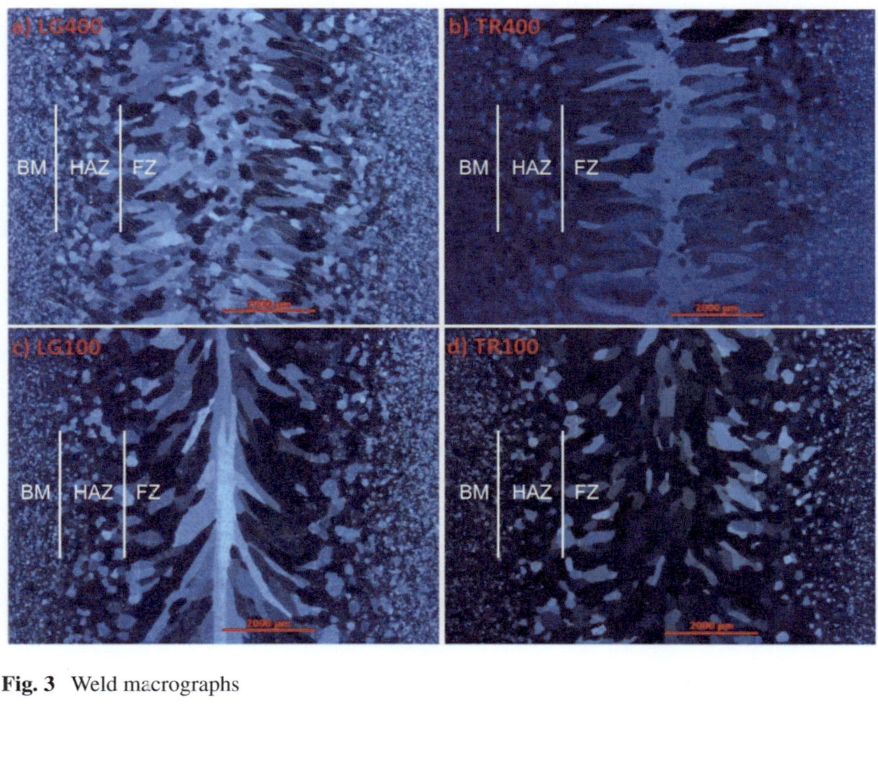

Fig. 3 Weld macrographs

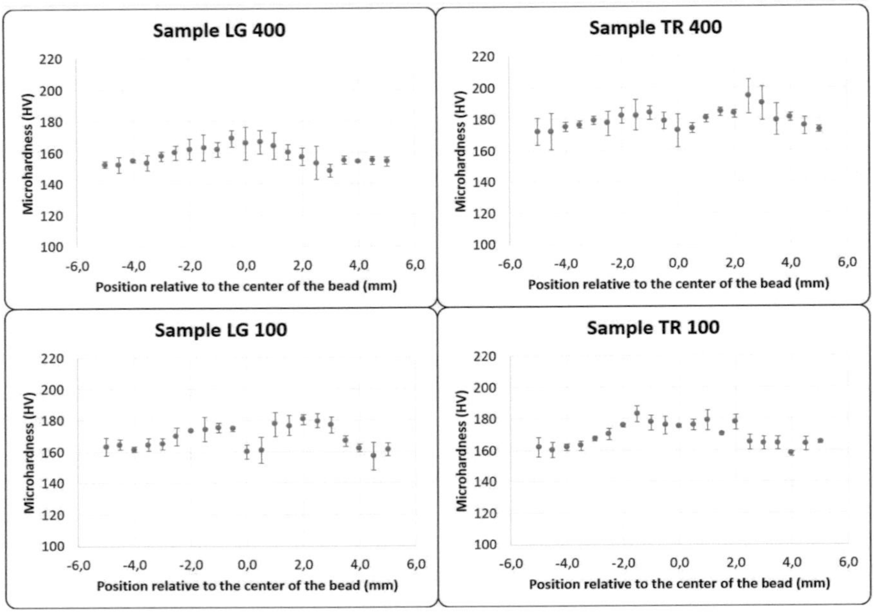

Fig. 4 Microhardness profiles of weld beads

3.2 Corrosion Behavior

The stabilization of the open circuit potential (OCP) for the welds and the BM in the 3.5% NaCl aqueous solution is shown in Fig. 5. The final values obtained after the stabilization period are also shown. For both samples with the welded bead longitudinally in relation to the rolling direction (LG100 and LG400), the OCP values were more negative (indicating a more chemically active surface) compared to the BM, in contrast to the samples with the welded bead transversely in relation to the rolling direction (TR100 and TR400), which showed less negative values (indicating a less chemically active surface).

With the stabilized OCP values indicated in Fig. 5, electrochemical impedance spectroscopy (EIS) measurements were performed, the results of which are presented in the form of Nyquist diagrams in Fig. 6, for each welded sample and BM. The scattered points refer to the measured values and the line drawn refers to the equivalent circuit model generated by the IviumSoft software. The model was chosen which represents the theoretical electrical properties of the electric double layer (McCafferty 2010). The electrochemical parameters obtained from the EIS analyses are presented in Table 2, and the equivalent circuit model used for fitting the experimental data is shown as an inset in Fig. 6.

The results obtained in the EIS analyses alloy elucidation of the corrosion mechanisms of welds and BM. It can be observed that, in both cases, the Nyquist diagrams

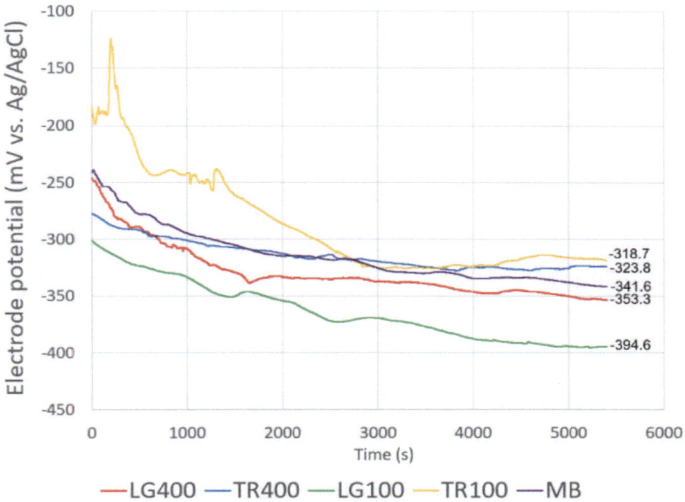

Fig. 5 Stabilization of the open circuit potential of each sample

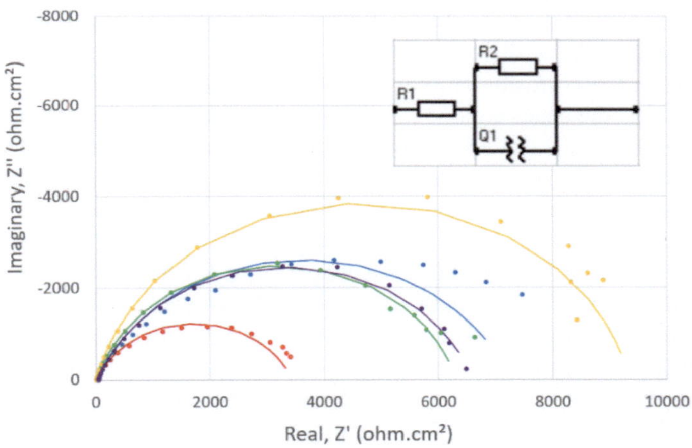

Fig. 6 EIS test results and corresponding electrical circuit model

exhibit a single depressed capacitive arc, modeled by a constant phase element (CPE) identified as Q in the equivalent circuit shown in Fig. 6. The impedance behavior (ZCPE) of this element is calculated by Eq. (1), given by (Stoynov 1990):

$$Z_{CPE} = \left[Q(j\omega)^n\right]^{-1} \tag{1}$$

where Q is a constant and n is an adjustable factor that can vary from -1 (when the element acts as an inductor) to 1 (when it behaves as an ideal capacitor). In

Table 2 Main electrochemical parameters with fitting error values obtained from the analysis of the EIS equivalent circuit model (Rs—solution resistance; Rp—polarization resistance; *Q*—constant phase element; *n*—constant phase element time exponent)

Sample ID	Rs (Ohm.cm^2)	Rp (Ohm.cm^2)	Q (× 10^{-5} s^(n/Ohm)/cm^2)	n
LG400	10.1 ± 0.2	3414 ± 81	21.6 ± 0.6	0.79 ± 0.006
TR400	10.4 ± 0.4	7243 ± 326	17.8 ± 0.7	0.79 ± 0.008
LG100	11.3 ± 0.1	6293 ± 87	11.2 ± 0.1	0.85 ± 0.003
TR100	11.0 ± 0.2	9329 ± 188	7.6 ± 0.1	0.88 ± 0.005
MB	39.1 ± 0.5	6555 ± 130	14.6 ± 0.3	0.86 ± 0.005

both tests, n was approximately 0.8, indicating that the electrochemical interface formed between each sample and the electrolyte exhibited predominantly capacitive behavior.

Regarding the values determined for the CPE in each sample, it is possible to note that, for both the welds and the BM, the interface capacitance is in the order of 10-5 s^(n/Ohm)/cm^2, which is typical of an exposed metal surface, that is, in the absence of a passive oxide layer (Orazem and Tribollet 2017). Finally, it is important to analyze the relative values obtained for the polarization resistance (Rp) of the different samples. Rp is directly related to corrosion resistance, as it is inversely proportional to the corrosion current density and the corrosion rate. In this study, it was noted that the beads welded longitudinally in relation to the rolling direction showed lower Rp values (indicating lower corrosion resistance) compared to the BM, in contrast to the beads welded transversely in relation to the rolling direction, which showed higher Rp values (indicating greater corrosion resistance).

To further evaluate the corrosion behavior of AISI 409 steel (BM and welds), potentiodynamic polarization analyses were performed. The results are presented in Fig. 7. It can be observed that, in the BM and AISI 409 welds, a well-defined active region was found in the polarization diagrams, which indicates absence of passivation of the metal surface, in agreement with the EIS tests. Based on the polarization test results, the Tafel extrapolation model was applied using IviumSoft software, generating the results shown in Table 3, as well as the corrosion rates calculated through the Faraday equation. The results demonstrate greater corrosion resistance of the welded beads transversely compared to the rolling direction, which presented corrosion rates of 4.9 and 6.2 μm/year for the samples welded at 400 and 100 mm/min, respectively. Concurrently, the BM presented a corrosion rate of 24.7 μm/year. This contrasts with the values for the welded beads longitudinally compared to the rolling direction, which exhibited corrosion rates of 57.7 and 40.8 μm/year for the samples welded at 400 and 100 mm/min, respectively, indicating lower corrosion resistance compared to the BM and an influence of the welding direction relative to the materials rolling direction on corrosion resistance.

After the electrochemical tests, the exposed surfaces of the analyzed samples were observed by optical microscopy and it was found that different corrosion patterns

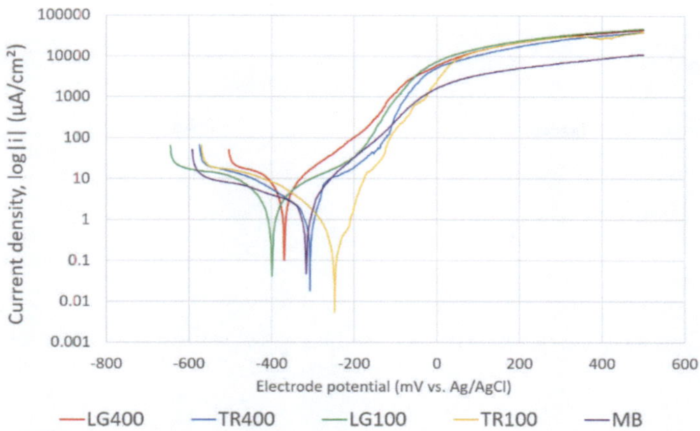

Fig. 7 Potentiodynamic polarization tests of welds and BM

Table 3 Electrochemical parameters from Tafel extrapolation (Ecorr—Corrosion potential; Icorr—Corrosion current; ba—Tafel anode coefficient; bc—Tafel cathodic coefficient, and W—Corrosion rate)

Sample ID	Ecorr (V)	Icorr (μA/cm^2)	Ba (V/dec)	Bc (V/dec)	*W* (μm/year)
LG400	– 0.374	4.97	0.13	0.12	57.7
TR400	– 0.306	0.43	0.03	0.04	4.9
LG100	– 0.403	3.52	0.22	0.19	40.8
TR100	– 0.245	0.54	0.06	0.12	6.2
MB	– 0.320	2.13	0.10	0.29	24.7

emerged, as shown in Fig. 8. The samples welded with higher welding speeds (LG400 and TR400), as well as the BM, exhibited mostly uniform corrosion, with some localized material loss which is spread randomly along the surface, e.g., the white spots seen in Fig. 8e. In contrast, the samples produced with lower welding speeds (LG100 and TR100) exhibited significant localized corrosion in the central region of the FZ, with a pattern that suggests preferential crystallographic dissolution in the welds Figs. 3c and 3d.

4 Conclusions

This work analyzed the influence of the welding parameters of the GTAW process on the formation of the microstructure formed in the weld bead and how this profile affects the corrosion resistance of AISI 409 stainless steel compared to the base metal in a 3.5% (by weight) NaCl in water solution. It was concluded that:

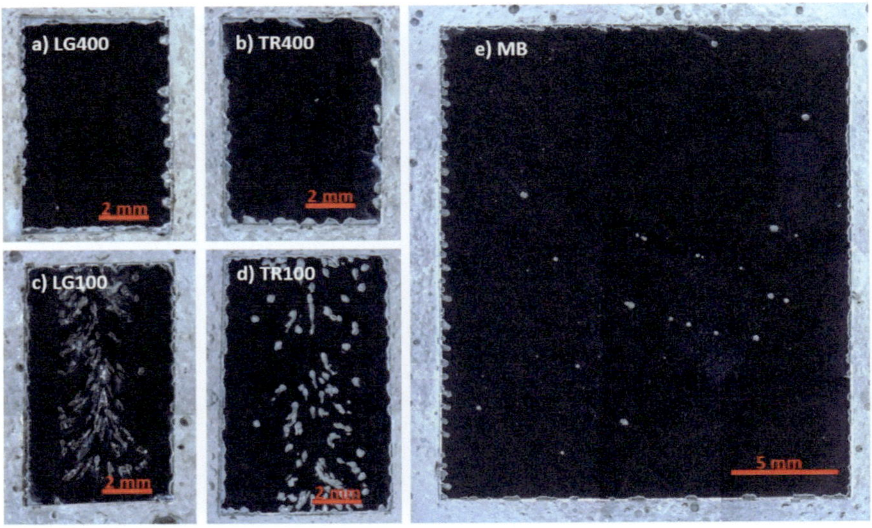

Fig. 8 Macrograph of welded samples and BM

- For the same heat input, welding speed significantly influences the grain profile. At higher speeds, elongated grains develop parallel to the direction of the weld bead; at lower speeds, the grains are coarser and have a profile that converges from the transverse direction to the weld bead at the ends of the fusion zone to the center, parallel to the weld bead.
- Impedance spectrometry and potentiodynamic polarization analyses indicated that in the tested electrolyte, the material does not form a passive layer. Samples welded transversely to the sheet metal rolling direction had superior corrosion resistance compared to the base metal, unlike samples welded longitudinally.
- The corrosion pattern of the samples welded at the lower speed of 100 mm/min was significantly different that observed on the welds obtained with higher welding speeds (400 mm/min), presenting a preferential corrosion profile in the center of the bead which follows the coarse grain morphology of the fusion zone.

Acknowledgements This work was carried out with the support of the Coordenação de Aperfeiçoamento de Pessoal de Nível Superior—Brasil (CAPES)—Financing Code 001.

Responsibility Notice The authors are the only responsible for the printed material included in this paper.

References

Anttila S, Karjalainen P, Lantto S (2013) Mechanical properties of ferritic stainless steel welds in using type 409 and 430 filler metals. Weld World 57:335–347

Huttunen-Saarivirta EV, Kuokkala T, Pohjanne P (2014) Thermally grown oxide films and corrosion performance of ferritic stainless steels under simulated exhaust gas condensate conditions. Corros Sci 87:344–365

Inoue Y, Kikuchi M (2003) Present and future trends of stainless steel for automotive exhaust system. Nippon Steel Tech Rep 88:62–69

Loto RT 2019 Effect of SO42—and Cl—anionic attack on the localized corrosion resistance and morphology of 409 ferritic stainless steel. Results Phys 12:738–742

McCafferty E (2010) Introduction to Corrosion science. Springer Science and Business Media

Menezes MR, Andrade MS, Junqueira RMR (2010) Avaliação da resistência à corrosão de aços inoxidáveis ferríticos em solução aquosa com adição de álcool e de sais contendo cloreto. In: Proceedings of the 2020 Intercorr, Fortaleza, Brazil.

Oliver DC, Sephton M (2003) External corrosion resistance of steel and ferritic stainless steel exhaust systems. J South Afr Inst Min Metall 103:93–100

Orazem ME, Tribollet B (2017) Electrochemical impedance spectroscopy. Wiley, New Jersey

Stoynov Z (1990) Impedance modeling and data processing: structural and parametrical estimation. Electrochim Acta 35:1493–1499

Van Niekerk CJ, Du Toit M, Erwee MW (2012) Sensitization of AISI 409 ferritic stainless steel during low heat input arc welding. Weld World 56:54–64

Villafuerte JC, Kerr HC (1990) Electromagnetic stirring and grain refinement in stainless steel GTA welds. Weld J 69:S1–S13

MIX
Papier aus verantwortungsvollen Quellen
Paper from responsible sources
FSC® C105338

If you have any concerns about our products,
you can contact us on
ProductSafety@springernature.com

In case Publisher is established outside the EU,
the EU authorized representative is:
Springer Nature Customer Service Center GmbH
Europaplatz 3, 69115 Heidelberg, Germany

Printed by Libri Plureos GmbH
in Hamburg, Germany